Surveys in Representation Theory of Algebras

CONTEMPORARY MATHEMATICS

716

Surveys in Representation Theory of Algebras

Maurice Auslander Distinguished Lectures
and International Conference
Woods Hole Oceanographic Institute, Woods Hole, MA
April 26–May 1, 2017
April 29–May 4, 2015
April 18–23, 2013

Alex Martsinkovsky
Kiyoshi Igusa
Gordana Todorov
Editors

AMS AMERICAN MATHEMATICAL SOCIETY

2010 *Mathematics Subject Classification.* Primary 16E30, 18E30, 18E10, 16G99, 18G25, 16G20, 16G70.

Library of Congress Cataloging-in-Publication Data

Names: Maurice Auslander Distinguished Lectures and International Conference. | Martsinkovsky, A. (Alex), editor. | Igusa, Kiyoshi, 1949- editor. | Todorov, G. (Gordana), editor.

Title: Surveys in representation theory of algebras : Maurice Auslander Distinguished Lectures and International Conference, April 26-May 1, 2017, April 29-May 4, 2015, April 18-23, 2013, Woods Hole Oceanographic Institute, Woods Hole, MA / Alex Martsinkovsky, Kiyoshi Igusa, Gordana Todorov, editors.

Description: Providence, Rhode Island : American Mathematical Society, [2018] | Series: Contemporary mathematics ; volume 716 | Includes bibliographical references.

Identifiers: LCCN 2018018535 | ISBN 9781470436797 (alk. paper)

Subjects: LCSH: Representations of algebras–Congresses. | AMS: Associative rings and algebras – Homological methods – Homological functors on modules (Tor, Ext, etc.). msc | Category theory; homological algebra – Abelian categories – Derived categories, triangulated categories. msc | Category theory; homological algebra – Abelian categories – Exact categories, abelian categories. msc | Associative rings and algebras – Representation theory of rings and algebras – None of the above, but in this section. msc | Category theory; homological algebra – Homological algebra – Relative homological algebra, projective classes. msc | Associative rings and algebras – Representation theory of rings and algebras – Representations of quivers and partially ordered sets. msc | Associative rings and algebras – Representation theory of rings and algebras – Auslander-Reiten sequences (almost split sequences) and Auslander-Reiten quivers. msc

Classification: LCC QA176 .M385 2018 | DDC 512/.6–dc23 LC record available at https://lccn.loc.gov/2018018535

Contemporary Mathematics ISSN: 0271-4132 (print); ISSN: 1098-3627 (online)

DOI: https://doi.org/10.1090/conm/716

---

10 9 8 7 6 5 4 3 2 1 23 22 21 20 19 18

Contents

Preface

This volume is an outgrowth of some expository lectures from the annual Maurice Auslander Distinguished Lectures and International Conference. The latest event took place April 26 - May 1, 2017 at the Woods Hole Oceanographic Institute in Falmouth, MA. However, some of the presented material goes back to earlier years due to an extended editorial process which we hope has improved the contents of this volume.

The continued – and gratefully acknowledged – support from the National Science Foundation and Bernice Auslander results in a large number of graduate students and young researchers attending the conference. Their interest and unabating enthusiasm is *raison d'être* for this series of expository works (for the previous collections, see Contemporary Mathematics, vols. 607 and 673).

The selected expository lectures in this volume cover a wide area of research. Lidia Angeleri-Hügel gives a quick introduction to the recent concepts of silting and cosilting modules and their applications. Peter Jørgensen's brief survey of co-t-structures in triangulated categories comes with insightful philosophical observations. The reader who thinks they know what a module is may be surprised to learn about a more general point of view, as expounded in Mike Prest's paper. Chrysostomos Psaroudakis presents a comprehensive survey of recollements of abelian categories. The subject of relative homological algebra is brought back to life in Øyvind Solberg's survey. Hugh Thomas explains connections between Coxeter groups and representations of quivers. Jan Trlifaj's paper on tree modules explains recent striking results on approximation theory.

As we mentioned above, the primary audience for this volume are graduate students and young researchers working in representation theory. It is to be hoped that a much wider audience may also find this book useful.

The Editors

Contemporary Mathematics
Volume **716**, 2018
https://doi.org/10.1090/conm/716/14424

On the abundance of silting modules

Lidia Angeleri Hügel

ABSTRACT. Silting modules are abundant. Indeed, they parametrise the definable torsion classes over a noetherian ring, and the hereditary torsion pairs of finite type over a commutative ring. Also the universal localisations of a hereditary ring, or of a finite dimensional algebra of finite representation type, can be parametrised by silting modules. In these notes, we give a brief introduction to the fairly recent concepts of silting and cosilting module, and we explain the classification results mentioned above.

1. Introduction

The notion of a (compact) silting complex was introduced by Keller and Vossieck [**36**], and it was later rediscovered in work of Aihara-Iyama, Keller-Nicolàs, König-Yang, Mendoza-Sáenz-Santiago-Souto Salorio, and Wei [**2, 35, 37, 45, 55**]. This renewed interest was motivated by cluster theory, and also by the interplay with certain torsion pairs in triangulated categories. In fact, silting complexes are closely related with t-structures and co-t-structures in the derived category. This connection was recently extended to non-compact silting complexes, and more generally, to silting objects in triangulated categories [**5, 48, 50**].

But silting is also closely related to torsion pairs and localisation of abelian categories, as well as to ring theoretic localisation. The aim of these notes is to review some of these connections. We will see that localisation techniques provide constructions of silting objects and lead to classification results over several important classes of rings. There is also an interesting interaction with combinatorial aspects of silting. Certain posets studied in cluster theory have a ring theoretic interpretation which sheds new light on their structure. On the other hand, silting theory yields a new approach to general questions on homological properties of ring epimorphisms.

We will consider silting modules, that is, the modules that arise as zero cohomologies of (not necessarily compact) 2-term silting complexes over an arbitrary ring. These modules were introduced in [**5**]. They provide a generalisation of (not necessarily finitely generated) tilting modules. Moreover, over a finite dimensional algebra, the finitely generated silting modules are precisely the support τ-tilting modules introduced in [**1**] and studied in cluster theory.

In representation theory, one usually studies finite dimensional tilting or support τ-tilting modules up to isomorphism and multiplicities. Similarly, in the infinite dimensional case, it is convenient to study silting classes rather than modules.

The silting class $\operatorname{Gen} T$ given by a silting module T consists of all T-generated modules, and it determines the additive closure $\operatorname{Add} T$ of T. It has several useful closure properties, making it a definable torsion class. A silting class thus provides every module M with a (minimal) right $\operatorname{Gen} T$-approximation given by the trace of T in M, and with a left $\operatorname{Gen} T$-approximation. Furthermore, $\operatorname{Gen} T$ - even when T is not finitely generated - satisfies an important finiteness condition: it is determined by a set Σ of morphisms between finitely generated projective modules [**44**]. This "finite type" result extends the analogous result for tilting modules from [**13**] stating that every tilting class is determined by a set of finitely presented modules of projective dimension at most one.

It turns out that in many respects the dual concept of a cosilting module studied in [**16, 17, 57**] is more accessible. Cosilting modules are pure-injective, and cosilting classes are definable torsionfree classes, in fact, they coincide with the definable torsionfree classes. Notice that the dual result is not true in general: not every definable torsion class is silting. Indeed, there are more cosilting than silting classes, because in general cosilting modules are not of "cofinite type", that is, they need not be determined by a set Σ as above. Also this phenomenon was already known for cotilting modules [**12**], a cosilting example is exhibited in Example 4.10.

However, over a noetherian ring, every torsion pair with definable torsionfree class is generated by a set of finitely presented modules. This allows to show that all cosilting modules are of cofinite type. Then we obtain dually that the silting classes coincide with the definable torsion classes [**4**]. This can be regarded as a "large" analog of [**1**, Theorem 2.7] stating that over a finite dimensional algebra A, there is a bijection between isomorphism classes of basic support τ-tilting modules and functorially finite torsion classes in mod-A.

Similarly, one can show that over a commutative ring, silting modules correspond bijectively to hereditary torsion pairs of finite type, and if the ring is commutative noetherian, they are parametrised by subsets of the spectrum closed under specialisation [**4**].

Silting is thus often related to Gabriel localisation of module categories. But it is also related to ring theoretic localisation, in particular to universal localisation of rings in the sense of [**52**]. In fact, by [**6**] every partial silting module over a ring A induces a ring epimorphism $A \to B$ with nice homological properties, and it is proved in [**44**] that every universal localisation of A arises in this way from some partial silting module. Further, in certain cases, the universal localisations of A, or the homological ring epimorphisms starting in A, are parametrised by silting modules.

More precisely, one obtains a parametrisation by minimal silting modules, that is, silting modules satisfying a condition which ensures the existence of a minimal left $\operatorname{Gen} T$-approximation $f : A \to T_0$ for the regular module A. For example, when A is a finite dimensional algebra, every finite dimensional silting module is minimal. Moreover, the cokernel of the approximation f is a partial silting module uniquely determined by T, so one can associate to T a ring epimorphism $A \to B$, which turns out to be a universal localisation of A. If A is of finite representation type, or more generally, τ-tilting finite in the sense of [**26**], this assignment yields a bijection between silting modules and universal localisations of A, see [**43**].

A similar result holds true when A is a hereditary ring. As shown in [**6**], the minimal silting A-modules are in bijection with the universal localisations of A, which by [**39**] coincide with the homological ring epimorphisms. If A is a finite dimensional hereditary algebra, this correspondence yields a bijection between finite dimensional support tilting modules and functorially finite wide subcategories of mod-A, recovering results from [**32, 41**]. The combinatorial interpretation of finite dimensional support tilting modules in terms of noncrossing partitions or clusters can then be translated into ring theoretic terms by considering the poset of all universal localisations of A. Notice that universal localisations of A can have infinite dimension. Leaving the finite dimensional world, however, has the advantage that the poset becomes a lattice, while the same is not true if we restrict to ring epimorphisms with finite dimensional target, see e.g. Example 6.2(1).

The paper is organised as follows. Section 2 is devoted to the notion of a silting module. In Section 3, we collect some results on cosilting modules scattered in the literature. The interplay between silting and cosilting modules is discussed in Section 4, which also contains the classification results over noetherian or commutative rings mentioned above. In Section 5, we review the connections with ring epimorphisms, and we explain the parametrisation of all universal localisations of a hereditary ring. Finally, in Section 6, we focus on finite dimensional algebras and on the poset of their universal localisations.

NOTATION. Throughout the paper, A will denote a ring, Mod-A the category of all right A-modules, and mod-A the category of all finitely presented right A-modules. The corresponding categories of left A-modules are denoted by A-Mod and A-mod. The unbounded derived category of Mod-A is denoted by $\mathrm{D}(A)$. We further denote by Proj-A and proj-A the full subcategory of Mod-A consisting of all projective and all finitely generated projective right A-modules, respectively, and we write $^* = \mathrm{Hom}_A(-, A)$.

Given a subcategory $\mathcal{C}$ of Mod-A, we denote

$$\mathcal{C}^o = \{M \in \text{Mod-}A \mid \mathrm{Hom}_A(\mathcal{C}, M) = 0\},$$

$$\mathcal{C}^{\perp_1} = \{M \in \text{Mod-}A \mid \mathrm{Ext}^1_A(\mathcal{C}, M) = 0\}.$$

The classes ${}^o\mathcal{C}$ and ${}^{\perp_1}\mathcal{C}$ are defined dually. Moreover, $\mathrm{Add}\,\mathcal{C}$ and $\mathrm{Prod}\,\mathcal{C}$ are the subcategories of Mod-A formed by the modules that are isomorphic to a direct summand of a coproduct of modules in $\mathcal{C}$, or of a product of such modules, respectively. $\mathrm{Gen}\,\mathcal{C}$ is the subcategory of the $\mathcal{C}$-generated modules, i. e. the epimorphic images of modules in $\mathrm{Add}\,\mathcal{C}$, and $\mathrm{Cogen}\,\mathcal{C}$ is defined dually. When $\mathcal{C}$ just consists of a single module M, we write $\mathrm{Add}\,M, \mathrm{Prod}\,M, \mathrm{Gen}\,M, \mathrm{Cogen}\,M$.

Furthermore, we denote by $\mathrm{Mor}(\mathcal{C})$ the class of all morphisms in Mod-A between objects in $\mathcal{C}$. Given a set of morphisms $\Sigma \subset \mathrm{Mor}(\mathcal{C})$, we consider the classes

$$\begin{aligned}
\mathcal{D}_\Sigma &= \{X \in \text{Mod-}A \mid \mathrm{Hom}_A(\sigma, X) \text{ is surjective for all } \sigma \in \Sigma\};\\
\mathcal{X}_\Sigma &= \{X \in \text{Mod-}A \mid \mathrm{Hom}_A(\sigma, X) \text{ is bijective for all } \sigma \in \Sigma\};\\
\mathcal{C}_\Sigma &= \{X \in \text{Mod-}A \mid \mathrm{Hom}_A(X, \sigma) \text{ is surjective for all } \sigma \in \Sigma\};\\
\mathcal{F}_\Sigma &= \{X \in A\text{-Mod} \mid \sigma \otimes_A X \text{ is injective for all } \sigma \in \Sigma\}.
\end{aligned}$$

Again, when $\Sigma = \{\sigma\}$, we write $\mathcal{D}_\sigma, \mathcal{X}_\sigma, \mathcal{C}_\sigma, \mathcal{F}_\sigma$.

2. Silting modules

We start out by briefly reviewing the notion of a (not necessarily compact) silting complex.

DEFINITION 2.1. [**55**] A bounded complex of projective A-modules σ is said to be *silting* if

(1) $\mathrm{Hom}_{\mathrm{D}(A)}(\sigma, \sigma^{(I)}[i]) = 0$, for all sets I and $i > 0$.
(2) the smallest triangulated subcategory of $\mathrm{D}(A)$ containing $\mathrm{Add}\,\sigma$ is $\mathrm{K^b}(\text{Proj-}A)$.

A prominent role is played by the silting complexes of length two which were studied in [**1**] in connection with cluster mutation. Their endomorphism ring has interesting properties investigated in [**18–20**]. Another important feature is the fact that 2-term silting complexes (which we always assume concentrated in cohomological degrees -1 and 0) are determined by their zero cohomology. In fact, identifying a complex $\ldots 0 \to P_{-1} \xrightarrow{\sigma} P_0 \to 0 \ldots$ with the morphism σ in $\mathrm{Mor}(\text{Proj-}A)$, one obtains the following result which goes back to work of Hoshino-Kato-Miyachi [**29**].

PROPOSITION 2.2. *Let σ be 2-term complex in $\mathrm{K^b}(\text{Proj-}A)$ and $T = H^0(\sigma)$. Then σ is a silting complex if and only if the class $\mathrm{Gen}\,T$ of T-generated modules coincides with $\mathcal{D}_\sigma$.*

We are interested in the modules that occur in this way.

DEFINITION 2.3. A right A-module T is *silting* if it admits a projective presentation $P_{-1} \xrightarrow{\sigma} P_0 \to T \to 0$ such that $\mathrm{Gen}\,T = \mathcal{D}_\sigma$. The class $\mathrm{Gen}\,T$ is then called a *silting class.*

We say that two silting modules T, T' are *equivalent* if they generate the same silting class. By [**5**, Section 3], this is equivalent to $\mathrm{Add}\,T = \mathrm{Add}\,T'$.

EXAMPLES 2.4. (1) A module T is silting with respect to a monomorphic projective presentation $\sigma : P_{-1} \hookrightarrow P_0$ if and only if $\mathrm{Gen}\,T = \mathrm{Ker}\,\mathrm{Ext}^1_A(T, -)$, i. e. T is a *tilting* module (of projective dimension at most one, not necessarily finitely generated).

(2) If A is a finite dimensional algebra over a field, and $T \in \text{mod-}A$, then T is silting if and only if it is *support τ-tilting* in the sense of [**1**].

(3) Let A be the path algebra of the quiver Q having two vertices, 1 and 2, and countably many arrows from 1 to 2. Let $P_i = e_i A$ be the indecomposable projective A-module for $i = 1, 2$. Then $T := S_2$ with the projective presentation

$$0 \longrightarrow P_1^{(\mathbb{N})} \xrightarrow{\ \sigma\ } P_2 \longrightarrow T \longrightarrow 0,$$

is a silting module (of projective dimension one, not finitely presented) which is not tilting.

Indeed, it is not tilting as the class $\mathrm{Gen}\,T$ consists precisely of the semisimple injective A-modules, so $\mathrm{Gen}\,T = \mathrm{Ker}\,\mathrm{Hom}_A(P_1, -) \subsetneq \mathrm{Ker}\,\mathrm{Ext}^1_A(T, -)$. But T is silting with respect to the projective presentation γ of T obtained as the direct sum of σ with the trivial map $P_1 \to 0$, since $\mathcal{D}_\gamma = \mathrm{Ker}\,\mathrm{Ext}^1_A(T, -) \cap \mathrm{Ker}\,\mathrm{Hom}_A(P_1, -) = \mathrm{Ker}\,\mathrm{Hom}_A(P_1, -) = \mathrm{Gen}\,T$.

Notice that every class of the form $\mathcal{D}_\sigma$ for a morphism $\sigma \in \mathrm{Mor}(\text{Proj-}A)$ is closed under extensions, epimorphic images, and direct products. In the silting case, $\mathcal{D}_\sigma = \mathrm{Gen}\,T$ is also closed under coproducts, and it is therefore a torsion class. Moreover, even though σ need not belong to $\mathrm{Mor}(\text{proj-}A)$, the silting class $\mathcal{D}_\sigma$ is determined by a set Σ of morphisms in $\mathrm{Mor}(\text{proj-}A)$, that is, $\mathcal{D}_\sigma = \mathcal{D}_\Sigma$. This property is obtained in [**44**] as a consequence of the analogous result for tilting modules proved in [**13**]; an alternate proof in [**4**] uses that $\mathcal{D}_\sigma$ is definable (as shown in [**5**, 3.5 and 3.10]). Recall that a class of modules is said to be *definable* if it is closed under direct limits, direct products, and pure submodules.

In fact, these properties characterise silting classes.

THEOREM 2.5. [**44**] *The following statements are equivalent for a full subcategory $\mathcal{D}$ of* Mod-A.

(i) *$\mathcal{D}$ is a silting class;*
(ii) *there is $\sigma \in$ Mor(Proj-A) such that $\mathcal{D} = \mathcal{D}_\sigma$, and $\mathcal{D}$ is closed for coproducts;*
(iii) *there is a set $\Sigma \subset$ Mor(proj-A) such that $\mathcal{D} = \mathcal{D}_\Sigma$.*

Moreover, $\mathcal{D}$ is a tilting class if and only if $\mathcal{D} = \mathcal{D}_\Sigma$ for a set Σ of monomorphisms in Mor(proj-A).

A silting class $\mathcal{D}$, being a torsion class, provides any module M with a minimal right $\mathcal{D}$-approximation $d(M) \hookrightarrow M$ where d denotes the torsion radical corresponding to $\mathcal{D}$. Moreover, being definable, $\mathcal{D}$ also provides M with a left $\mathcal{D}$-approximation. In fact, the existence of special left approximations can be used to detect the torsion classes which are generated by a tilting module.

THEOREM 2.6. [**10**] *A torsion class $\mathcal{T}$ in* Mod-A *is of the form $\mathcal{T} = \operatorname{Gen} T$ for some tilting module T if and only if for every A-module M (or equivalently, for $M = A$) there is a short exact sequence $0 \to M \to B \to C \to 0$ with $B \in \mathcal{T}$ and $C \in {}^{\perp_1}\mathcal{T}$.*

At the end of the next section, we will discuss how this result extends to silting modules.

3. Cosilting modules

When dealing with classification results for silting modules, it turns out that the dual concept of a cosilting module, which was introduced by Breaz and Pop in [**16**], is often more accessible. In order to study the interplay between the two notions, it will be convenient to consider right silting and left cosilting modules.

DEFINITION 3.1. A left A-module C is *cosilting* if it admits an injective copresentation $0 \to C \longrightarrow E_0 \stackrel{\omega}{\longrightarrow} E_1$ such that the class $\operatorname{Cogen} C$ of C-cogenerated modules coincides with the class $\mathcal{C}_\omega$. The class $\operatorname{Cogen} C$ is then called a *cosilting class*.

Two cosilting modules C, C' are said to be *equivalent* if they cogenerate the same class $\operatorname{Cogen} C = \operatorname{Cogen} C'$. It will follow from Remark 3.4 below that this is equivalent to $\operatorname{Prod} C = \operatorname{Prod} C'$.

Dually to Example 2.4(1), we have that C is cosilting with respect to an epimorphic injective copresentation $E_0 \stackrel{\omega}{\twoheadrightarrow} E_1$ if and only if $\operatorname{Cogen} C = \operatorname{Ker} \operatorname{Ext}^1_A(-, C)$, i. e. C is a *cotilting* module (of injective dimension at most one, not necessarily finitely generated).

Further, dually to the silting case, we see that every cosilting class $\mathcal{C}_\omega = \operatorname{Cogen} C$ is a torsionfree class. How to describe the torsionfree classes that arise in this way? For cotilting modules there is the following characterisation in terms of the existence of approximations.

THEOREM 3.2. [**10**] *A torsionfree class* $\mathcal{F}$ *in* A-Mod *is of the form* $\mathcal{F} = \operatorname{Cogen} C$ *for some cotilting module* C *if and only if for every left* A*-module* M *(or equivalently, for an injective cogenerator* E *of* A-Mod*) there is a short exact sequence* $0 \to K \to L \to M \to 0$ *with* $L \in \mathcal{F}$ *and* $K \in \mathcal{F}^{\perp_1}$.

An extension of this theorem to cosilting modules was first obtained by Zhang and Wei [**57**], as a result of their comparison of several notions generalizing the definition of a cotilting module. Here we focus only on the arguments relevant to the notion of a cosilting module, which we collect below for the reader's convenience. We start with the following observations.

LEMMA 3.3. *Let* C *be a left* A*-module with injective copresentation* $0 \to C \longrightarrow E_0 \overset{\omega}{\longrightarrow} E_1$.

(1) *A module* X *belongs to the class* $\mathcal{C}_\omega$ *if and only if* $\operatorname{Hom}_{\mathrm{D}(A)}(X, \omega[1]) = 0$, *if and only if* $\operatorname{Hom}_{\mathrm{D}(A)}(\sigma, \omega[1]) = 0$ *for any injective copresentation* $0 \to X \longrightarrow I_0 \overset{\sigma}{\longrightarrow} I_1$.
(2) *Assume that* $0 \to C \longrightarrow E_0 \overset{\omega}{\longrightarrow} E_1$ *is a minimal injective copresentation. Then* $\operatorname{Cogen} C \subset {}^{\perp_1}C$ *if and only if* $\operatorname{Hom}_{\mathrm{D}(A)}(\omega^I, \omega[1]) = 0$ *for any set* I, *if and only if* $\operatorname{Hom}_{\mathrm{D}(A)}(X, \omega[1]) = 0$ *for all* $X \in \operatorname{Cogen} C$.

PROOF. (1) is shown by standard arguments. Moreover, using (1), the statement in (2) can be rephrased as follows: $\operatorname{Cogen} C \subset {}^{\perp_1}C$ if and only if all products of copies of C are in $\mathcal{C}_\omega$, if and only if $\operatorname{Cogen} C \subset \mathcal{C}_\omega$. Now, since $\mathcal{C}_\omega$ is always closed under submodules, and $\mathcal{C}_\omega \subset {}^{\perp_1}C$, the second condition means $\operatorname{Cogen} C \subset \mathcal{C}_\omega$, which in turn entails the first condition $\operatorname{Cogen} C \subset {}^{\perp_1}C$. So it remains to show that the first condition implies the second. We sketch an argument from [**57**, Lemma 4.13]. Consider a cardinal κ and a map $f : C^\kappa \to E_1$. The map ω factors as $\omega = e \circ \omega'$ with $e : \operatorname{Im} \omega \hookrightarrow E_1$ being an injective envelope. Setting $Z = \operatorname{Coker} \omega$, we obtain the following commutative diagram with exact rows

$$\begin{array}{ccccccccccc} & & & & 0 & \longrightarrow & K & \overset{i}{\longrightarrow} & C^\kappa & \overset{gf}{\longrightarrow} & Z & & \\ & & & & & & \downarrow{\scriptstyle h} & & \downarrow{\scriptstyle f} & & \| & & \\ 0 & \longrightarrow & C & \longrightarrow & & & E_0 & \overset{\omega}{\longrightarrow} & E_1 & \overset{g}{\longrightarrow} & Z & \longrightarrow & 0 \end{array} \tag{3.1}$$

where h is constructed by first taking the map $f' : K \to \operatorname{Im} \omega$ induced by f, and then lifting f' to h thanks to the fact that $K \in \operatorname{Cogen} C \subset {}^{\perp_1}C$. Now the injectivity of E_0, E_1 yields maps $s_0 : C^\kappa \to E_0$ and $s_1 : Z \to E_1$ such that $f = \omega s_0 + s_1(gf)$, and one verifies $\operatorname{Im}(f - \omega s_0) \cap \operatorname{Im} \omega = 0$. Since $\operatorname{Im} \omega$ is an essential submodule of E_1, it follows $f = \omega s_0$, as required. □

REMARK 3.4. A module M in a torsionfree class $\mathcal{F}$ is said to be *Ext-injective* in $\mathcal{F}$ if $\operatorname{Ext}^1_A(\mathcal{F}, M) = 0$. Notice that the condition $\operatorname{Cogen} C \subset {}^{\perp_1}C$ implies that $({}^{o}C, \operatorname{Cogen} C)$ is a torsion pair, and it can be rephrased by saying that C is an Ext-injective object in the torsionfree class $\operatorname{Cogen} C$.

Now assume that C is a cosilting module with respect to an injective copresentation ω. Since $\mathcal{C}_\omega \subset {}^{\perp_1}C$, it follows that C is Ext-injective in $\operatorname{Cogen} C$. More precisely, $\operatorname{Prod} C$ is the class of all Ext-injective modules in $\operatorname{Cogen} C$. Indeed, for every module $M \in \operatorname{Cogen} C$ there is an embedding $f : M \to C'$ with $C' \in \operatorname{Prod} C$

and Coker $f \in \operatorname{Cogen} C$. If M is Ext-injective in $\operatorname{Cogen} C$, then f is a split monomorphism and $M \in \operatorname{Prod} C$. For details we refer to [**5**, Lemma 2.3 and 3.3, Proposition 3.10] where the dual statements are proved.

The following result is inspired by the proof of [**57**, Proposition 4.13].

PROPOSITION 3.5. *Let C be a left A-module which is Ext-injective in* $\operatorname{Cogen} C$. *Assume there are an injective cogenerator E of A-*Mod*, a cardinal κ, and a right* $\operatorname{Prod} C$*-approximation $g : C^\kappa \to E$. Then* $\operatorname{Ker} g \oplus C$ *is a cosilting module with cosilting class* $\operatorname{Cogen} C$.

PROOF. Take a minimal injective copresentation $0 \to C \longrightarrow E_0 \stackrel{\gamma}{\longrightarrow} E_1$. Then γ^κ is an injective copresentation of C^κ. The map $g : C^\kappa \to E$ induces a map $\tilde{g} : {E_0}^\kappa \to E$, which can be viewed as a map of complexes $\tilde{g} : \gamma^\kappa \to E^\bullet$ where $E^\bullet$ is the complex with E concentrated in degree 0. Note that $\tilde{g}$ has zero cohomology g. Considering the mapping cone $K_{\tilde{g}} : {E_0}^\kappa \stackrel{(\tilde{g},\gamma^\kappa)}{\to} E \oplus {E_1}^\kappa$ and setting $\omega = K_{\tilde{g}}[-1]$, we obtain a triangle in $\mathrm{D}(A)$

$$\omega \to \gamma^\kappa \stackrel{\tilde{g}}{\to} E^\bullet \to$$

whose zero cohomologies give rise to the exact sequence

$$0 \to C_1 \to C^\kappa \stackrel{g}{\to} E$$

with $C_1 = \operatorname{Ker} g = H^0(\omega)$.

We claim that $\operatorname{Cogen} C = \mathcal{C}_\omega$. Indeed, for a module X, applying $\operatorname{Hom}_{\mathrm{D}(A)}(X, -)$ to the triangle above yields a long exact sequence

$$\operatorname{Hom}_{\mathrm{D}(A)}(X, \gamma^\kappa) \stackrel{\operatorname{Hom}_{\mathrm{D}(A)}(X,\tilde{g})}{\longrightarrow} \operatorname{Hom}_{\mathrm{D}(A)}(X, E^\bullet) \to \operatorname{Hom}_{\mathrm{D}(A)}(X, \omega[1]) \to \operatorname{Hom}_{\mathrm{D}(A)}(X, \gamma^\kappa[1]) \to 0.$$

Now, since $g : C^\kappa \to E$ is a right $\operatorname{Prod} C$-approximation of an injective cogenerator, $X \in \operatorname{Cogen} C$ if and only if $\operatorname{Hom}_A(X, g)$ is surjective, which amounts to $\operatorname{Hom}_{\mathrm{D}(A)}(X, \tilde{g})$ being surjective. Further, recall from Lemma 3.3(1) that $X \in \mathcal{C}_\omega$ if and only if $\operatorname{Hom}_{\mathrm{D}(A)}(X, \omega[1]) = 0$. This immediately gives the inclusion $\operatorname{Cogen} C \supset \mathcal{C}_\omega$. For the reverse inclusion, use that $\operatorname{Hom}_{\mathrm{D}(A)}(X, \gamma[1]) = 0$ for all $X \in \operatorname{Cogen} C$ by assumption and Lemma 3.3(2).

In order to prove that $\tilde{C} = C_1 \oplus C$ is a cosilting module, we now consider its injective copresentation $\tilde{\omega} = \omega \oplus \gamma$. By construction $\mathcal{C}_\omega \subset \mathcal{C}_{\gamma^\kappa} = \mathcal{C}_\gamma$. Therefore $\mathcal{C}_{\tilde{\omega}} = \mathcal{C}_\omega \cap \mathcal{C}_\gamma = \mathcal{C}_\omega = \operatorname{Cogen} C = \operatorname{Cogen} \tilde{C}$. □

We know from [**1**] that τ-tilting modules are "non-faithful tilting" modules. The same holds true for silting modules, as shown in [**5**]. Here is the dual case.

THEOREM 3.6. [**57**] *The following statements are equivalent for an A-module C.*

(1) *C is a cosilting module.*
(2) $\operatorname{Cogen} C$ *is a torsionfree class, and C is cotilting over $\overline{A} = A/\operatorname{Ann}(C)$.*
(3) *C is Ext-injective in* $\operatorname{Cogen} C$*, and there are an injective cogenerator E of A-*Mod *and an exact sequence $0 \to C_1 \to C_0 \stackrel{g}{\to} E$ such that $C_1, C_0 \in$* $\operatorname{Prod} C$ *and g is a right* $\operatorname{Cogen} C$*-approximation.*

PROOF. First of all, notice that $\mathrm{Ann}(C) = \mathrm{Ann}(\mathrm{Cogen}\, C)$ and $\mathrm{Cogen}_{\overline{A}}C = \mathrm{Cogen}_{A}C$. Moreover, if $\mathrm{Cogen}\, C$ is extension closed (which holds true in all three statements), then $\mathrm{Ext}^1_{\overline{A}}(\mathrm{Cogen}\, C, C) = 0$ if and only if $\mathrm{Ext}^1_A(\mathrm{Cogen}\, C, C) = 0$. Finally, note also that $\mathrm{Ann}(C) = \bigcap_{h\in \mathrm{Hom}_A(A,C)} \mathrm{Ker}\, h$, so the class $\mathrm{Cogen}_{\overline{A}}C$ contains $\overline{A}$ and all projective $\overline{A}$-modules.

(1) $\Rightarrow$ (2): Every cosilting module satisfies $\mathrm{Cogen}\, C \subset {}^{\perp_1}C$, thus $\mathrm{Cogen}_{\overline{A}}C \subset \mathrm{Ker}\,\mathrm{Ext}^1_{\overline{A}}(-, C)$. To verify that ${}_{\overline{A}}C$ is cotilting, it remains to show the reverse inclusion. Pick a left $\overline{A}$-module X with $\mathrm{Ext}^1_{\overline{A}}(X, C) = 0$. Since $\mathrm{Cogen}\, C$ contains all projective $\overline{A}$-modules, there is a short exact sequence $0 \to K \to L \to X \to 0$ in $\overline{A}$-Mod with $K, L \in \mathrm{Cogen}\, C$. Now one proves (dually to the proof of [**5**, Proposition 3.10]) that the diagonal map $K \hookrightarrow C^I$ with $I = \mathrm{Hom}_A(K, C)$ has cokernel $M \in \mathrm{Cogen}\, C$. By assumption on X, the map $\mathrm{Hom}_{\overline{A}}(L, C^I) \to \mathrm{Hom}_{\overline{A}}(K, C^I)$ is surjective, yielding a map f and a commutative diagram

$$\begin{array}{ccccccccc} 0 & \longrightarrow & K & \longrightarrow & L & \longrightarrow & X & \longrightarrow & 0 \\ & & \| & & \downarrow{\scriptstyle f} & & \downarrow{\scriptstyle g} & & \\ 0 & \longrightarrow & K & \longrightarrow & C^I & \longrightarrow & M & \longrightarrow & 0. \end{array} \tag{3.2}$$

Since $L, M \in \mathrm{Cogen}\, C$, we infer $\mathrm{Ker}\, g \cong \mathrm{Ker}\, f \in \mathrm{Cogen}\, C$, and $\mathrm{Im}\, g \in \mathrm{Cogen}\, C$. But $\mathrm{Cogen}\, C$ is closed under extensions, so also $X \in \mathrm{Cogen}\, C$.

(2) $\Rightarrow$ (3): If E is an injective cogenerator of A-Mod, then $E' = \{x \in E \mid \mathrm{Ann}(C)\cdot x = 0\}$ is an injective cogenerator of $\overline{A}$-Mod. By Theorem 3.2 there is a short exact sequence of $\overline{A}$-modules $0 \to C_1 \to C_0 \overset{g'}{\to} E' \to 0$ with $C_0 \in \mathrm{Cogen}\, C$ and $C_1 \in \mathrm{Ker}\,\mathrm{Ext}^1_{\overline{A}}(\mathrm{Cogen}\, C, -)$. Then g' is a right $\mathrm{Cogen}\, C$-approximation. Further, since $E' \in \mathrm{Ker}\,\mathrm{Ext}^1_{\overline{A}}(\mathrm{Cogen}\, C, -)$, both C_1 and C_0 belong to $\mathrm{Cogen}\, C \cap \mathrm{Ker}\,\mathrm{Ext}^1_{\overline{A}}(\mathrm{Cogen}\, C, -) = \mathrm{Prod}\, C$.

Keeping in mind that every module in $\mathrm{Cogen}\, C$ is also a $\overline{A}$-module, we conclude that the map $g : C_0 \overset{g'}{\to} E' \subset E$ is a right $\mathrm{Cogen}\, C$-approximation with the stated properties. Finally, C is Ext-injective in $\mathrm{Cogen}\, C$ by the first paragraph of the proof.

(3) $\Rightarrow$ (1): We can assume w.l.o.g. that $C_0 = C^\kappa$ for some cardinal κ. By Proposition 3.5, there is an injective copresentation $\tilde{\omega}$ of the module $\tilde{C} = C_1 \oplus C$ such that $\mathcal{C}_{\tilde{\omega}} = \mathrm{Cogen}\, \tilde{C} = \mathrm{Cogen}\, C$. Take minimal injective copresentations α and γ of C_1 and C, respectively. Then $\alpha \oplus \gamma$ is a minimal injective copresentation of $\tilde{C}$, hence there are injective modules I, I' such that $\tilde{\omega} = (\alpha \oplus \gamma) \oplus (0 \to I) \oplus (I' \overset{\mathrm{id}}{\to} I')$. Further, it is easy to see that $C_1 \in \mathrm{Prod}\, C$ implies $\mathcal{C}_\gamma \subset \mathcal{C}_\alpha$, so $\mathcal{C}_{\alpha\oplus\gamma} = \mathcal{C}_\gamma$. We infer $\mathcal{C}_{\tilde{\omega}} = \mathcal{C}_\gamma \cap \mathcal{C}_{(0\to I)}$. So the injective copresentation $\omega = \gamma \oplus (0 \to I)$ of C satisfies $\mathcal{C}_\omega = \mathcal{C}_{\tilde{\omega}} = \mathrm{Cogen}\, C$. This completes the proof. $\square$

It was shown in [**11**] that every cotilting module is pure-injective, and cotilting classes are definable. The characterisation given in Theorem 3.6(2) now allows to deduce the same properties for cosilting modules.

COROLLARY 3.7. [**16, 57**] *Every cosilting module is pure-injective, every cosilting class is definable.*

We are ready for a generalisation of Theorem 3.2 and [**12**, Theorem 6.1].

THEOREM 3.8. [**17**,**57**] *The following statements are equivalent for a torsionfree class $\mathcal{F}$ in A-Mod.*

(1) $\mathcal{F} = \operatorname{Cogen} C$ *for some cosilting module* C.
(2) *For every left A-module M there is an exact sequence* $0 \to K \to L \xrightarrow{g} M$ *such that g is a right $\mathcal{F}$-approximation and K is Ext-injective in $\mathcal{F}$.*
(3) *Every left A-module admits a minimal right $\mathcal{F}$-approximation.*

Let us sketch the proof. If $\mathcal{F}$ is definable, then every module admits a minimal right $\mathcal{F}$-approximation, see [**12**, Corollary 2.6] and the references therein. Moreover, Wakamatsu's Lemma ensures that every minimal right $\mathcal{F}$-approximation has an Ext-injective kernel. Therefore (1) $\Rightarrow$ (3) $\Rightarrow$ (2). The implication (2) $\Rightarrow$ (1) is shown by looking at the special case when $M = E$ is an injective cogenerator of A-Mod. In this case $\mathcal{F} = \operatorname{Cogen} L = \operatorname{Cogen} C$ for $C = K \oplus L$, and by Theorem 3.6, the module C is cosilting provided it is Ext-injective in $\mathcal{F}$. So it remains to verify Ext-injectivity of L. To this end, one works over $\overline{A} = \operatorname{Ann}(C) = \operatorname{Ann}(\mathcal{F})$, where the injective cogenerator $E' = \{x \in E \mid \operatorname{Ann}(C) \cdot x = 0\}$ admits a short exact sequence $0 \to K \to L \to E' \to 0$, and L is in $\operatorname{Ker} \operatorname{Ext}^1_{\overline{A}}(\mathcal{F}, -)$ since so are K and E' (cf. the proof of Theorem 3.6).

COROLLARY 3.9. *The assignment* $C \mapsto \operatorname{Cogen} C$ *defines a bijection between*

(i) *equivalence classes of cosilting left A-modules;*
(ii) *definable torsionfree classes in* A-Mod.

Notice that the statements above rely on the existence of *minimal* injective copresentations. Breaz and Žemlička show in [**17**] that the dual version of Theorem 3.8 holds true over perfect or hereditary rings. In general, however, one only has that the conditions dual to statements (2) and (3) in Theorem 3.6 are equivalent [**5**, Proposition 3.2] and are satisfied by every silting module. In [**5**, Proposition 3.10] this is phrased by saying that every silting module is *finendo quasitilting.* But the converse is not true. A counterexample will be given in Example 4.10.

Moreover, there is a further asymmetry. Indeed, definable classes give rise to minimal right approximations, and also to left approximations, but not necessarily to *minimal* left approximations. So not all silting (nor tilting) classes satisfy the dual of condition (3) in Theorem 3.8. Moreover, not all definable torsion classes are generated by a finendo quasitilting module, an example is given in [**12**, Proposition 7.2]. However, we will see in the forthcoming section that definable torsion classes coincide with silting classes over noetherian rings.

4. Duality

For a better understanding of the asymmetries discussed above, we need to review the interplay between definable subcategories of Mod-A and A-Mod. Let us briefly recall that a subcategory $\mathcal{D}$ of Mod-A is definable if and only if it is the intersection of the kernels of a set of coherent functors Mod-$A \to$ Ab, that is, of additive functors commuting with direct limits and direct products. Since the coherent functors are precisely the cokernels of morphisms of functors $\operatorname{Hom}_A(\sigma, -)$: $\operatorname{Hom}_A(M, -) \longrightarrow \operatorname{Hom}_A(N, -)$ induced by morphisms $\sigma : N \to M$ in mod-A, one obtains that $\mathcal{D}$ is definable if and only if it is of the form $\mathcal{D} = \mathcal{D}_\Sigma$ for a set Σ of morphisms between finitely presented right A-modules, see e.g. [**25**, Sections 2.1 and 2.3].

Further, denoting by (mod-A, Ab) and (A-mod, Ab) the categories of additive covariant functors on mod-A and A-mod, respectively, one has an assignment

$$(\text{mod-}A, \text{Ab}) \to (A\text{-mod}, \text{Ab}), F \mapsto F^\vee$$

where $F^\vee$ is defined on a module N in A-mod by

$$F^\vee(N) = \text{Hom}(F, - \otimes N).$$

Notice that Hom-functors $F = \text{Hom}_A(M, -)$ given by $M \in \text{mod-}A$ are mapped to $\otimes$-functors $F^\vee = - \otimes_A M$, see e.g. [**49**, Example 10.3.1].

This assignment was first studied by Auslander and by Gruson and Jensen. It induces a duality between the finitely presented objects in (mod-A, Ab) and (A-mod, Ab), and it allows to associate to every definable category $\mathcal{D}$ in Mod-A a *dual definable category* $\mathcal{D}^\vee$ in A-Mod. Indeed, $\mathcal{D}^\vee = \mathcal{F}_\Sigma$ where Σ is the maximal collection of morphisms in mod-A such that $\mathcal{D} = \mathcal{D}_\Sigma$. In this way one obtains a bijection between definable subcategories of Mod-A and A-Mod interchanging definable torsion classes with definable torsionfree classes. For details we refer to [**12**, Section 5] and the references therein.

We now review some results from [**4**] explaining how the assignment $\mathcal{D} \mapsto \mathcal{D}^\vee$ acts on the definable torsion or torsionfree classes given by silting and cosilting modules, respectively. We fix a commutative ring k such that A is a k-algebra, and given an A-module M, we denote by M^+ its dual with respect to an injective cogenerator of Mod-k. For example, take $k = \mathbb{Z}$ and M^+ the character dual of M.

PROPOSITION 4.1. *Let $\sigma \in \text{Mor}(\text{Proj-}A)$.*

1. *σ^+ is a morphism between injective left A-modules with $\mathcal{C}_{\sigma^+} = \mathcal{F}_\sigma$.*
2. *If $\mathcal{D}_\sigma$ is a silting class, then ${\mathcal{D}_\sigma}^\vee = \mathcal{F}_\sigma$.*
3. *If T is a silting module with respect to σ, then T^+ is a cosilting module with respect to σ^+.*

DEFINITION 4.2. A cosilting left A-module C (or the cosilting class Cogen C) is said to be *of cofinite type* if there is a set $\Sigma \subset \text{Mor}(\text{proj-}A)$ such that $\text{Cogen}\, C = \mathcal{F}_\Sigma$.

COROLLARY 4.3. *The assignment $\mathcal{D} \mapsto \mathcal{D}^\vee$ defines a bijection between silting classes in* Mod-A *and cosilting classes of cofinite type in* A-Mod.

If we restrict to tilting classes, that is, to classes $\mathcal{D} = \mathcal{D}_\Sigma$ with Σ consisting of monomorphisms in Mor(proj-A), then we recover the bijection between tilting classes and cotilting classes of cofinite type established in [**12**]. Indeed, in this case $\mathcal{F}_\Sigma = \text{Ker}\,\text{Tor}_1^A(\mathcal{S}, -)$ where $\mathcal{S}$ is the set of finitely presented right A-modules of projective dimension at most one that arise as cokernels of the monomorphisms in Σ. So Definition 4.2 agrees with the definition of a cotilting class of cofinite type in [**12**].

Here is a useful criterion for a torsionfree class to be cosilting of cofinite type.

LEMMA 4.4. *A torsion pair $(\mathcal{T}, \mathcal{F})$ in* A-Mod *is generated by a set of finitely presented left A-modules if and only if $\mathcal{F}$ is a cosilting class of cofinite type.*

PROOF. Let $\mathcal{U}$ be a set in A-mod such that $\mathcal{F} = \{M \in A\text{-Mod} \mid \text{Hom}_A(U, M) = 0 \text{ for all } U \in \mathcal{U}\}$. Choosing a projective presentation $\alpha_U \in \text{Mor(A-proj)}$ for each $U \in \mathcal{U}$ and applying $^* = \text{Hom}_A(-, A)$ on it, we obtain a set $\Sigma = \{{\alpha_U}^* \mid U \in \mathcal{U}\} \subset \text{Mor}(\text{proj-}A)$ such that $\mathcal{F} = \mathcal{F}_\Sigma$. The other implication is proven similarly. □

Now recall that over a left noetherian ring every torsion pair $(\mathcal{T}, \mathcal{F})$ in A-Mod restricts to a torsion pair $(\mathcal{U}, \mathcal{V})$ in A-mod with $\mathcal{U} = \mathcal{T} \cap A$-mod and $\mathcal{V} = \mathcal{F} \cap A$-mod, and moreover, taking direct limit closures, the latter torsion pair $(\mathcal{U}, \mathcal{V})$ extends to a torsion pair $(\varinjlim \mathcal{U}, \varinjlim \mathcal{V})$ in A-Mod, see [**24**, Lemma 4.4]. So, if $\mathcal{F}$ is definable, that is, closed under direct limits, we see that $(\mathcal{T}, \mathcal{F}) = (\varinjlim \mathcal{U}, \varinjlim \mathcal{V})$ is generated by $\mathcal{U}$. Using Lemma 4.4, we conclude that every definable torsionfree class in A-Mod is a cosilting class of cofinite type. This also implies that every definable torsion class in Mod-A is a silting class.

THEOREM 4.5. [**4**, Corollary 3.8] *If A is a left noetherian ring, the assignment $\mathcal{D} \mapsto \mathcal{D}^\vee$ defines a bijection between silting classes in* Mod-A *and cosilting classes in* A-Mod, *and there are bijections between*

(i) *equivalence classes of silting right A-modules;*
(ii) *equivalence classes of cosilting left A-modules;*
(iii) *definable torsion classes in* Mod-A*;*
(iv) *definable torsionfree classes in* A-Mod.

The bijection (i) $\to$ (iii) is given by the assignment $T \mapsto \operatorname{Gen} T$, the bijection (ii) $\to$ (iv) is defined dually.

The theorem above can be regarded as a "large" version of the following result for finite dimensional algebras due to Adachi-Iyama-Reiten.

THEOREM 4.6. [**1**, Theorem 2.7] *If A is a finite dimensional algebra over a field, there are bijections between*

(i) *isomorphism classes of basic support τ-tilting (i.e. finite dimensional silting) right A-modules;*
(ii) *isomorphism classes of basic finite dimensional cosilting left A-modules;*
(iii) *functorially finite torsion classes in* mod-A*;*
(iv) *functorially finite torsionfree classes in* A-mod.

The bijection (i) $\to$ (iii) is given by the assignment $T \mapsto \operatorname{gen} T = \operatorname{Gen} T \cap \text{mod-}A$, the bijection (ii) $\to$ (iv) is defined dually.

Next, we apply the criterion in Lemma 4.4 to hereditary torsion pairs. Recall that a torsion pair $(\mathcal{T}, \mathcal{F})$ is *hereditary* if the torsion class $\mathcal{T}$ is closed under submodules, or equivalently, the torsionfree class $\mathcal{F}$ is closed under injective envelopes. Moreover, $(\mathcal{T}, \mathcal{F})$ has *finite type* if $\mathcal{F}$ is closed under direct limits.

Hereditary torsion pairs are in bijection with the Gabriel topologies on A. We refer to [**54**, Ch.VI] or [**49**, Sections 11.1.1 and 11.1.2] for details. Here we only mention that every hereditary torsion pair $(\mathcal{T}, \mathcal{F})$ in A-Mod is associated to a *Gabriel filter* $\mathcal{G}$ which is formed by the left ideals I of A such that $A/I \in \mathcal{T}$. When the torsion pair is of finite type, then $\mathcal{G}$ has a basis of finitely generated ideals, that is, every ideal in $\mathcal{G}$ contains a finitely generated ideal from $\mathcal{G}$. This implies that the torsion pair $(\mathcal{T}, \mathcal{F})$ is generated by the set $\mathcal{U} = \{A/I \mid I \in \mathcal{G} \text{ finitely generated }\} \subset A$-mod, and so $\mathcal{F}$ is a cosilting class of cofinite type. The dual definable category $\mathcal{F}^\vee$ is then given by the right A-modules with $M \otimes A/I = 0$ for all finitely generated ideals $I \in \mathcal{G}$, and one easily checks that this amounts to M being *I-divisible*, i.e. $MI = M$, for all such I. We will say that $\mathcal{F}^\vee$ is the *class of divisibility* by these ideals.

COROLLARY 4.7. *The assignment $\mathcal{D} \mapsto \mathcal{D}^\vee$ restricts to a bijection between the silting classes occurring as classes of divisibility by sets of finitely generated left ideals, and the torsionfree classes in hereditary torsion pairs of finite type.*

Over a commutative ring, every cosilting class of cofinite type arises from a hereditary torsion pair of finite type, because it turns out that every torsionfree class of the form $\mathcal{F}_\sigma$ for some $\sigma \in \mathrm{Mor}(\text{proj-}A)$ is closed under injective envelopes [**4**, Lemma 4.2]. This yields the following classification result.

THEOREM 4.8. [**4**, Theorems 4.7 and 5.1] *If A is commutative ring, there is a bijection between*

(i) *equivalence classes of silting A-modules,*
(ii) *hereditary torsion pairs of finite type in* Mod-A.

If A is commutative noetherian, there are further bijections with

(iii) *equivalence classes of cosilting A-modules,*
(iv) *subsets $P \subseteq \mathrm{Spec}(A)$ closed under specialisation.*

The bijection between (ii) and (iv) is well known, see [**54**, Chapter VI, §6.6]. An explicit construction of a silting module in (i) and a cosilting module in (iii) is provided in [**4**].

We close this section with an example of a cosilting module not of cofinite type.

LEMMA 4.9. *Let A be a ring, and S a simple module such that S is a finitely generated module over* $\mathrm{End}_A S$ *and* $\mathrm{Ext}^1_A(S,S)=0$. *Then S satisfies condition (3) in Theorem* 3.6 *and its dual. In particular, S is a cosilting module.*

PROOF. Since $\mathrm{End}_A S$ is a skew-field, the module S has finite length over $\mathrm{End}_A S$, and therefore $\mathrm{Add}\, S = \mathrm{Prod}\, S$, see [**38**]. By assumption, $\mathrm{Gen}\, S = \mathrm{Add}\, S = \mathrm{Cogen}\, S$ are contained both in $S^{\perp_1}$ and ${}^{\perp_1}S$. Moreover, if E is an injective cogenerator of Mod-A, the codiagonal map $S^{(I)} \to E$ given by $I = \mathrm{Hom}_A(S,E)$ is a right $\mathrm{Cogen}\, S$-approximation with kernel in $\mathrm{Prod}\, S$, yielding condition (3) in Theorem 3.6. For the dual condition take the diagonal map $A \to S^J$ given by $J = \mathrm{Hom}_A(A,S)$. □

EXAMPLE 4.10. [**4**, Example 5.4] Let $(A,\mathfrak{m})$ be a valuation domain whose maximal ideal $\mathfrak{m} = \mathfrak{m}^2$ is idempotent and non-zero. Then $S = A/\mathfrak{m}$ satisfies the conditions of the Lemma above (indeed, we will see in Example 5.7(2) and Theorem 5.3(2) that the idempotency of $\mathfrak{m}$ implies $\mathrm{Ext}^1_A(S,S)=0$), so it is a cosilting module with $\mathrm{Cogen}\, S = \mathrm{Gen}\, S = \mathrm{Add}\, S = \{M \in \text{Mod-}A \mid M\mathfrak{m} = 0\}$.

Notice that $\mathrm{Cogen}\, S$ does not contain the injective envelope of S, and so it does not arise from a hereditary torsion pair. It follows from Theorem 4.8 that S is not a cosilting module of cofinite type. Moreover, despite the fact that S satisfies the statement dual to condition (3) in Theorem 3.6, it is not a silting module. This follows again from Theorem 4.8, because if $\mathrm{Gen}\, S$ were a silting class, it would be the class of divisibility by a set of finitely generated left ideals. But the only ideal I with $SI = S$ is $I = A$, which would entail $\mathrm{Gen}\, S = \text{Mod-}A$, a contradiction.

5. Ring epimorphisms

We have seen above that silting modules are closely related with localisation at Gabriel topologies, and silting classes are often given by divisibility conditions.

In this section, we discuss the connections between silting and ring theoretic localisation. We first recall the relevant terminology.

DEFINITION 5.1. [**54**, Ch.XI][**28**] A ring homomorphism $f : A \to B$ is a *ring epimorphism* if it is an epimorphism in the category of rings with unit, or equivalently, if the functor given by restriction of scalars $f_* : \text{Mod-}B \to \text{Mod-}A$ is a full embedding.

Further, f is a *homological ring epimorphism* if it is a ring epimorphism and $\text{Tor}_i^A(B, B) = 0$ for all $i > 0$, or equivalently, the functor given by restriction of scalars $f_* : \text{D}(B) \to \text{D}(A)$ is a full embedding.

Finally, f is a *(right) flat ring epimorphism* if it is a ring epimorphism and B is a flat right A-module.

Two ring epimorphisms $f_1 : A \to B_1$ and $f_2 : A \to B_2$ are *equivalent* if there is a ring isomorphism $h : B_1 \to B_2$ such that $f_2 = h \circ f_1$. We then say that they lie in the same *epiclass* of A.

DEFINITION 5.2. [**28**],[**33**, Theorem 1.6.3] A full subcategory $\mathcal{X}$ of Mod-A is called *bireflective* if the inclusion functor $\mathcal{X} \hookrightarrow \text{Mod-}A$ admits both a left and right adjoint, or equivalently, if it is closed under products, coproducts, kernels and cokernels.

Moreover, a full subcategory $\mathcal{W}$ of mod-A is said to be *wide* if it is an abelian subcategory of mod-A closed under extensions.

THEOREM 5.3. *(1)* [**27**] *There is a bijection between*

(i) *epiclasses of ring epimorphisms;*
(ii) *bireflective subcategories of* Mod-A.

It assigns to a ring epimorphism $f : A \to B$ *the essential image* $\mathcal{X}_B$ *of the restriction functor* $f_* : \text{Mod-}B \hookrightarrow \text{Mod-}A$.

(2) [**15**][**52**, Theorem 4.8] *The following statements are equivalent for a ring epimorphism* $f : A \to B$*:*

(1) $\mathcal{X}_B$ *is closed under extensions in* Mod-A*;*
(2) $\text{Tor}_1^A(B, B) = 0$*;*
(3) *the functors* Ext_A^1 *and* Ext_B^1 *coincide on* B*-modules;*
(4) *the functors* Tor_1^A *and* Tor_1^B *coincide on* B*-modules.*

(3) [**27**, **28**], [**33**, Theorem 1.6.1] *If* A *is a finite dimensional algebra, the assignment* $f \mapsto \mathcal{X}_B \cap \text{mod-}A$ *defines a bijection between*

(i) *epiclasses of ring epimorphisms* $f : A \to B$ *with* B *finite dimensional and* $\text{Tor}_1^A(B, B) = 0$*;*
(ii) *functorially finite wide subcategories of* mod-A.

The following result provides a large supply of ring epimorphisms with the properties in Theorem 5.3(2).

THEOREM 5.4. [**52**, Theorem 4.1] *Let* A *be a ring and* Σ *be a set of morphisms in* Mor(proj-A). *Then there is a ring homomorphism* $f : A \to A_\Sigma$*, called* universal localisation *of* A *at* Σ*, such that*

(i) f *is* Σ-inverting, *i.e.* $\sigma \otimes_A A_\Sigma$ *is an isomorphism for every* σ *in* Σ*, and*
(ii) f *is* universal Σ-inverting, *i.e. for any* Σ*-inverting morphism* $f' : A \to B$ *there exists a unique ring homomorphism* $g : A_\Sigma \to B$ *such that* $g \circ f = f'$.

Moreover, $f\colon A \to A_\Sigma$ is a ring epimorphism with $Tor_1^A(A_\Sigma, A_\Sigma) = 0$.

Notice that A_Σ coincides with the universal localisation at $\Sigma^* = \{\sigma^* \mid \sigma \in \Sigma\} \subset \mathrm{Mor}(\text{A-proj})$. Moreover, the essential image of the restriction functor along $f\colon A \to A_\Sigma$ consists of the right A-modules such that $X \otimes_A \sigma^*$ is an isomorphism for every $\sigma \in \Sigma$, and it equals

$$\mathcal{X}_\Sigma = \{X \in \text{Mod-}A \mid \mathrm{Hom}_A(\sigma, X) \text{ is bijective for all } \sigma \in \Sigma\}.$$

We are going to discuss how classes of the latter shape are related with silting modules.

Definition 5.5. A module T_1 (the reason for this notation will become clear later) is called a *partial silting module* if it admits a projective presentation $P_{-1} \xrightarrow{\sigma} P_0 \to T_1 \to 0$ such that $\mathcal{D}_\sigma$ is a torsion class containing T_1.

Extending [**1**, Theorem 2.10], it was shown in [**5**, Theorem 3.12] that every partial silting module T_1 can be completed to a silting module T which is called a *Bongartz completion* of T_1 and generates the same torsion class $\mathcal{D}_\sigma$. Moreover, $\mathrm{Gen}\, T_1 \subset \mathcal{D}_\sigma \subset T_1{}^{\perp_1}$, and $(\mathrm{Gen}\, T_1, T_1{}^o)$ is a torsion pair. Then

$$\mathcal{X}_\sigma = \mathcal{D}_\sigma \cap (\mathrm{Coker}\,\sigma)^o = \mathrm{Gen}\, T \cap T_1{}^o.$$

As shown in [**6**, Proposition 3.3], the class $\mathcal{X}_\sigma$ is bireflective and extension closed. Thus it can be realised as $\mathcal{X}_B$ for some ring epimorphism $f : A \to B$ as in Theorem 5.3(2), and by [**6**, Theorem 3.5] the ring B can be described as an idempotent quotient of $\mathrm{End}_A\, T$.

A ring epimorphism arising from a partial silting module as above will be called *silting ring epimorphism.*

Theorem 5.6. [**44**] *Every universal localisation is a silting ring epimorphism.*

Examples 5.7. (1) A module T_1 is partial silting with respect to an monomorphic projective presentation $\sigma : P_{-1} \hookrightarrow P_0$ if and only if it is partial tilting, and in this case $\mathcal{X}_\sigma = T_1^{\perp_1} \cap T_1{}^o$ is the perpendicular category of T_1 studied in [**22**, **28**].

(2) For any ideal I of A, the canonical surjection $f : A \to \overline{A} = A/I$ is a ring epimorphism. Moreover, since $I/I^2 \cong \mathrm{Tor}_1^A(\overline{A}, \overline{A})$, the ideal $I = I^2$ is idempotent if and only if $\mathcal{X}_{\overline{A}}$ is closed under extensions, in which case $\mathcal{X}_{\overline{A}} = I^o$.

An important example of an idempotent ideal is provided by the trace ideal $I = \tau_P(A)$ of a projective right A-module P. Notice that every idempotent ideal I has the form $I = AeA$ for some $e = e^2 \in A$ whenever A is a (one-sided) perfect ring [**46**, Proposition 2.1], or when A is commutative and I is finitely generated, in which case f is even a split epimorphism [**40**, Lemma 2.43]. Moreover, every idempotent ideal I with ${}_AI$ being finitely generated is the trace ideal of a countably generated projective right A-module, see [**56**].

(3) Let us focus on the case when $I = \tau_P(A)$ is the trace ideal of a projective right A-module P, and $f : A \to \overline{A} = A/I$. In this case $\mathcal{X}_{\overline{A}}$ consists of the modules M with $\tau_P(M) = MI = 0$, so it is the perpendicular category of the partial tilting module P, and f is therefore a silting ring epimorphism. In fact, f is even a universal localisation. This is clear when P is finitely generated (then f is the universal localisation at $\Sigma = \{0 \to P\}$). For the general case, one uses the following argument due to Pavel Příhoda. First of all, keeping in mind that every projective module is a direct sum of countably generated projectives by a celebrated result

of Kaplansky, we can assume w.l.o.g. that P is countably generated. Then P can be written as a direct limit of a direct system of finitely generated free modules $F_1 \overset{\alpha_1}{\to} F_2 \overset{\alpha_2}{\to} F_3 \overset{\alpha_3}{\to} \dots$ where each map α_i is given by multiplication with a matrix X_i having its entries in the trace ideal I, and moreover, for each $i > 1$ there is a map $\beta_i : F_{i+1} \to F_i$ such that $\beta_i\alpha_i\alpha_{i-1} = \alpha_{i-1}$, see [**56**, Theorem 1.9], [**34**, Proposition 1.4]. Now we set $\Sigma = \{1_{F_i} - \beta_i\alpha_i \mid i > 1\}$. Since f is Σ-inverting, the universal property in Theorem 5.4 implies that $\mathcal{X}_{\overline{A}} \subset \mathcal{X}_\Sigma$. Conversely, the fact that $A \to A_\Sigma$ is Σ-inverting entails $\alpha_i \otimes_A A_\Sigma = 0$ for all $i \geq 1$, hence $P \otimes_A A_\Sigma = 0$. But then $\mathrm{Hom}_A(P, A_\Sigma) \cong \mathrm{Hom}_{A_\Sigma}(P \otimes_A A_\Sigma, A_\Sigma) = 0$, that is, A_Σ belongs to $\mathcal{X}_{\overline{A}}$, and so do all A_Σ-modules.

(4) If A is a semihereditary ring, then all flat ring epimorphisms $A \to B$ are universal localisations [**3**, Proposition 5.3], and the converse is true if A is also commutative [**14**, Theorem 7.8]. Moreover, Tor_i^A vanishes for $i \geq 2$ (see e.g. [**40**, Theorem 4.67]), hence every epimorphism as in Theorem 5.3(2) is homological, and its kernel is an idempotent ideal by [**14**, Lemma 4.5]. On the other hand, for a semihereditary commutative ring, the fact that every homological epimorphism is a universal localisation amounts to the validity of the Telescope Conjecture for $\mathrm{D}(A)$, and it fails in general, see [**14**, Section 8] and the example below.

(5) Let $(A, \mathfrak{m})$ be as in Example 4.10. Then $f : A \to \overline{A} = A/\mathfrak{m}$ is a homological ring epimorphism. But f is not a silting ring epimorphism, and thus not a universal localisation. Indeed, if there were a partial silting module T_1 with Bongartz completion T such $\mathcal{X}_{\overline{A}} = \mathrm{Gen}\, T \cap T_1{}^o$, then by Theorem 4.8 the silting class $\mathrm{Gen}\, T$ would be the class of divisibility by a set of finitely generated ideals of A, and then also by all ideals in the corresponding Gabriel filter $\mathcal{G}$. Since $\mathfrak{m}$ is the unique maximal ideal and $\overline{A} \in \mathrm{Gen}\, T$ is certainly not $\mathfrak{m}$-divisible, we infer that $\mathcal{G}$ can only contain the ideal A. But then $\mathrm{Gen}\, T = \text{Mod-}A$, and T and T_1 are projective, hence free, thus $\mathrm{Gen}\, T \cap T_1{}^o = 0$ cannot coincide with $\mathcal{X}_{\overline{A}}$.

(6) If A is a hereditary ring, then homological ring epimorphisms and universal localisations of A coincide [**39**]. Moreover, by [**53**, Theorem 2.3], there is a bijection between wide subcategories of mod-A and universal localisations of A, which maps a wide subcategory $\mathcal{W}$ to the universal localisation at (projective resolutions of the modules in) $\mathcal{W}$. Conversely, every $A \to A_\Sigma$ is associated to the wide subcategory $\mathcal{W}$ formed by the Σ-*trivial* modules, that is, the modules $M \in$ mod-A admitting a projective resolution $\sigma : P_{-1} \hookrightarrow P_0$ with $\sigma \otimes_A A_\Sigma$ being an isomorphism.

(7) It follows from (6) that silting ring epimorphisms and universal localisations coincide for hereditary rings. The same holds true for commutative noetherian rings of Krull-dimension at most one, but it fails already in Krull-dimension two, see [**8**, **44**]. In particular, there is no analog of Theorem 2.5 for classes $\mathcal{X}_\sigma$: in general, the map $\sigma \in \mathrm{Mor}(\text{Proj-}A)$ cannot be replaced by a set $\Sigma \subset \mathrm{Mor}(\text{proj-}A)$. However, it is shown in [**44**] that one can always find a set Σ of morphisms between *countably* generated projective modules such that $\mathcal{X}_\sigma = \mathcal{X}_\Sigma$.

Our next aim is to investigate the relationship between silting modules and silting ring epimorphisms. We will need the following construction of silting modules which is dual to Proposition 3.5.

Proposition 5.8. [**7**] *Let T be a module with a projective presentation $P_{-1} \overset{\sigma}{\longrightarrow} P_0$ such that $\mathrm{Hom}_{\mathrm{D}(A)}(\sigma, \sigma^{(I)}[1]) = 0$ for any set I. Assume there are a cardinal κ and a left $\mathrm{Add}\, T$-approximation $f : A \longrightarrow T^{(\kappa)}$. Then $T \oplus \mathrm{Coker}\, f$ is a silting*

A-module with silting class $\operatorname{Gen} T$*. Moreover, if* $A \longrightarrow \sigma^{(\kappa)} \longrightarrow \omega \longrightarrow$ *is the triangle in* $\mathsf{D}(A)$ *induced by* f*, then* $\operatorname{Coker} f$ *is a partial silting module with respect to* ω *and* $\operatorname{Gen} T = \mathcal{D}_\omega$.

Now, if T is a silting module with respect to a projective presentation σ, then by [**55**] there is a triangle

$$A \xrightarrow{\phi} \sigma_0 \longrightarrow \sigma_1 \longrightarrow \tag{5.1}$$

in $\mathsf{D}(A)$, where σ_0 and σ_1 lie in $\operatorname{Add} \sigma$ and ϕ is a left $\operatorname{Add} \sigma$-approximation of A. Applying the functor $H^0(-)$ to this triangle, we obtain a left $\operatorname{Gen} T$-approximation sequence for A in Mod-A

$$A \xrightarrow{f} T_0 \longrightarrow T_1 \longrightarrow 0 \tag{5.2}$$

where $T_0, T_1 \in \operatorname{Add} T$ and T_0 satisfies the assumptions of Proposition 5.8, so T_1 is a partial silting module with respect to σ_1. Moreover, f is left minimal if so is ϕ.

Definition 5.9. Let T be a silting module with respect to σ. If the map ϕ in the triangle (5.1) above can be chosen left-minimal, then T is said to be a *minimal* silting module.

Examples 5.10. (1) If A is (right) hereditary or perfect, then any approximation sequence as in (5.2) can be lifted to a triangle as in (5.1), and a silting module is minimal if and only if the map f in (5.2) can be chosen left-minimal, compare with [**6**, Definition 5.4].

(2) Every finite dimensional silting module over a finite dimensional algebra is minimal.

(3) The following example extends a construction of tilting modules from [**9**]. If $f : A \longrightarrow B$ is a homological ring epimorphism such that B is an A-module of projective dimension at most one, then $B \oplus \operatorname{Coker} f$ is a minimal silting A-module. This follows immediately from Proposition 5.8 keeping in mind that f can always be regarded as a minimal left $\operatorname{Add} B$-approximation of A, in fact, it is the $\mathcal{X}_B$-reflection of A. Moreover, $\operatorname{Coker} f$ is partial silting with respect to a projective presentation ω such that $\operatorname{Gen} B = \mathcal{D}_\omega$, and f is the corresponding silting ring epimorphism, because $\mathcal{X}_\omega = \operatorname{Gen} B \cap (\operatorname{Coker} f)^o = \mathcal{X}_B$.

The importance of minimal silting modules is due to the fact that an approximation triangle (5.1) with ϕ being left minimal is unique up to isomorphism, and so is the module T_1. We can thus associate to T a uniquely determined silting ring epimorphism $f : A \to B$ with $\mathcal{X}_B = \mathcal{X}_{\sigma_1}$.

Notice that the class $\mathcal{X}_{\sigma_1}$ can also be described in a different way. For any torsion class $\mathcal{T}$ in an abelian category $\mathcal{A}$, we consider the subcategory of $\mathcal{A}$

$$\mathfrak{a}(\mathcal{T}) := \{X \in \mathcal{T} : \text{ if } (g : Y \to X) \in \operatorname{Mor}(\mathcal{T}), \text{ then } \operatorname{Ker}(g) \in \mathcal{T}\} \tag{5.3}$$

studied in [**32**]. It turns out that in our situation $\mathcal{X}_{\sigma_1} = \mathfrak{a}(\operatorname{Gen} T)$, see [**6**, Remark 5.7]. In summary:

PROPOSITION 5.11. [**7**] *There is a commutative diagram*

$$\begin{array}{ccc}
\left\{\begin{array}{c}\textit{equivalence classes}\\ \textit{of minimal}\\ \textit{silting } A\textit{-modules}\end{array}\right\} & \xrightarrow{\quad\alpha\quad} & \left\{\begin{array}{c}\textit{epiclasses of ring}\\ \textit{epimorphisms } A\to B\\ \textit{with } \operatorname{Tor}_1^A(B,B)=0\end{array}\right\} \\
 & {\searrow^{\mathfrak{a}} \qquad \swarrow_{\epsilon}} & \\
 & \left\{\begin{array}{c}\textit{bireflective}\\ \textit{extension-closed}\\ \textit{subcategories}\\ \textit{of } \text{Mod-}A\end{array}\right\} &
\end{array}$$

where α assigns to a silting module T the associated silting ring epimorphism, the map ϵ is the bijection from Theorem 5.3*(1) and $\mathfrak{a}$ is the map defined above.*

Furthermore, if A is hereditary, then T_0 is a projective generator of $\mathfrak{a}(\operatorname{Gen} T)$, and the category of projective B-modules is equivalent to $\operatorname{Add} T_0$, see [**6**, Proposition 5.6]. This shows that the epiclass of f determines $\operatorname{Add} T_0$ and thus also the silting class $\operatorname{Gen} T_0 = \operatorname{Gen} T$. Hence α is an injective map. On the other hand, we also have a reverse assignment. According to Example 5.10(3), to every homological ring epimorphism $f : A \to B$ we can associate the minimal silting module $T = B \oplus \operatorname{Coker} f$, and taking the silting ring epimorphism corresponding to T we recover f.

Combined with the results from [**39**, **53**] explained in Example 5.7(6), we obtain

THEOREM 5.12. [**6**] *Let A be a hereditary ring. There is a bijection between*

(i) *equivalence classes of minimal silting A-modules;*
(ii) *epiclasses of homological ring epimorphisms of A;*
(ii') *epiclasses of universal localisations of A;*
(iii) *wide subcategories of* mod-A.

Under this bijection, epiclasses of injective homological ring epimorphisms of A correspond to minimal tilting A-modules.

EXAMPLE 5.13. If A is a commutative hereditary ring, then every silting module is minimal. Indeed, by Theorem 4.8 every silting module T corresponds to a hereditary torsion pair of finite type, that is, to a Gabriel filter $\mathcal{G}$ with a basis of finitely generated ideals. Since all ideals are projective, $\mathcal{G}$ is a *perfect* Gabriel topology according to [**54**, Chapter XI, Proposition 3.3 and Theorem 3.4], and it induces a flat ring epimorphism $f : A \to B$ satisfying the assumptions of Example 5.10(3). Then one can show as in [**30**, Theorem 5.4] that T is equivalent to the minimal silting module $B \oplus \operatorname{Coker} f$.

In particular, if A is a Dedekind domain, there is a bijection between equivalence classes of tilting modules and sets of maximal ideals, and the trivial module 0 is the only silting module that is not tilting, see [**9**, Corollary 6.12].

6. The lattice of ring epimorphisms

The partial order on bireflective subcategories given by inclusion corresponds under the bijection in Theorem 5.3(1) to the partial order on the epiclasses of A defined by setting

$$f_1 \geq f_2$$

whenever there is a commutative diagram of ring homomorphisms

$$\begin{array}{ccc} A & \xrightarrow{\quad f_2 \quad} & B_2 \\ & {\scriptstyle f_1}\searrow \qquad \nearrow{\scriptstyle g} & \\ & B_1 & \end{array}$$

Since bireflective subcategories are determined by closure properties, the poset induced by $\geq$ is a lattice, and the ring epimorphisms $A \to B$ with $\mathrm{Tor}_1^A(B, B) = 0$ form a sublattice in it by Theorem 5.3(2). We now compare these posets with other posets recently studied in cluster theory.

First of all, observe that over a finite dimensional algebra, Theorem 5.12 has also a "small" version.

Corollary 6.1. [**32**, Section 2],[**41**, Theorem 4.2] *If A is a finite dimensional hereditary algebra, there are bijections between*

(i) *isomorphism classes of basic support tilting (i.e. finite dimensional silting) modules;*
(ii) *epiclasses of homological ring epimorphisms $A \to B$ with B finite dimensional;*
(iii) *functorially finite wide subcategories of* mod-A.

In [**31**, **32**, **51**] further bijections are established, providing a combinatorial interpretation of finitely generated silting modules in terms of noncrossing partitions, clusters, or antichains. Observe that the poset of noncrossing partitions corresponds to the poset given by $\geq$ on the epiclasses in condition (ii), and it does not form a lattice in general (unless we relax the condition that B is finite dimensional, as discussed above).

Examples 6.2. (1) Let A be a tame hereditary finite dimensional algebra with a tube $\mathcal{U}$ of rank 2, and let S_1, S_2 be the simple regular modules in $\mathcal{U}$. The universal localisation $f_i : A \to A_{\{S_i\}}$ of A at (a projective resolution of) S_i has a finite dimensional codomain $A_{\{S_i\}}$ for $i = 1, 2$. The meet of f_1 and f_2, however, is the universal localisation $A \to A_{\{S_1,S_2\}}$ of A at $\mathcal{U}$ and $A_{\{S_1,S_2\}}$ is an infinite dimensional algebra by [**23**, Theorem 4.2].

(2) Let A be the Kronecker algebra, i.e. the path algebra of the quiver $\bullet \rightrightarrows \bullet$ over an algebraically closed field $\mathbb{K}$. The lattice of universal localisations has the

following shape

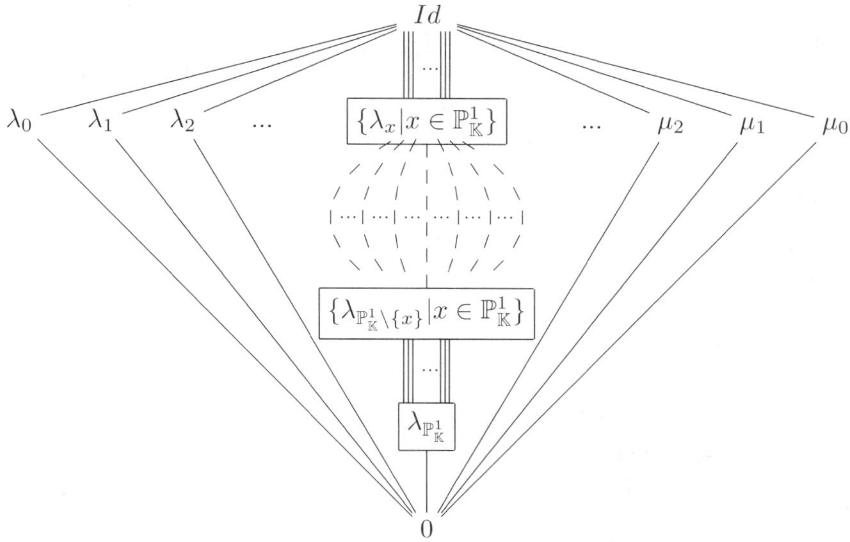

Here the λ_i are the homological ring epimorphisms corresponding to preprojective silting modules, the μ_i correspond to preinjective silting modules, and the ring epimorphisms in frames are those with infinite dimensional codomain, that is, those given by universal localisation at (projective resolutions of) simple regular modules. The interval between Id and $\lambda_{\mathbb{P}^1_{\mathbb{K}}}$ represents the dual poset of subsets of the projective line $\mathbb{P}^1_{\mathbb{K}}$ over $\mathbb{K}$.

Up to equivalence, there is just one additional silting module which is not minimal and thus does not appear in the lattice above. It is called Lukas tilting module and it generates the class of all modules without preprojective summands. More details are given in [**6**, Examples 5.10 and 5.18].

We have seen in Section 5 that over a hereditary ring the assignment $\alpha : T \mapsto f$ from Proposition 5.11 defines a bijective correspondence between minimal silting modules and universal localisations. Another important case where α plays a similar role is established by Marks and Stovicek in [**43**].

PROPOSITION 6.3. [**43**] *If A is a finite dimensional algebra, there is a commutative diagram of injections*

$$\begin{array}{ccc}
\left\{\begin{array}{c}\textit{equivalence classes of}\\ \textit{finite dimensional}\\ \textit{silting } A\textit{-modules}\end{array}\right\} & \xrightarrow{\quad\alpha\quad} & \left\{\begin{array}{c}\textit{epiclasses of}\\ \textit{universal localisations}\\ f : A \to B \textit{ with } \dim B < \infty\end{array}\right\} \\
& {\scriptstyle \mathfrak{a}}\searrow \qquad \swarrow{\scriptstyle \epsilon} & \\
& \left\{\begin{array}{c}\textit{functorially finite wide}\\ \textit{subcategories of } \text{mod-}A\end{array}\right\} &
\end{array}$$

where α assigns to a silting module T the associated silting ring epimorphism, ϵ is the assignment $f \mapsto \mathcal{X}_B \cap$ mod-A, and $\mathfrak{a}$ is the map defined in (5.3) for the abelian category mod-A.

PROOF. If T is a finite dimensional silting module, then it is minimal by Example 5.10(2), and it is associated to a partial silting module T_1 with projective presentation $\sigma_1 \in \mathrm{Mor}(\text{proj-}A)$. By [**43**, Lemma 3.8] the functorially finite torsion class $\mathrm{gen}\,T$ gives rise to the functorially finite wide subcategory $\mathfrak{a}(\mathrm{gen}\,T) = \mathrm{gen}\,T \cap T_1{}^o = \mathcal{X}_{\sigma_1} \cap \text{mod-}A$, which is precisely the wide subcategory corresponding to the silting ring epimorphism f associated to T_1. Notice that f is the universal localisation of A at σ_1, and it has a finite dimensional codomain by Theorem 5.3(3). Hence α is well-defined, and the diagram commutes. Finally, ϵ is obviously injective, and $\mathfrak{a}$ is injective by [**43**, Proposition 3.9]. □

The image of the map $\mathfrak{a}$ in the proposition above is determined in [**43**, Theorem 3.10]. It consists of the functorially finite wide subcategories $\mathcal{W} \subset \text{mod-}A$ for which the smallest torsion class in mod-A containing $\mathcal{W}$ is functorially finite. In particular, $\mathfrak{a}$ is bijective whenever all torsion classes are functorially finite. Let us turn to algebras with such property.

DEFINITION 6.4. [**1**, **26**] Let A be a finite dimensional algebra, and denote by τ the Auslander-Reiten transpose. A module $M \in \text{mod-}A$ is *τ-rigid* if $\mathrm{Hom}_A(M, \tau M) = 0$, and it is *$\tau$-tilting* if, in addition, the number of non-isomorphic indecomposable direct summands of M coincides with the number of isomorphism classes of simple A-modules. The algebra A is *τ-tilting finite* if there are only finitely many isomorphism classes of basic τ-tilting A-modules.

We list some characterisations of τ-tilting finite algebras from [**26**], and we add two new conditions from [**7**].

THEOREM 6.5. *The following statements are equivalent for a finite dimensional algebra A.*

(1) *A is τ-tilting finite.*
(2) *There are only finitely many finite dimensional silting A-modules up to equivalence.*
(3) *Every torsion class in* mod-A *is functorially finite.*
(4) *Every silting A-module is finite dimensional up to equivalence.*
(5) *There are only finitely many epiclasses of ring epimorphisms $A \to B$ with* $\mathrm{Tor}_1^A(B, B) = 0$.

In particular, if A is τ-tilting finite, all ring epimorphisms $A \to B$ with $\mathrm{Tor}_1^A(B, B) = 0$ *are universal localisations with a finite dimensional codomain.*

THEOREM 6.6. [**43**] *Let A be a finite dimensional τ-tilting finite algebra. Then there are bijections between*

(i) *isomorphism classes of basic support τ-tilting (i.e. silting) modules;*
(ii) *epiclasses of ring epimorphisms $A \to B$ with* $\mathrm{Tor}_1^A(B, B) = 0$;
(ii') *epiclasses of universal localisations of A;*
(iii) *wide subcategories of* mod-A.

The theorem above applies in particular to all algebras of finite representation type, but also to many representation-infinite algebras.

EXAMPLES 6.7. (1) [**41**, Lemma 3.5] If A is a local finite dimensional algebra, then A and 0 are the only silting modules, up to equivalence.

(2) Let A be a preprojective algebra of Dynkin type. Then the collections in Theorem 6.6 are further in bijection with the elements of the Weyl group of the

underlying Dynkin quiver [**47**]. The interplay between combinatorial and ring theoretic data is used in [**42**] to describe the algebras arising as universal localisations of A and to determine the homological ring epimorphisms.

(3) The combinatorics of universal localisations is also used in [**41**] to classify silting modules over Nakayama algebras.

(4) Finally, let us remark that the maps α and $\mathfrak{a}$ in Proposition 6.3 do not preserve the poset structure, see e.g. [**41**, Example 4.8] or [**6**, Example 5.18].

We close this survey with some open questions.

QUESTION 1. What is the image of the map α in Proposition 6.3? Recall that all universal localisations are silting ring epimorphisms by Theorem 5.6, but in general we do not know how finite dimensional silting modules interact with silting ring epimorphisms having a finite dimensional codomain.

QUESTION 2. Let T be a silting module such that A has a minimal left $\operatorname{Add} T$-approximation $A \to T_0$. Is the module T_1 in the sequence $A \to T_0 \to T_1 \to 0$ always partial silting?

QUESTION 3. In Theorems 4.8, 5.12, 6.6, we have seen several instances of the interplay between silting modules and localisations. Is there a suitable notion of localisation with a general statement encompassing all these cases?

Acknowledgments

These notes arose from my talk at the "Maurice Auslander Distinguished Lectures and Conference 2015" in Woods Hole, and from lectures delivered at the University of Verona. I would like to thank the organisers for the invitation to a very nice and interesting conference. Furthermore, I thank Simion Breaz, Frederik Marks, Jan Šťovíček, and Jorge Vitória for many valuable comments on a first version of the manuscript, and I thank Pavel Příhoda for providing the argument in Example 5.7(3).

References

[1] Takahide Adachi, Osamu Iyama, and Idun Reiten, *τ-tilting theory*, Compos. Math. **150** (2014), no. 3, 415–452, DOI 10.1112/S0010437X13007422. MR3187626

[2] Takuma Aihara and Osamu Iyama, *Silting mutation in triangulated categories*, J. Lond. Math. Soc. (2) **85** (2012), no. 3, 633–668, DOI 10.1112/jlms/jdr055. MR2927802

[3] Lidia Angeleri Hügel and Maria Archetti, *Tilting modules and universal localization*, Forum Math. **24** (2012), no. 4, 709–731, DOI 10.1515/form.2011.080. MR2949120

[4] Lidia Angeleri Hügel and Michal Hrbek, *Silting modules over commutative rings*, Int. Math. Res. Not. IMRN **13** (2017), 4131–4151, DOI 10.1093/imrn/rnw147. MR3671512

[5] Lidia Angeleri Hügel, Frederik Marks, and Jorge Vitória, *Silting modules*, Int. Math. Res. Not. IMRN **4** (2016), 1251–1284, DOI 10.1093/imrn/rnv191. MR3493448

[6] Lidia Angeleri Hügel, Frederik Marks, and Jorge Vitória, *Silting modules and ring epimorphisms*, Adv. Math. **303** (2016), 1044–1076, DOI 10.1016/j.aim.2016.08.035. MR3552543

[7] L. Angeleri Hügel, F. Marks, J. Vitória, A characterisation of τ-tilting finite algebras, `https://arxiv.org/abs/1801.04312`.

[8] L. Angeleri Hügel, F. Marks, J. Šťovíček, R. Takahashi, J. Vitória, Flat ring epimorphisms and universal localisations of commutative rings, in preparation.

[9] Lidia Angeleri Hügel and Javier Sánchez, *Tilting modules arising from ring epimorphisms*, Algebr. Represent. Theory **14** (2011), no. 2, 217–246, DOI 10.1007/s10468-009-9186-x. MR2776783

[10] Lidia Angeleri Hügel, Alberto Tonolo, and Jan Trlifaj, *Tilting preenvelopes and cotilting precovers*, Algebr. Represent. Theory **4** (2001), no. 2, 155–170, DOI 10.1023/A:1011485800557. MR1834843

[11] S. Bazzoni, *Cotilting modules are pure-injective*, Proc. Amer. Math. Soc. **131** (2003), no. 12, 3665–3672, DOI 10.1090/S0002-9939-03-06938-7. MR1998172

[12] Silvana Bazzoni, *When are definable classes tilting and cotilting classes?*, J. Algebra **320** (2008), no. 12, 4281–4299, DOI 10.1016/j.jalgebra.2008.08.028. MR2464106

[13] Silvana Bazzoni and Dolors Herbera, *One dimensional tilting modules are of finite type*, Algebr. Represent. Theory **11** (2008), no. 1, 43–61, DOI 10.1007/s10468-007-9064-3. MR2369100

[14] Silvana Bazzoni and Jan Šťovíček, *Smashing localizations of rings of weak global dimension at most one*, Adv. Math. **305** (2017), 351–401, DOI 10.1016/j.aim.2016.09.028. MR3570139

[15] George M. Bergman and Warren Dicks, *Universal derivations and universal ring constructions*, Pacific J. Math. **79** (1978), no. 2, 293–337. MR531320

[16] Simion Breaz and Flaviu Pop, *Cosilting modules*, Algebr. Represent. Theory **20** (2017), no. 5, 1305–1321, DOI 10.1007/s10468-017-9688-x. MR3707917

[17] S. Breaz, J. Žemlička, Torsion classes generated by silting modules, `http://arxiv.org/abs/1601.06655`.

[18] Aslak Bakke Buan and Yu Zhou, *A silting theorem*, J. Pure Appl. Algebra **220** (2016), no. 7, 2748–2770, DOI 10.1016/j.jpaa.2015.12.009. MR3457990

[19] Aslak Bakke Buan and Yu Zhou, *Silted algebras*, Adv. Math. **303** (2016), 859–887, DOI 10.1016/j.aim.2016.07.004. MR3552540

[20] Aslak Bakke Buan and Yu Zhou, *Endomorphism algebras of 2-term silting complexes*, Algebr. Represent. Theory **21** (2018), no. 1, 181–194, DOI 10.1007/s10468-017-9709-9. MR3748359

[21] P. M. Cohn, *Free rings and their relations*, 2nd ed., London Mathematical Society Monographs, vol. 19, Academic Press, Inc. [Harcourt Brace Jovanovich, Publishers], London, 1985. MR800091

[22] Riccardo Colpi, Alberto Tonolo, and Jan Trlifaj, *Perpendicular categories of infinite dimensional partial tilting modules and transfers of tilting torsion classes*, J. Pure Appl. Algebra **211** (2007), no. 1, 223–234, DOI 10.1016/j.jpaa.2007.01.004. MR2333768

[23] W. W. Crawley-Boevey, *Regular modules for tame hereditary algebras*, Proc. London Math. Soc. (3) **62** (1991), no. 3, 490–508, DOI 10.1112/plms/s3-62.3.490. MR1095230

[24] William Crawley-Boevey, *Locally finitely presented additive categories*, Comm. Algebra **22** (1994), no. 5, 1641–1674, DOI 10.1080/00927879408824927. MR1264733

[25] William Crawley-Boevey, *Infinite-dimensional modules in the representation theory of finite-dimensional algebras*, Algebras and modules, I (Trondheim, 1996), CMS Conf. Proc., vol. 23, Amer. Math. Soc., Providence, RI, 1998, pp. 29–54. MR1648602

[26] L. Demonet, O. Iyama, G. Jasso, τ-tilting finite algebras, bricks, and g-vectors, to appear in Int. Math. Res. Notices IMRN `https://doi.org/10.1093/imrn/rnx135`.

[27] P. Gabriel and J. A. de la Peña, *Quotients of representation-finite algebras*, Comm. Algebra **15** (1987), no. 1-2, 279–307, DOI 10.1080/00927878708823421. MR876981

[28] Werner Geigle and Helmut Lenzing, *Perpendicular categories with applications to representations and sheaves*, J. Algebra **144** (1991), no. 2, 273–343, DOI 10.1016/0021-8693(91)90107-J. MR1140607

[29] Mitsuo Hoshino, Yoshiaki Kato, and Jun-Ichi Miyachi, *On t-structures and torsion theories induced by compact objects*, J. Pure Appl. Algebra **167** (2002), no. 1, 15–35, DOI 10.1016/S0022-4049(01)00012-3. MR1868115

[30] Michal Hrbek, *One-tilting classes and modules over commutative rings*, J. Algebra **462** (2016), 1–22, DOI 10.1016/j.jalgebra.2016.05.014. MR3519496

[31] Kiyoshi Igusa and Ralf Schiffler, *Exceptional sequences and clusters*, J. Algebra **323** (2010), no. 8, 2183–2202, DOI 10.1016/j.jalgebra.2010.02.003. MR2596373

[32] Colin Ingalls and Hugh Thomas, *Noncrossing partitions and representations of quivers*, Compos. Math. **145** (2009), no. 6, 1533–1562, DOI 10.1112/S0010437X09004023. MR2575093

[33] O. Iyama, Rejective subcategories of artin algebras and orders, `http://arxiv:math/0311281`.

[34] Dolors Herbera and Pavel Příhoda, *Reconstructing projective modules from its trace ideal*, J. Algebra **416** (2014), 25–57, DOI 10.1016/j.jalgebra.2014.06.010. MR3232793

[35] Bernhard Keller and Pedro Nicolás, *Weight structures and simple dg modules for positive dg algebras*, Int. Math. Res. Not. IMRN **5** (2013), 1028–1078, DOI 10.1093/imrn/rns009. MR3031826

[36] B. Keller and D. Vossieck, *Aisles in derived categories*, Bull. Soc. Math. Belg. Sér. A **40** (1988), no. 2, 239–253. Deuxième Contact Franco-Belge en Algèbre (Faulx-les-Tombes, 1987). MR976638

[37] Steffen Koenig and Dong Yang, *Silting objects, simple-minded collections, t-structures and co-t-structures for finite-dimensional algebras*, Doc. Math. **19** (2014), 403–438. MR3178243

[38] H. Krause, M. Saorín, On minimal approximations of modules, In: Trends in the representation theory of finite dimensional algebras (ed. by E. L. Green and B. Huisgen-Zimmermann), Contemp. Math. **229** (1998) 227-236.

[39] Henning Krause and Jan Šťovíček, *The telescope conjecture for hereditary rings via Ext-orthogonal pairs*, Adv. Math. **225** (2010), no. 5, 2341–2364, DOI 10.1016/j.aim.2010.04.027. MR2680168

[40] T. Y. Lam, *Lectures on modules and rings*, Graduate Texts in Mathematics, vol. 189, Springer-Verlag, New York, 1999. MR1653294

[41] Frederik Marks, *Universal localisations and tilting modules for finite dimensional algebras*, J. Pure Appl. Algebra **219** (2015), no. 7, 3053–3088, DOI 10.1016/j.jpaa.2014.10.003. MR3313518

[42] Frederik Marks, *Homological embeddings for preprojective algebras*, Math. Z. **285** (2017), no. 3-4, 1091–1106, DOI 10.1007/s00209-016-1739-1. MR3623741

[43] Frederik Marks and Jan Šťovíček, *Torsion classes, wide subcategories and localisations*, Bull. Lond. Math. Soc. **49** (2017), no. 3, 405–416, DOI 10.1112/blms.12033. MR3723626

[44] F. Marks, J. Šťovíček, Universal localisations via silting, to appear in Proc. Roy. Soc. Edinburgh Sect. A. `http://arxiv.org/abs/1605.04222`

[45] Octavio Mendoza Hernández, Edith Corina Sáenz Valadez, Valente Santiago Vargas, and María José Souto Salorio, *Auslander-Buchweitz context and co-t-structures*, Appl. Categ. Structures **21** (2013), no. 5, 417–440, DOI 10.1007/s10485-011-9271-2. MR3097052

[46] Gerhard Michler, *Idempotent ideals in perfect rings*, Canad. J. Math. **21** (1969), 301–309, DOI 10.4153/CJM-1969-031-0. MR0238885

[47] Yuya Mizuno, *Classifying τ-tilting modules over preprojective algebras of Dynkin type*, Math. Z. **277** (2014), no. 3-4, 665–690, DOI 10.1007/s00209-013-1271-5. MR3229959

[48] P. Nicolás, M. Saorín, A. Zvonareva, Silting theory in triangulated categories with coproducts. `http://arxiv.org/abs/1512.04700`

[49] Mike Prest, *Purity, spectra and localisation*, Encyclopedia of Mathematics and its Applications, vol. 121, Cambridge University Press, Cambridge, 2009. MR2530988

[50] Chrysostomos Psaroudakis and Jorge Vitória, *Realisation functors in tilting theory*, Math. Z. **288** (2018), no. 3-4, 965–1028, DOI 10.1007/s00209-017-1923-y. MR3778987

[51] Claus Michael Ringel, *The Catalan combinatorics of the hereditary Artin algebras*, Recent developments in representation theory, Contemp. Math., vol. 673, Amer. Math. Soc., Providence, RI, 2016, pp. 51–177, DOI 10.1090/conm/673/13490. MR3546710

[52] A. H. Schofield, *Representation of rings over skew fields*, London Mathematical Society Lecture Note Series, vol. 92, Cambridge University Press, Cambridge, 1985. MR800853

[53] A. H. Schofield, Universal localisations of hereditary rings, `http://arxiv.org/0708.0257`.

[54] Bo Stenström, *Rings of quotients*, Springer-Verlag, New York-Heidelberg, 1975. Die Grundlehren der Mathematischen Wissenschaften, Band 217; An introduction to methods of ring theory. MR0389953

[55] Jiaqun Wei, *Semi-tilting complexes*, Israel J. Math. **194** (2013), no. 2, 871–893, DOI 10.1007/s11856-012-0093-1. MR3047094

[56] James M. Whitehead, *Projective modules and their trace ideals*, Comm. Algebra **8** (1980), no. 19, 1873–1901, DOI 10.1080/00927878008822551. MR588450

[57] Peiyu Zhang and Jiaqun Wei, *Cosilting complexes and AIR-cotilting modules*, J. Algebra **491** (2017), 1–31, DOI 10.1016/j.jalgebra.2017.07.022. MR3699086

Dipartimento di Informatica - Settore di Matematica, Università degli Studi di Verona, Strada le Grazie 15 - Ca' Vignal, I-37134 Verona, Italy

Email address: `lidia.angeleri@univr.it`

Contemporary Mathematics
Volume **716**, 2018
https://doi.org/10.1090/conm/716/14425

Co-t-structures: The first decade

Peter Jørgensen

ABSTRACT. Co-t-structures were introduced about ten years ago as a type of mirror image of t-structures. Like t-structures, they permit us to divide an object in a triangulated category $\mathscr{T}$ into a "left part" and a "right part", but there are crucial differences. For instance, a bounded t-structure gives rise to an abelian subcategory of $\mathscr{T}$, while a bounded co-t-structure gives rise to a so-called silting subcategory.

This brief survey will emphasise three philosophical points. First, bounded t-structures are akin to the canonical example of "soft" truncation of complexes in the derived category. Secondly, bounded co-t-structures are akin to the canonical example of "hard" truncation of complexes in the homotopy category.

Thirdly, a triangulated category $\mathscr{T}$ may be skewed towards t-structures or co-t-structures, in the sense that one type of structure is more useful than the other for studying $\mathscr{T}$. In particular, we think of derived categories as skewed towards t-structures, and of homotopy categories as skewed towards co-t-structures.

0. Introduction

The notion of co-t-structure in a triangulated category $\mathscr{T}$ was introduced independently by Bondarko and Pauksztello, see Definition 2.1. It is a mirror image of the classic notion of t-structure due to Beilinson, Bernstein, and Deligne, see Definition 1.1.

Given an object $t \in \mathscr{T}$, both types of structure give a way to divide t into a "left part" and a "right part". This is exemplified by dividing a complex of modules into a left and a right part by "soft" or "hard" truncation, see Figures 1 and 2.

Each case gives a triangle $u \to t \to v$. Crucially, for a t-structure, u is the left part of t and v the right part; for a co-t-structure, vice versa. This reversal leads to a number of differences, and the theories of t-structures and co-t-structures are far from being simple mirrors of each other. For instance, while a bounded t-structure induces an abelian subcategory of $\mathscr{T}$, a bounded co-t-structure induces a so-called silting subcategory; see Definition 2.3.

2010 *Mathematics Subject Classification.* Primary 18E30.

Key words and phrases. Abelian subcategory, Auslander–Reiten quiver, co-heart, co-t-structure, complex of modules, derived category, heart, homotopy category, silting mutation, silting quiver, silting subcategory, simple minded collection, t-structure, triangulated category, truncation.

$$
\begin{array}{ccccccccccccccc}
a & = & \cdots \to & M^{-2} & \to & M^{-1} & \to & \operatorname{Ker} d^0 & \longrightarrow & 0 & \longrightarrow & 0 & \longrightarrow & 0 & \to \cdots \\
\downarrow & & & \downarrow & & \downarrow & & \downarrow & & \downarrow & & \downarrow & & \downarrow & \\
t & = & \cdots \to & M^{-2} & \to & M^{-1} & \longrightarrow & M^0 & \xrightarrow{d^0} & M^1 & \longrightarrow & M^2 & \to & M^3 & \to \cdots \\
\downarrow & & & \downarrow & & \downarrow & & \downarrow & & \downarrow & & \downarrow & & \downarrow & \\
b & = & \cdots \longrightarrow & 0 & \longrightarrow & 0 & \longrightarrow & 0 & \longrightarrow & \operatorname{Coker} d^0 & \to & M^2 & \to & M^3 & \to \cdots
\end{array}
$$

FIGURE 1. A complex $t = \cdots \to M^{-1} \to M^0 \xrightarrow{d^0} M^1 \to M^2 \to \cdots$ has the soft truncations a and b. There is a triangle $a \to t \to b$ in the derived category. Each vertical module homomorphism is either the identity, a canonical inclusion or surjection, or zero.

In the decade since their inception, the theory of co-t-structures has grown considerably. This brief survey is far from encyclopedic, but has the goal of communicating three philosophical points:

(i) Bounded t-structures are akin to soft truncation in the bounded derived category of an abelian category.
(ii) Bounded co-t-structures are akin to hard truncation in the bounded homotopy category of an additive category.
(iii) A triangulated category $\mathscr{T}$ may be skewed in the direction of t-structures or co-t-structures, in the sense that one type of structure is more useful than the other for studying $\mathscr{T}$.

We explain these points in the next three subsections. Throughout, $\mathscr{T}$ is a triangulated category with Hom-spaces $\mathscr{T}(-,-)$ and suspension functor Σ (sometimes called the shift functor and denoted $t \mapsto t[1]$). If $\mathscr{M}$ is an abelian category then $\mathscr{D}^{\mathrm{b}}(\mathscr{M})$ is the derived category of bounded complexes over $\mathscr{M}$, and if $\mathscr{P}$ is an additive category then $\mathscr{K}^{\mathrm{b}}(\mathscr{P})$ is the homotopy category of bounded complexes over $\mathscr{P}$.

(i) Bounded t-structures are akin to soft truncation in the bounded derived category. Let R be a ring, $\mathscr{M} = \operatorname{Mod} R$ the category of left modules over R. Each complex $t \in \mathscr{D}^{\mathrm{b}}(\mathscr{M})$ has soft truncations a and b as shown in Figure 1, and there is a triangle $a \to t \to b$ in $\mathscr{D}^{\mathrm{b}}(\mathscr{M})$. The full subcategories $\mathscr{A}$ and $\mathscr{B}$ consisting of complexes isomorphic to such truncations satisfy Definition 1.1 and hence form a bounded t-structure, sometimes known as the standard t-structure, see Example 1.2.

Up to isomorphism, the heart $\mathscr{H} = \mathscr{A} \cap \Sigma\mathscr{B}$ consists of complexes concentrated in degree 0. The heart is an abelian subcategory of $\mathscr{D}^{\mathrm{b}}(\mathscr{M})$ which is equivalent to $\mathscr{M}$. Each $t \in \mathscr{D}^{\mathrm{b}}(\mathscr{M})$ permits a "tower" as shown in Proposition 3.1 due to Beilinson, Bernstein, and Deligne. This expresses how to build t from objects of the form $\Sigma^i h$ with $h \in \mathscr{H}$. The objects can be taken to be $\Sigma^i \mathrm{H}^{-i}(t)$ and then the tower shows how t is built from its cohomology modules, see Example 3.3.

In general, if $(\mathscr{A}, \mathscr{B})$ is a bounded t-structure in a triangulated category $\mathscr{T}$, then each $t \in \mathscr{T}$ still permits a triangle $a \to t \to b$ with $a \in \mathscr{A}$, $b \in \mathscr{B}$. The heart $\mathscr{H}$ is still an abelian subcategory of $\mathscr{T}$ by Theorem 1.3 due to Beilinson, Bernstein, and Deligne, and each $t \in \mathscr{T}$ still has the tower in Proposition 3.1.

$$
\begin{array}{ccccccccccccccc}
x & = & \cdots & \longrightarrow & 0 & \longrightarrow & 0 & \longrightarrow & 0 & \longrightarrow & P^0 & \longrightarrow & P^1 & \longrightarrow & P^2 \longrightarrow \cdots \\
\downarrow & & & & \downarrow & & \downarrow & & \downarrow & & \downarrow & & \downarrow & & \downarrow \\
t & = & \cdots & \longrightarrow & P^{-3} & \longrightarrow & P^{-2} & \longrightarrow & P^{-1} & \longrightarrow & P^0 & \longrightarrow & P^1 & \longrightarrow & P^2 \longrightarrow \cdots \\
\downarrow & & & & \downarrow & & \downarrow & & \downarrow & & \downarrow & & \downarrow & & \downarrow \\
y & = & \cdots & \longrightarrow & P^{-3} & \longrightarrow & P^{-2} & \longrightarrow & P^{-1} & \longrightarrow & 0 & \longrightarrow & 0 & \longrightarrow & 0 \longrightarrow \cdots
\end{array}
$$

FIGURE 2. A complex $t = \cdots \to P^{-2} \to P^{-1} \to P^0 \to P^1 \to \cdots$ has the hard truncations x and y. There is a triangle $x \to t \to y$ in the homotopy category. Each vertical module homomorphism is the identity or zero.

Working in such a setup is akin to working with soft truncation in $\mathscr{D}^{\mathrm{b}}(\mathscr{M})$.

(ii) Bounded co-t-structures are akin to hard truncation in the bounded homotopy category. Let R be a ring, $\mathscr{P} = \operatorname{Prj} R$ the category of projective left modules over R. Each complex $t \in \mathscr{K}^{\mathrm{b}}(\mathscr{P})$ has hard truncations x and y as shown in Figure 2, and there is a triangle $x \to t \to y$ in $\mathscr{K}^{\mathrm{b}}(\mathscr{P})$. The full subcategories $\mathscr{X}$ and $\mathscr{Y}$ consisting of complexes isomorphic to such truncations satisfy Definition 2.1 and hence form a bounded co-t-structure, sometimes known as the standard co-t-structure, see Example 2.2.

Up to isomorphism, the co-heart $\mathscr{C} = \mathscr{X} \cap \Sigma^{-1}\mathscr{Y}$ consists of complexes concentrated in degree 0. The co-heart is an additive subcategory of $\mathscr{K}^{\mathrm{b}}(\mathscr{P})$ which is equivalent to $\mathscr{P}$.

The co-heart is not in general abelian, but it does have the strong property of being a so-called silting subcategory of $\mathscr{K}^{\mathrm{b}}(\mathscr{P})$, see Definition 2.3 and Theorem 2.4 due to Mendoza Hernández et.al. Such subcategories are structurally important.

Each $t = \cdots \to P^{-3} \to P^{-2} \to P^{-1} \to P^0 \to P^1 \to P^2 \to \cdots$ in $\mathscr{K}^{\mathrm{b}}(\mathscr{P})$ permits a "tower" as shown in Proposition 3.2 due to Bondarko. This expresses how to build t from objects of the form $\Sigma^i c$ with $c \in \mathscr{C}$. The objects can be taken to be $\Sigma^i P^{-i}$ and then the tower shows how t is built from its constituent modules, see Example 3.4.

In general, if $(\mathscr{X}, \mathscr{Y})$ is a bounded co-t-structure in a triangulated category $\mathscr{T}$, then each $t \in \mathscr{T}$ still permits a triangle $x \to t \to y$ with $x \in \mathscr{X}$, $y \in \mathscr{Y}$. The co-heart $\mathscr{C}$ is still a silting subcategory of $\mathscr{T}$ by Theorem 2.4, and each $t \in \mathscr{T}$ still has the tower in Proposition 3.2.

Working in such a setup is akin to working with hard truncation in $\mathscr{K}^{\mathrm{b}}(\mathscr{P})$.

(iii) Triangulated categories skewed in the direction of t-structures or co-t-structures. When studying a given triangulated category, bounded co-t-structures may be more useful than bounded t-structures, simply because there are none of the latter. See Section 4 for a toy example.

Section 5 shows a subtler skewing phenomenon. Let Λ be a finite dimensional $\mathbb{C}$-algebra, $\mathscr{M} = \operatorname{mod} \Lambda$ the category of finite dimensional left modules over Λ, and $\mathscr{P} = \operatorname{prj} \Lambda$ the category of finite dimensional projective left modules over Λ. Theorem 5.2, due to König and Yang, shows a bijection between all bounded co-t-structures in $\mathscr{K}^{\mathrm{b}}(\mathscr{P})$ and the bounded t-structures in $\mathscr{D}^{\mathrm{b}}(\mathscr{M})$ whose hearts are length categories (these are, in a sense, the "algebraic" t-structures; for instance, they permit a nice mutation theory).

In itself, this does not imply that either category has more t-structures than co-t-structures or vice versa, but we think of it as indicating that when going from derived to homotopy categories, the role played by t-structures is taken over by co-t-structures.

Summing up, Sections 1 and 2 show the definitions of t-structures and co-t-structures, emphasising the similarities with soft and hard truncation of complexes of modules. Section 3 shows the "towers" whereby general objects can be built from objects in the (co-)heart. Sections 4 and 5 illustrate how a triangulated category may be skewed towards t-structures or co-t-structures.

Section 6 shows the silting mutation of Aihara and Iyama which mutates between bounded co-t-structures. It permits the definition of the so-called silting quiver which is a combinatorial picture of how the bounded co-t-structures fit inside a fixed triangulated category.

1. t-structures

The following definition was made in [**4**, def. 1.3.1].

DEFINITION 1.1 (Beilinson, Bernstein, and Deligne). A *t-structure* in the triangulated category $\mathscr{T}$ is a pair $(\mathscr{A},\mathscr{B})$ of full subcategories, closed under isomorphisms, direct sums, and direct summands, which satisfy the following conditions.

(i) $\Sigma\mathscr{A} \subseteq \mathscr{A}$ and $\Sigma^{-1}\mathscr{B} \subseteq \mathscr{B}$.

(ii) $\mathscr{T}(\mathscr{A},\mathscr{B}) = 0$.

(iii) For each object $t \in \mathscr{T}$ there is a triangle $a \to t \to b$ with $a \in \mathscr{A}$, $b \in \mathscr{B}$.

The *heart* is $\mathscr{H} = \mathscr{A} \cap \Sigma\mathscr{B}$.

The t-structure is called *bounded* if

$$\bigcup_{i\in\mathbb{Z}} \Sigma^i\mathscr{A} = \bigcup_{i\in\mathbb{Z}} \Sigma^i\mathscr{B} = \mathscr{T}. \qquad \square$$

The objects a and b in Definition 1.1(iii) depend functorially on t. The resulting functor $t \mapsto a$ is a right-adjoint to the inclusion $\mathscr{A} \hookrightarrow \mathscr{T}$. Similarly, $t \mapsto b$ is a left adjoint to the inclusion $\mathscr{B} \hookrightarrow \mathscr{T}$. See [**4**, prop. 1.3.3].

The following is the canonical example of a t-structure.

EXAMPLE 1.2 (The standard t-structure). Let R be a ring, $\mathscr{M} = \operatorname{Mod} R$ the category of left modules over R. Let $\mathscr{A}$ and $\mathscr{B}$ be the isomorphism closures in the bounded derived category $\mathscr{D}^{\mathrm{b}}(\mathscr{M})$ of the subsets

$$\{\cdots \to M^{-2} \to M^{-1} \to M^0 \longrightarrow 0 \longrightarrow 0 \longrightarrow 0 \longrightarrow \cdots \,|\, M^i \in \mathscr{M},\ M^i = 0 \text{ for } i \ll 0\},$$

$$\{\cdots \longrightarrow 0 \longrightarrow 0 \longrightarrow 0 \longrightarrow M^1 \to M^2 \to M^3 \to \cdots \,|\, M^i \in \mathscr{M},\ M^i = 0 \text{ for } i \gg 0\}.$$

We will show that if $\mathscr{A}$ and $\mathscr{B}$ are viewed as full subcategories, then $(\mathscr{A},\mathscr{B})$ is a bounded t-structure with heart $\mathscr{H}$ equivalent to $\mathscr{M}$.

It is easy to show

$$\begin{aligned} \mathscr{A} &= \{\, M \in \mathscr{D}^{\mathrm{b}}(\mathscr{M}) \mid \mathrm{H}^i(M) = 0 \text{ for } i \geqslant 1 \,\}, \\ \mathscr{B} &= \{\, M \in \mathscr{D}^{\mathrm{b}}(\mathscr{M}) \mid \mathrm{H}^i(M) = 0 \text{ for } i \leqslant 0 \,\}. \end{aligned} \tag{1.1}$$

This description clearly implies that $\mathscr{A}$ and $\mathscr{B}$ are closed under isomorphisms, direct sums, and direct summands.

We next check the conditions in Definition 1.1. Condition (i) is immediate. Condition (ii) requires that if objects

$$a = \cdots \to M^{-2} \to M^{-1} \to M^0 \longrightarrow 0 \longrightarrow 0 \longrightarrow 0 \longrightarrow \cdots,$$

$$b = \cdots \longrightarrow 0 \longrightarrow 0 \longrightarrow 0 \longrightarrow M^1 \to M^2 \to M^3 \to \cdots$$

in $\mathscr{A}$ and $\mathscr{B}$ are given, then $\operatorname{Hom}_{\mathscr{D}^{\mathrm{b}}(\mathscr{M})}(a,b) = 0$. This can be shown by noting that a has a projective resolution

$$p = \cdots \to P^{-2} \to P^{-1} \to P^0 \to 0 \to 0 \to 0 \to \cdots,$$

and that

$$\operatorname{Hom}_{\mathscr{D}^{\mathrm{b}}(\mathscr{M})}(a,b) = \operatorname{Hom}_{\mathscr{K}(\mathscr{M})}(p,b) = 0.$$

Here $\mathscr{K}(\mathscr{M})$ is the homotopy category of complexes over $\mathscr{M}$, and the second $=$ holds because in each degree, either the complex p or the complex b is zero. The triangle in Definition 1.1(iii) can be obtained by soft truncation of the object $t = \cdots \to M^{-1} \to M^0 \xrightarrow{d^0} M^1 \to M^2 \to \cdots$ in $\mathscr{D}^{\mathrm{b}}(\mathscr{M})$, see Figure 1 in the introduction.

It is immediate that the t-structure $(\mathscr{A}, \mathscr{B})$ is bounded.

Finally, it follows from Equation (1.1) that the heart $\mathscr{H} = \mathscr{A} \cap \Sigma\mathscr{B}$ is

$$\mathscr{H} = \{\, M \in \mathscr{D}^{\mathrm{b}}(\mathscr{M}) \mid \mathrm{H}^i(M) = 0 \text{ for } i \neq 0 \,\},$$

and this subcategory of $\mathscr{D}^{\mathrm{b}}(\mathscr{M})$ is equivalent to $\mathscr{M}$. □

The following pivotal result is one of the motivations for the definition of t-structures. It was proved in [**4**, thm. 1.3.6].

THEOREM 1.3 (Beilinson, Bernstein, and Deligne). *Let $(\mathscr{A}, \mathscr{B})$ be a t-structure in $\mathscr{T}$. Then the heart $\mathscr{H} = \mathscr{A} \cap \Sigma\mathscr{B}$ is an abelian subcategory of $\mathscr{T}$.*

EXAMPLE 1.4. Let $\Lambda = \mathbb{C}A_2$ be the path algebra of the quiver

$$A_2 \;=\; 1 \to 2 \tag{1.2}$$

and let $\mathscr{M} = \operatorname{mod} \Lambda$ be the category of finite dimensional left modules over Λ.

There is a bounded t-structure $(\mathscr{A}, \mathscr{B})$ in $\mathscr{D}^{\mathrm{b}}(\mathscr{M})$ defined by Equation (1.1); this is shown by the same method as in Example 1.2.

There are three isomorphism classes of indecomposable objects in $\mathscr{M}$ given by the following representations of the quiver A_2.

$$0 \to \mathbb{C} \;,\; \mathbb{C} \xrightarrow{\mathrm{id}} \mathbb{C} \;,\; \mathbb{C} \to 0$$

They induce isomorphism classes x_0, x_1, x_2 of indecomposable objects in $\mathscr{D}^{\mathrm{b}}(\mathscr{M})$, and we define further isomorphism classes recursively by $\Sigma x_i = x_{i+3}$ for $i \in \mathbb{Z}$.

The Auslander–Reiten quiver of $\mathscr{D}^{\mathrm{b}}(\mathscr{M})$ looks as follows, where $x_0, x_1, x_2, \ldots$ are in $\mathscr{A}$ and $\ldots, x_{-3}, x_{-2}, x_{-1}$ are in $\mathscr{B}$.

(1.3)

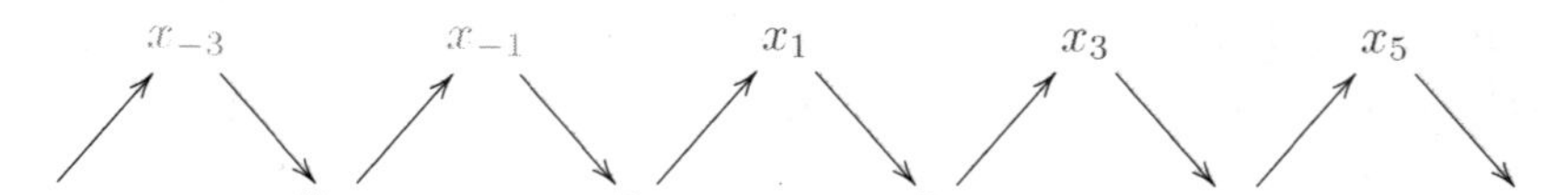

Note that if $t \in \mathscr{D}^{\mathrm{b}}(\mathscr{M})$ is indecomposable, then t is in $\mathscr{A}$ or in $\mathscr{B}$, so the triangle in Definition 1.1(iii) is trivial in the sense that it reads $t \to t \to 0$ or $0 \to t \to t$.

The heart $\mathscr{H} = \mathscr{A} \cap \Sigma\mathscr{B}$ is determined by

$$\mathscr{H} = \operatorname{add}(x_0 \oplus x_1 \oplus x_2). \qquad \square$$

2. Co-t-structures

The following definition was made in [**5**, def. 1.1.1] and [**13**, def. 2.4].

DEFINITION 2.1 (Bondarko and Pauksztello). A *co-t-structure* in $\mathscr{T}$ is a pair $(\mathscr{X}, \mathscr{Y})$ of full subcategories, closed under isomorphisms, direct sums, and direct summands, which satisfy the following conditions.

(i) $\Sigma^{-1}\mathscr{X} \subseteq \mathscr{X}$ and $\Sigma\mathscr{Y} \subseteq \mathscr{Y}$.
(ii) $\mathscr{T}(\mathscr{X}, \mathscr{Y}) = 0$.
(iii) For each object $t \in \mathscr{T}$ there is a triangle $x \to t \to y$ with $x \in \mathscr{X}$, $y \in \mathscr{Y}$.

The *co-heart* is $\mathscr{C} = \mathscr{X} \cap \Sigma^{-1}\mathscr{Y}$.

The co-t-structure is called *bounded* if

$$\bigcup_{i \in \mathbb{Z}} \Sigma^i \mathscr{X} = \bigcup_{i \in \mathbb{Z}} \Sigma^i \mathscr{Y} = \mathscr{T}. \qquad \square$$

In contrast to t-structures, the objects x and y in Definition 2.1(iii) do *not* in general depend functorially on t, see [**5**, rmk. 1.2.2].

The following is the canonical example of a co-t-structure.

EXAMPLE 2.2 (The standard co-t-structure). Let R be a ring, $\mathscr{P} = \operatorname{Prj} R$ the category of projective left-modules over R. Let $\mathscr{X}$ and $\mathscr{Y}$ be the isomorphism closures in the bounded homotopy category $\mathscr{K}^{\mathrm{b}}(\mathscr{P})$ of the subsets

$$(2.1)\qquad \begin{aligned} &\{\cdots \to 0 \longrightarrow 0 \longrightarrow 0 \to P^0 \rightarrowtail P^1 \rightarrowtail P^2 \rightarrowtail \cdots \mid P^i \in \mathscr{P},\ P^i = 0 \text{ for } i \gg 0\}, \\ &\{\cdots \rightarrowtail P^{-3} \rightarrowtail P^{-2} \rightarrowtail P^{-1} \to 0 \to 0 \to 0 \to \cdots \mid P^i \in \mathscr{P},\ P^i = 0 \text{ for } i \ll 0\}. \end{aligned}$$

We will show that if $\mathscr{X}$ and $\mathscr{Y}$ are viewed as full subcategories, then $(\mathscr{X}, \mathscr{Y})$ is a bounded co-t-structure with co-heart $\mathscr{C}$ equivalent to $\mathscr{P}$.

Recall that $\mathscr{X}$ and $\mathscr{Y}$ are required to be closed under isomorphisms, direct sums, and direct summands. The two former properties are immediate, and the latter follows from the results in [**14**, secs. 3 and 4].

We next check the conditions in Definition 2.1. Conditions (i) and (ii) are clear. The triangle in Definition 2.1(iii) can be obtained by hard truncation of the object $t = \cdots \to P^{-2} \to P^{-1} \to P^0 \to P^1 \to \cdots$ in $\mathscr{K}^{\mathrm{b}}(\mathscr{P})$, see Figure 2 in the introduction.

It is immediate that the co-t-structure $(\mathscr{X}, \mathscr{Y})$ is bounded.

Finally, it follows from [**14**, cor. 4.11] that the coheart $\mathscr{C} = \mathscr{X} \cap \Sigma^{-1}\mathscr{Y}$ is equivalent to $\mathscr{P}$. $\square$

The following definition was made in [**1**, def. 2.1]; see also [**9**].

DEFINITION 2.3. A *silting subcategory* $\mathscr{C}$ of $\mathscr{T}$ is a full subcategory, closed under isomorphisms, direct sums, and direct summands, which satisfies

(i) $\mathscr{T}(\mathscr{C}, \Sigma^{>0}\mathscr{C}) = 0$.
(ii) Each object in $\mathscr{T}$ can be obtained from $\mathscr{C}$ by taking finitely many (de)suspensions, triangles, and direct summands.

A *silting object* s of $\mathscr{T}$ is an object such that $\operatorname{add}(s)$ is a silting subcategory. $\square$

The following was proved in [**12**, cor. 5.9].

THEOREM 2.4 (Mendoza Hernández et.al.). *The map*

$$(\mathscr{X},\mathscr{Y}) \mapsto \mathscr{C} = \mathscr{X} \cap \Sigma^{-1}\mathscr{Y}$$

is a bijection between bounded co-t-structures and silting subcategories of $\mathscr{T}$.

REMARK 2.5. The inverse map sends a silting subcategory $\mathscr{C}$ to a pair $(\mathscr{X},\mathscr{Y})$ where $\mathscr{X}$ is the smallest full subcategory, closed under isomorphisms, direct sums, and direct summands, which is closed under Σ^{-1} and contains $\mathscr{C}$. Similarly, $\mathscr{Y}$ is the smallest full subcategory, closed under isomorphisms, direct sums, and direct summands, which is closed under Σ and contains $\Sigma\mathscr{C}$. □

EXAMPLE 2.6. We continue Example 1.4, so $\Lambda = \mathbb{C}A_2$ is the path algebra of the quiver A_2 from Equation (1.2) and $\mathscr{P} = \operatorname{prj} \Lambda$ is the category of finite dimensional projective left modules over Λ.

There is a bounded co-t-structure $(\mathscr{X},\mathscr{Y})$ in $\mathscr{K}^{\mathrm{b}}(\mathscr{P})$ where $\mathscr{X}$ and $\mathscr{Y}$ are the isomorphism closures in $\mathscr{K}^{\mathrm{b}}(\mathscr{P})$ of the subsets in Equation (2.1); this is shown by the same method as in Example 2.2.

Recall that Λ also has a bounded derived category $\mathscr{D}^{\mathrm{b}}(\mathscr{M})$, see Example 1.4. Since Λ has global dimension 1, the triangulated categories $\mathscr{K}^{\mathrm{b}}(\mathscr{P})$ and $\mathscr{D}^{\mathrm{b}}(\mathscr{M})$ are equivalent, so $\mathscr{K}^{\mathrm{b}}(\mathscr{P})$ has the Auslander–Reiten quiver shown in Equation (1.3). We draw the quiver again, noting that $\ldots$, x_{-1}, x_0, x_1 are in $\mathscr{X}$ while x_3, x_4, x_5, $\ldots$ are in $\mathscr{Y}$.

$$\begin{array}{ccccccccccccccccccccc} & & x_{-3} & & & & x_{-1} & & & & x_1 & & & & x_3 & & & & x_5 & & \\ & \nearrow & & \searrow & & \nearrow & & \searrow & & \nearrow & & \searrow & & \nearrow & & \searrow & & \nearrow & & \searrow & \\ \cdots & & & & x_{-2} & & & & x_0 & & & & x_2 & & & & x_4 & & & & \cdots \end{array} \tag{2.2}$$

Note that x_2 is neither in $\mathscr{X}$ nor $\mathscr{Y}$. Indeed, if we abuse notation to confuse isomorphism classes with individual objects, then we can set $t = x_2$ and the triangle in Definition 2.1(iii) becomes $x_1 \to x_2 \to x_3$.

The coheart $\mathscr{C} = \mathscr{X} \cap \Sigma^{-1}\mathscr{Y}$ is determined by

$$\mathscr{C} = \operatorname{add}(x_0 \oplus x_1).$$

Theorem 2.4 implies that $\mathscr{C}$ is a silting subcategory of $\mathscr{K}^{\mathrm{b}}(\mathscr{P})$. The corresponding isomorphism class of silting objects is $x_0 \oplus x_1$. □

3. Towers which build an arbitrary object from objects of the (co-)heart

The following two results were proved in [**4**, p. 34] and [**5**, prop. 1.5.6]. A wavy arrow $s \rightsquigarrow t$ denotes a morphism $s \to \Sigma t$.

PROPOSITION 3.1 (Beilinson, Bernstein, and Deligne). *Let $(\mathscr{A},\mathscr{B})$ be a bounded t-structure in $\mathscr{T}$ with heart $\mathscr{H}$. For each object $t \in \mathscr{T}$, there is an integer $n \geqslant 1$ and a diagram consisting of triangles,*

$$\begin{array}{ccccccccccc} 0 = t_0 & \longrightarrow & t_1 & \longrightarrow & t_2 & \longrightarrow \cdots \longrightarrow & t_{n-1} & \longrightarrow & t_n = t, \\ & \nwarrow\!\!\!\sim \quad \swarrow & & \nwarrow\!\!\!\sim \quad \swarrow & & & & \nwarrow\!\!\!\sim \quad \swarrow & \\ & \Sigma^{i_1}h_1 & & \Sigma^{i_2}h_2 & & & & \Sigma^{i_n}h_n & \end{array}$$

with $h_m \in \mathscr{H}$ for each m and $i_1 > i_2 > \cdots > i_n$.

PROPOSITION 3.2 (Bondarko). *Let $(\mathscr{X}, \mathscr{Y})$ be a bounded co-t-structure in $\mathscr{T}$ with co-heart $\mathscr{C}$. For each object $t \in \mathscr{T}$, there is an integer $n \geqslant 1$ and a diagram consisting of triangles,*

$$\begin{array}{ccccccccccc} 0 = t_0 & \longrightarrow & t_1 & \longrightarrow & t_2 & \longrightarrow & \cdots & \longrightarrow & t_{n-1} & \longrightarrow & t_n = t, \\ & \nwarrow \; \swarrow & & \nwarrow \; \swarrow & & & & & & \nwarrow \; \swarrow & \\ & \Sigma^{i_1} c_1 & & \Sigma^{i_2} c_2 & & & & & & \Sigma^{i_n} c_n & \end{array}$$

with $c_m \in \mathscr{C}$ for each m and $i_1 < i_2 < \cdots < i_n$.

EXAMPLE 3.3. Consider the t-structure in Example 1.2. If $t \in \mathscr{D}^{\mathrm{b}}(\mathscr{M})$ is given, then there is a diagram as in Proposition 3.1 where the objects $\Sigma^{i_m} h_m$ are of the form $\Sigma^i \mathrm{H}^{-i}(t)$. The diagram expresses that t can be built from its cohomology modules $\mathrm{H}^{-i}(t)$. □

EXAMPLE 3.4. Consider the co-t-structure in Example 2.2. If $t = \cdots \to P^{-2} \to P^{-1} \to P^0 \to P^1 \to \cdots$ in $\mathscr{K}^{\mathrm{b}}(\mathscr{P})$ is given, then there is a diagram as in Proposition 3.2 where the objects $\Sigma^{i_m} c_m$ are of the form $\Sigma^i P^{-i}$. The diagram expresses that t can be built from its constituent modules P^{-i}. □

4. Categories skewed towards t- or co-t-structures

In this section, d is a fixed integer.

DEFINITION 4.1. If $\mathscr{T}$ is $\mathbb{C}$-linear, then an object $s \in \mathscr{T}$ is called *d-spherical* if there is an isomorphism

$$\mathscr{T}(s, \Sigma^* s) \cong \mathbb{C}[X]/(X^2)$$

of graded algebras where $\deg X = d$. □

REMARK 4.2. If $d \neq 0$, then s is d-spherical if and only if there are isomorphisms of $\mathbb{C}$-vector spaces

$$\mathscr{T}(s, \Sigma^i s) \cong \begin{cases} \mathbb{C} & \text{for } i \in \{0, d\}, \\ 0 & \text{otherwise.} \end{cases}$$ □

The following was proved in [**10**, thm. 2.1]. Note that the result does hold for an arbitrary integer d; it is not necessary to assume d positive. Recall that suspension is sometimes called shift.

THEOREM 4.3 (Keller, Yang, and Zhou). *There is a triangulated category $\mathscr{S}_d$ which is algebraic, $\mathbb{C}$-linear with finite dimensional* Hom*-spaces, and contains a d-spherical object s such that each object in $\mathscr{S}_d$ can be obtained from s by taking finitely many (de)suspensions, triangles, and direct summands.*

Up to triangulated equivalence, $\mathscr{S}_d$ is unique.

The following was proved in [**7**, thm. A]. It shows that if $d \leqslant 0$, then bounded co-t-structures are more useful than bounded t-structures for the study of $\mathscr{S}_d$.

THEOREM 4.4 (Holm, J, and Yang). *If $d \leqslant 0$ then $\mathscr{S}_d$ has no bounded t-structures. It has one familiy of bounded co-t-structures, all of which are (de)suspensions of a canonical one.*

If $d \geqslant 1$ then $\mathscr{S}_d$ has no bounded co-t-structures. It has one familiy of bounded t-structures, all of which are (de)suspensions of a canonical one.

5. The bijections of König and Yang

DEFINITION 5.1 (Rickard). A *simple minded collection* in $\mathscr{T}$ is a set $\{t_1, \ldots, t_n\}$ of objects of $\mathscr{T}$ with the following properties.

(i) $\mathscr{T}(t_i, \Sigma^{<0}t_j) = 0$ for all i and j.
(ii) $\mathscr{T}(t_i, t_i)$ is a division ring for each i and $\mathscr{T}(t_i, t_j) = 0$ when $i \neq j$.
(iii) Each object in $\mathscr{T}$ can be obtained from t_1, …, t_n by taking finitely many (de)suspensions, triangles, and direct summands. □

Let Λ be a finite dimensional $\mathbb{C}$-algebra, $\mathscr{M} = \operatorname{mod} \Lambda$ the category of finite dimensional left modules over Λ, and $\mathscr{P} = \operatorname{prj} \Lambda$ the category of finite dimensional projective left modules over Λ. The following was proved in [**11**, thm. (6.1)]. We interpret it as indicating that the role of t-structures in derived categories is taken over by co-t-structures in homotopy categories.

THEOREM 5.2. *There are bijections between the following sets.*

(i) *Bounded t-structures in $\mathscr{D}^{\mathrm{b}}(\mathscr{M})$ whose hearts are length categories ("length category" means that each object has finite length).*
(ii) *Bounded co-t-structures in $\mathscr{K}^{\mathrm{b}}(\mathscr{P})$.*
(iii) *Isomorphism classes of simple minded collections in $\mathscr{D}^{\mathrm{b}}(\mathscr{M})$.*
(iv) *Isomorphism classes of basic silting objects in $\mathscr{K}^{\mathrm{b}}(\mathscr{P})$ (basic means no repeated indecomposable summands).*

REMARK 5.3. There is an extensive body of work on the bijections of Theorem 5.2 which predates [**11**], see [**1**], [**2**], [**3**], [**5**], [**6**], [**8**], [**9**], and [**12**]. The contributions of these papers to Theorem 5.2 are explained in the introduction to [**11**]. □

REMARK 5.4. The proof of Theorem 5.2 occupies a significant part of [**11**], and we only show how some of the bijections are defined.

(i) to (iii): Let $(\mathscr{A}, \mathscr{B})$ be a bounded t-structure in $\mathscr{D}^{\mathrm{b}}(\mathscr{M})$ whose heart $\mathscr{H} = \mathscr{A} \cap \Sigma\mathscr{B}$ is a length category. Take a simple object from each isomorphism class of simple objects in $\mathscr{H}$. This gives a simple minded system in $\mathscr{D}^{\mathrm{b}}(\mathscr{M})$, see [**11**, sec. 5.3].

(ii) to (iv): Let $(\mathscr{X}, \mathscr{Y})$ be a bounded co-t-structure in $\mathscr{K}^{\mathrm{b}}(\mathscr{P})$. The co-heart $\mathscr{C} = \mathscr{X} \cap \Sigma^{-1}\mathscr{Y}$ is a silting subcategory by Theorem 2.4, and there is a silting object s such that $\mathscr{C} = \operatorname{add}(s)$, see [**11**, sec. 5.2].

(i) to (ii): Let $(\mathscr{X}, \mathscr{Y})$ be a bounded co-t-structure in $\mathscr{K}^{\mathrm{b}}(\mathscr{P})$. Set

$$\mathscr{A} = \{a \in \mathscr{D}^{\mathrm{b}}(\mathscr{M}) \mid \operatorname{Hom}_{\mathscr{D}^{\mathrm{b}}(\mathscr{M})}(x, a) = 0 \text{ for each } x \in \mathscr{X}\},$$

$$\mathscr{B} = \{b \in \mathscr{D}^{\mathrm{b}}(\mathscr{M}) \mid \operatorname{Hom}_{\mathscr{D}^{\mathrm{b}}(\mathscr{M})}(y, b) = 0 \text{ for each } y \in \mathscr{Y}\}$$

and view these two sets as full subcategories of $\mathscr{D}^{\mathrm{b}}(\mathscr{M})$. Then $(\mathscr{A}, \mathscr{B})$ is a bounded t-structure in $\mathscr{D}^{\mathrm{b}}(\mathscr{M})$ with length heart, see [**11**, sec. 5.7]. □

6. The silting mutation of Aihara and Iyama

Silting mutation is an operation which changes one silting subcategory into another. By virtue of Theorem 2.4, it can be viewed as changing one bounded co-t-structure into another. This leads to the definition of the so-called silting quiver of the triangulated category $\mathscr{T}$ which shows how silting subcategories, and hence co-t-structures, fit together inside $\mathscr{T}$.

In this section, $\mathscr{T}$ is $\mathbb{C}$-linear with finite dimensional Hom-spaces and split idempotents, and $m = m_0 \oplus m_1$ is a basic silting object of $\mathscr{T}$ with m_0 indecomposable. The following definition and theorem are special cases of [**1**, sec. 2.4].

DEFINITION 6.1 (Aihara and Iyama). Let $r \xrightarrow{\rho} m_0 \xrightarrow{\lambda} \ell$ be a minimal right add(m_1)-approximation and a minimal left add(m_1)-approximation of m_0. Complete these morphisms to triangles

$$m_0^{\sim} \to r \xrightarrow{\rho} m_0 \quad , \quad m_0 \xrightarrow{\lambda} \ell \to m_0^{\dagger}$$

in $\mathscr{T}$ and set

$$\mu^-(m, m_1) = m_0^{\sim} \oplus m_1 \quad , \quad \mu^+(m, m_1) = m_0^{\dagger} \oplus m_1.$$

These are called *right* and *left silting mutations of* m. □

THEOREM 6.2 (Aihara and Iyama). (i) *The silting mutations* $\mu^-(m, m_1)$ *and* $\mu^+(m, m_1)$ *are basic silting objects of* $\mathscr{T}$.

(ii) *Right and left silting mutations are inverse in the sense that*

$$\mu^-\big(\mu^+(m, m_1), m_1\big) \cong m \quad , \quad \mu^+\big(\mu^-(m, m_1), m_1\big) \cong m.$$

The following is a special case of [**1**, def. 2.41].

DEFINITION 6.3 (Aihara and Iyama). The *silting quiver* of $\mathscr{T}$ has a vertex for each isomorphism class of basic silting objects of $\mathscr{T}$, and an arrow $[m] \to [m^*]$ if m^* is a left silting mutation of m, where square brackets denote isomorphism class. □

REMARK 6.4. The silting quiver gives a picture of how silting mutation moves from one silting object to another, hence from one silting subcategory to another. By virtue of Theorem 2.4, it gives a picture of how silting mutation moves from one bounded co-t-structure to another. □

EXAMPLE 6.5. We continue Example 2.6, so $\Lambda = \mathbb{C}A_2$ is the path algebra of the quiver A_2 from Equation (1.2). The bounded homotopy category $\mathscr{K}^{\mathrm{b}}(\mathscr{P})$ has the Auslander–Reiten quiver shown in Equation (2.2).

Recall from Example 2.6 that $\mathscr{K}^{\mathrm{b}}(\mathscr{P})$ has the isomorphism class of silting objects $x_0 \oplus x_1$. Indeed, $x_0 \oplus x_1$ is a vertex in the silting quiver of $\mathscr{K}^{\mathrm{b}}(\mathscr{P})$. The full quiver was determined in [**1**, exa. 2.45], see Figure 3. As the quiver shows, there is a left silting mutation of $x_0 \oplus x_1$ which gives $x_0 \oplus x_4$. The isomorphism classes of silting objects $x_0 \oplus x_1$ and $x_0 \oplus x_4$ give rise to silting subcategories add$(x_0 \oplus x_1)$ and add$(x_0 \oplus x_4)$ which, under the bijection of Theorem 2.4, correspond to two bounded co-t-structures. The first of these is shown on the AR quiver of $\mathscr{K}^{\mathrm{b}}(\mathscr{P})$ in Equation (2.2). The second co-t-structure $(\mathscr{X}', \mathscr{Y}')$ can be shown as follows, where $\ldots$, x_{-1}, x_0, x_1 and x_4 are in $\mathscr{X}'$ while x_3 and x_6, x_7, x_8, $\ldots$ are in $\mathscr{Y}'$.

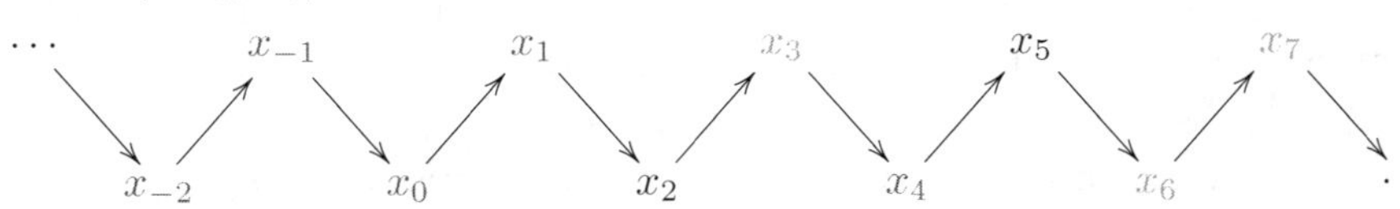

□

Acknowledgment

I thank Kiyoshi Igusa, Alex Martsinkovsky, and Gordana Todorov for inviting me to speak at the Maurice Auslander International Conference and to write this survey. I thank the anonymous referee for useful comments on an earlier version.

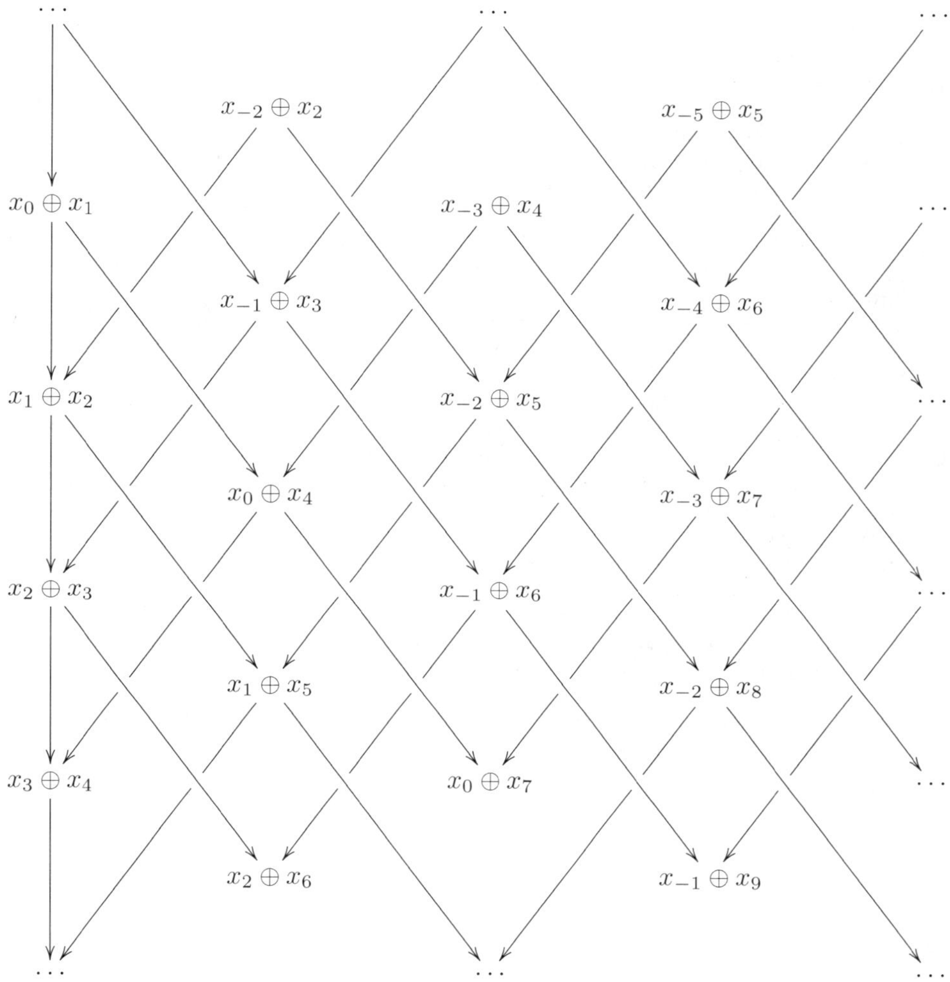

FIGURE 3. The silting quiver of the bounded homotopy category $\mathscr{K}^{\mathrm{b}}(\mathscr{P})$ where $\mathscr{P}$ is the category of finite dimensional projective modules over $\mathbb{C}A_2$.

References

[1] Takuma Aihara and Osamu Iyama, *Silting mutation in triangulated categories*, J. Lond. Math. Soc. (2) **85** (2012), no. 3, 633–668, DOI 10.1112/jlms/jdr055. MR2927802

[2] Salah Al-Nofayee, *Simple objects in the heart of a t-structure*, J. Pure Appl. Algebra **213** (2009), no. 1, 54–59, DOI 10.1016/j.jpaa.2008.05.014. MR2462984

[3] Ibrahim Assem, María José Souto Salorio, and Sonia Trepode, *Ext-projectives in suspended subcategories*, J. Pure Appl. Algebra **212** (2008), no. 2, 423–434, DOI 10.1016/j.jpaa.2007.06.002. MR2357343

[4] A. A. Beĭlinson, J. Bernstein, and P. Deligne, *Faisceaux pervers* (French), Analysis and topology on singular spaces, I (Luminy, 1981), Astérisque, vol. 100, Soc. Math. France, Paris, 1982, pp. 5–171. MR751966

[5] M. V. Bondarko, *Weight structures vs. t-structures; weight filtrations, spectral sequences, and complexes (for motives and in general)*, J. K-Theory **6** (2010), no. 3, 387–504, DOI 10.1017/is010012005jkt083. MR2746283

[6] Aslak Bakke Buan, Idun Reiten, and Hugh Thomas, *From m-clusters to m-noncrossing partitions via exceptional sequences*, Math. Z. **271** (2012), no. 3-4, 1117–1139, DOI 10.1007/s00209-011-0906-7. MR2945600

[7] Thorsten Holm, Peter Jørgensen, and Dong Yang, *Sparseness of t-structures and negative Calabi-Yau dimension in triangulated categories generated by a spherical object*, Bull. Lond. Math. Soc. **45** (2013), no. 1, 120–130, DOI 10.1112/blms/bds072. MR3033960

[8] Bernhard Keller and Pedro Nicolás, *Weight structures and simple dg modules for positive dg algebras*, Int. Math. Res. Not. IMRN **5** (2013), 1028–1078, DOI 10.1093/imrn/rns009. MR3031826

[9] B. Keller and D. Vossieck, *Aisles in derived categories*, Bull. Soc. Math. Belg. Sér. A **40** (1988), no. 2, 239–253. Deuxième Contact Franco-Belge en Algèbre (Faulx-les-Tombes, 1987). MR976638

[10] Bernhard Keller, Dong Yang, and Guodong Zhou, *The Hall algebra of a spherical object*, J. Lond. Math. Soc. (2) **80** (2009), no. 3, 771–784, DOI 10.1112/jlms/jdp054. MR2559128

[11] Steffen Koenig and Dong Yang, *Silting objects, simple-minded collections, t-structures and co-t-structures for finite-dimensional algebras*, Doc. Math. **19** (2014), 403–438. MR3178243

[12] Octavio Mendoza Hernández, Edith Corina Sáenz Valadez, Valente Santiago Vargas, and María José Souto Salorio, *Auslander-Buchweitz context and co-t-structures*, Appl. Categ. Structures **21** (2013), no. 5, 417–440, DOI 10.1007/s10485-011-9271-2. MR3097052

[13] David Pauksztello, *Compact corigid objects in triangulated categories and co-t-structures*, Cent. Eur. J. Math. **6** (2008), no. 1, 25–42, DOI 10.2478/s11533-008-0003-2. MR2379950

[14] O. M. Schnürer, Homotopy categories and idempotent completeness, weight structures and weight complex functors, preprint (2011). `math.CT/1107.1227v1`.

School of Mathematics, Statistics and Physics, Newcastle University, Newcastle upon Tyne NE1 7RU, United Kingdom

Email address: `peter.jorgensen@ncl.ac.uk`

URL: `http://www.staff.ncl.ac.uk/peter.jorgensen`

Contemporary Mathematics
Volume **716**, 2018
https://doi.org/10.1090/conm/716/14426

Modules as exact functors

Mike Prest

CONTENTS

1. Introduction

This mostly expository paper is about modules, by which I mean the following. A **module** is an exact functor from a small abelian category to the category, **Ab**, of abelian groups.

An **exact functor** is, recall, one which takes exact sequences to exact sequences.

Certainly, this is not the usual definition of a module. But it is a good alternative, in some senses better. And it gives a different perspective on modules. I will explain how this definition arises and we will see where it leads us. Of course, I should, and will, show that this definition is equivalent to the usual one.

The first step towards this definition of module is into a functor-category-theoretic framework, where we allow rings to be multi-sorted and where modules are just additive functors. One of the aims of the paper is to show that this all is very concrete and natural, especially through explicit computation of examples. Another is to show how model theory gives a convenient language for recognising and handling the multi-sorted structures that arise in this framework. Yet another aim is to reveal implicit definable structure and show how to make this explicit,

The author acknowledges the support of EPSRC grant EP/K022490/1 during the development of this work.

indeed, that is really the main step in reaching the rather strange-looking definition of "module" above.

I give most definitions, but the paper is really aimed at a reader who has at least a little background on rings, modules, quivers and representations and has seen some very basic category theory.

The "narrative" part of the paper proceeds as follows: modules as additive functors; the free abelian category generated by a ring (as a functor category and *via* systems of linear equations); the abelian category associated to a module (a quotient = localisation of a free abelian category); definable categories (additive categories which include, but are more general than, module categories) and the definable category generated by a module.

Most of the rest of the paper is filling-in/linking/examples/context, as well as explanation of concepts which are less likely to be familiar to all readers: Serre-localisation; many-sorted structures; definable structure/imaginary sorts; the idea of interpretation from model theory; interpretations regarded as functors; pp formulas; languages for modules. It is my hope that the reader will be able to dip in and out of the various sections and subsections.[1]

I'll start with an example which illustrates many of the main points and which we will gradually develop.

1.1. Finding modules within modules: an example. This is an example of how essentially the same module structure may be found over very different rings. By "essentially the same" I mean something which includes, but is much more general than, Morita equivalence between categories of modules.

Consider the quiver $\widetilde{A_1}$: $1 \overset{\alpha}{\underset{\beta}{\rightrightarrows}} 2$. Given a field K, a K-representation of $\widetilde{A_1}$ is given by two vector spaces V_1 and V_2 (one for each vertex of the quiver) and two K-linear maps t_α, t_β from V_1 to V_2 (corresponding to the arrows of the quiver) $V_1 \overset{t_\alpha}{\underset{t_\beta}{\rightrightarrows}} V_2$. One way to build such representations is to take a $K[T]$-module M - that is a K-vectorspace with a distinguished endomorphism (the action of T) and, from it, define the representation with $V_1 = V_2 = M$, $t_\alpha = 1_M$ and $t_\beta = T$ (meaning multiplication-by-T) $M \overset{1}{\underset{T}{\rightrightarrows}} M$. Note that t_α is an isomorphism.

Recall that the category of K-representations of a quiver, such as $\widetilde{A_1}$, is equivalent to the category of modules over the **path** algebra $K\widetilde{A_1}$. This algebra has a K-vectorspace basis consisting of all the paths in the quiver (including a lazy path at each vertex), in this case e_1, e_2, α, β. The multiplication is determined by defining the product of two paths to be their composition, written right to left, if defined, otherwise 0. Thus, for example, $e_1^2 = e_1$, $\alpha e_1 = \alpha = e_2\alpha$.

From a representation V of $\widetilde{A_1}$ we define the module over $K\widetilde{A_1}$ which has underlying set $V_1 \oplus V_2$ and with actions given by $e_1(v_1, v_2) = (v_1, 0)$, $e_2(v_1, v_2) = (0, v_2)$, $\alpha(v_1, v_2) = (0, t_\alpha v_1)$, $\beta(v_1, v_2) = (0, t_\beta v_1)$.

[1]Thanks to Mike Bushell, Lorna Gregory, Harry Gulliver and Sam Dean for comments on a preliminary version of this paper.

Conversely, from a $K\widetilde{A_1}$-module N we define the representation with vector spaces e_1N, e_2N (noting that $N = e_1N \oplus e_2N$ as K-vectorspaces) and with maps t_α, t_β given by α and β composed with the relevant injections and projections from and to these components. Combining with the process from the first paragraph, we have a map from $K[T]$-modules to $K\widetilde{A_1}$-modules given by taking M to $M \oplus M$ with the induced action of the path algebra $K\widetilde{A_1}$.

$$K[T]\text{-modules} \to K\text{-representations of } \widetilde{A_1} \to K\widetilde{A_1}\text{-modules}.$$

In the other direction, suppose we are given a $K\widetilde{A_1}$-module N where α restricts to a bijection from e_1N to e_2N, equivalently a representation V of $\widetilde{A_1}$ in which t_α is an isomorphism. Then the vector space e_1N, equipped with the endomorphism defined in terms of representations as $t_\alpha^{-1}t_\beta$, is a $K[T]$-module from which the original $K\widetilde{A_1}$-module may be recovered.

If one of these processes is applied and then the other, then the result is isomorphic to the module we began with. These processes extend to morphisms and are functorial. They give us an equivalence between the category of $K[T]$-modules and a large subcategory of the category of $K\widetilde{A_1}$-modules.

This is an equivalence which, like Morita equivalence, is very concrete and reversible: we start with a module, we construct a new module over a different ring and there is a construction in the other direction which recovers the original module. So, in some sense, the two modules - one over $K[T]$, the other over $K\widetilde{A_1}$ - are "implicitly the same" structure. We will make this explicit and precise. In particular, these will turn out to give isomorphic modules in the sense defined at the start of this paper.

2. Modules

By "the usual definition" of a module I mean: a **module** is an additive functor from a skeletally small preadditive category to **Ab**. Perhaps that is stretching the meaning of "usual" since it is a generalisation, but a mild one ([**26**]), of the first definition one meets - the case where the preadditive category has just one object. Let us start there, with a ring R which has an identity 1.

Rings as categories: We view R as a category which has a single object, $*_R$ say, with the elements of R as the arrows from $*_R$ to $*_R$. The addition in R gives an additive structure on this category, that is, the addition on the set $(*_R, *_R)$ of arrows, and the multiplication in R gives composition of morphisms (to the left). Thus we regard R as a **preadditive category** - a category where each hom set has an abelian group structure and where composition is bilinear.

EXAMPLE 2.1. Let K be a field and take R to be the ring $K[T]$ of polynomials in one variable with coefficients from K. In the corresponding 1-object preadditive category, $1 \in K[T]$ is the identity morphism $\mathrm{id}_{*_{K[T]}}$, T is an endomorphism of $*_{K[T]}$, T^2 is the composition of T with itself, indeed every morphism in $(*_{K[T]}, *_{K[T]})$ is a K-linear combination of 1 and powers of T. Note that, since $K[T]$ is a K-algebra, $(*_{K[T]}, *_{K[T]})$ is actually a K-vector space.

Modules as functors: An additive functor[2] M from R, regarded as a 1-object category, to **Ab** is given by an abelian group $M(*_R)$ and, for each $r \in R$, an endomorphism, which we call multiplication by r, of $M(*_R)$. One may check that the condition that M is an additive functor translates to this data defining a left R-module structure on $M(*_R)$. Moreover, a natural transformation from the functor M to another additive functor N is a morphism of abelian groups $M(*_R) \to N(*_R)$ which commutes with multiplication by r for every $r \in R$ - that is, it is an R-linear map from M to N. In this way, the category $(R, \mathbf{Ab})$ of (covariant) additive functors from R to **Ab** is precisely the category R-Mod of left R-modules. Similarly, the category Mod-R of right R-modules is the category $(R^{\mathrm{op}}, \mathbf{Ab})$ of contravariant functors. (All functors in this paper will be additive, so let us drop that adjective, except when used for emphasis, from now on.)

Example 2.2. Continuing Example 2.1, to give a functor M from $K[T]$, considered as a category, to **Ab**, it is enough to specify the abelian group $M(*_{K[T]})$, the K-vectorspace structure on $M(*_{K[T]})$ given by the scalar multiplications $M(\lambda)$ for $\lambda \in k$ and the action, $M(T)$, of T on $M(*_{K[T]})$. Which is exactly the data of a $K[T]$-module.

A general preadditive category $\mathcal{R}$ will have more objects but we assume that, up to isomorphism, there is just a set of these - that is, we assume that $\mathcal{R}$ is **skeletally small** ($\mathcal{R}$ is **small** if it has just a set of objects). As in the 1-object case, an (additive) functor from $\mathcal{R}$ to **Ab** may be referred to as a (left) $\mathcal{R}$-module. That is, $\mathcal{R}$-Mod $\simeq (\mathcal{R}, \mathbf{Ab})$ and Mod-$\mathcal{R} \simeq (\mathcal{R}^{\mathrm{op}}, \mathbf{Ab})$.

Example 2.3. Let K be a field and consider the preadditive (in fact, K-linear, see below) category $\mathcal{R}$ generated by two objects $*_1, *_2$ and a single arrow $\alpha : *_1 \to *_2$ from one to the other. Thus, $(*_1, *_1) = K.\mathrm{id}_{*_1}$, $(*_1, *_2) = K.\alpha$ and $(*_2, *_2) = K.\mathrm{id}_{*_2}$. This is the path category (see below) of the quiver A_2 which is the directed graph $1 \xrightarrow{\alpha} 2$.

An $\mathcal{R}$-module M is given by two abelian groups (indeed, as in 2.1, these will be K-vector spaces) $M(*_1)$, $M(*_2)$ and a (K-)linear map $M(\alpha) : M(*_1) \to M(*_2)$. That is, an $\mathcal{R}$-module is a K-representation of the quiver A_2. And a natural transformation from one module M to another N is precisely a morphism between the corresponding representations, that is, a pair of maps $f_1 : M(*_1) \to N(*_1)$, $f_2 : M(*_2) \to N(*_2)$, such that, for every $\lambda \in k$, the following diagram commutes.

$$\begin{array}{ccc} M(*_1) & \xrightarrow{M(\lambda\alpha)} & M(*_2) \\ \downarrow{\scriptstyle f_1} & & \downarrow{\scriptstyle f_2} \\ N(*_1) & \xrightarrow{N(\lambda\alpha)} & N(*_2) \end{array}$$

In this paper we will use the concept of sorts from model theory, see Section 9; this is well-illustrated in this example since any $\mathcal{R}$-module M is naturally a 2-sorted structure, the sorts being $M(*_1)$ and $M(*_2)$. Replacing M by the corresponding module over the path algebra would be to replace this two-sorted structure by a one-sorted structure (based on the single sort $M(*_1) \oplus M(*_2)$).

Like any preadditive category $\mathcal{R}$ with only finitely many objects (up to isomorphism), the category in the example above is essentially equivalent to a ring

[2]A functor F between preadditive categories is **additive** if $F(f+g) = Ff + Fg$ whenever f and g have the same domain and the same codomain.

R (namely the path algebra) in the sense that the functors from $\mathcal{R}$ to $\mathbf{Ab}$ are equivalent to the R-modules as usually defined. But preadditive categories include examples such as the category $\mathcal{R} = R$-mod of finitely presented modules over any ring (or small preadditive category) R. Then $\mathcal{R}$-Mod, the category $(R\text{-mod}, \mathbf{Ab})$ of functors from R-mod to $\mathbf{Ab}$, is almost never equivalent to the category of modules over any 1-object ring (exceptions being von Neumann regular rings and rings of finite representation type).

Functor categories: If $\mathcal{C}$, $\mathcal{D}$ are preadditive categories with $\mathcal{C}$ skeletally small (to avoid set-theoretic issues), we use the notation $(\mathcal{C}, \mathcal{D})$ for the category of additive functors from $\mathcal{C}$ to $\mathcal{D}$. The objects of this category are the functors; the arrows are the natural transformations between functors. This is itself a preadditive category: if $\sigma, \tau : F \to G$ are natural transformations from F to G, where $F, G : \mathcal{C} \to \mathcal{D}$ are additive functors, then we define $\sigma + \tau : F \to G$ to have, for its component $(\sigma + \tau)_C$ at $C \in \mathcal{C}$, the sum $\sigma_C + \tau_C$ of the components of σ and τ at C. We will always use this notation with $\mathcal{D}$ an **additive** category, meaning that $\mathcal{D}$ has a zero object as well as all finite direct sums of objects; in which case $(\mathcal{C}, \mathcal{D})$ also will be additive.

EXAMPLE 2.4. If $\mathcal{T}$ is a compactly generated triangulated category then, setting $\mathcal{T}^{c}$ to be the full subcategory of compact objects (this is skeletally small and additive), we may consider the functor category Mod-$\mathcal{T}^{c} = ((\mathcal{T}^{c})^{\mathrm{op}}, \mathbf{Ab})$ and the restricted Yoneda functor $\mathcal{T} \to$ Mod-$\mathcal{T}^{c}$ given by $T \mapsto (-, T) \upharpoonright_{\mathcal{T}^{c}}$ on objects. This is an abelian (Grothendieck) approximation of $\mathcal{T}$ and using it allows techniques and results from abelian categories to be applied in this non-abelian context (see, e.g., [**22**], [**3**]).

K-linear categories: A general point seen in examples above is that, if K is a field, or just a commutative ring, and $\mathcal{R}$ is a **K-linear** category, meaning that each morphism group (a, b) is a K-module with composition being (K, K)-bilinear, then every additive functor from $\mathcal{R}$ to $\mathbf{Ab}$ is K-linear. That is, directly from the definition of a functor, each group $M(a)$ has an induced K-module structure and each map $M(f) : M(a) \to M(b)$, where $f \in (a, b)$ is K-linear. Since every additive functor to $\mathbf{Ab}$ will automatically be a K-linear map to a K-module, the category $(\mathcal{R}, \mathbf{Ab})$ is, therefore, naturally equivalent to the category $(\mathcal{R}, K\text{-Mod})$ of K-linear functors to K-modules and so, in the definition of "$\mathcal{R}$-module", we may take the codomain category to be K-Mod.

Path categories: Generalising 2.3, let Q be a quiver and let K be a commutative ring. We define the **K-path category** $K\overrightarrow{Q}$ of Q to have, for objects, the vertices of Q and, given vertices s, t of Q, the set (s, t) of arrows is defined to be the free K-module on the set of **paths** (concatenations of arrows) from s to t. Composition is given by concatenation of paths when the end of the one path is the start of the next, 0 when they don't connect, as in the usual definition of the path algebra of a quiver. This path category is a small preadditive category, the modules over which are the K-representations of Q. If the number of vertices of Q is finite then $K\overrightarrow{Q}$ is morally equivalent to the path algebra KQ, which is a ring in the usual (one-object) sense, though these are not literally equivalent as categories unless Q has only one vertex. What are equivalent are their **idempotent-splitting=karoubian** additive completions. To get that, we add a 0 object and form finite direct sums of objects (that gives us the additive category generated by the initial preadditive category), then we add kernels and cokernels of idempotent endomorphisms.

One may check that none of these operations (additive completion, idempotent-splitting completion) on a preadditive category changes its category of modules, at least, not up to equivalence.

EXAMPLE 2.5. (additive and idempotent-splitting completion) For instance, if our initial preadditive category is a ring R, regarded as a 1-object category, then the additive category we form from this has finite (including empty) formal powers of the single object $*_R$ of R for its objects. The morphisms from $(*_R)^n$ to $(*_R)^m$ will be the $m \times n$ matrices of maps from $*_R$ to $*_R$, that is $m \times n$ matrices over R (the empty power is the zero object, with only the zero morphisms to and from it). This is equivalent, through the Yoneda embedding, to the opposite of the category of finitely generated free R-modules. On the other hand, the idempotent-splitting completion is equivalent to the category of finitely generated projective modules.

EXAMPLE 2.6. Consider the quiver A_3: $1 \xrightarrow{\alpha} 2 \xrightarrow{\beta} 3$. The corresponding K-path category has the three objects, labelled say 1, 2, 3, and 1-dimensional morphism spaces between each pair (i, j) of vertices with $i \leq j$, has basis the identity (if the vertices are equal), α (from 1 to 2), β (from 2 to 3) and $\beta\alpha$ (from 1 to 3). All other morphism spaces are zero. The corresponding path algebra can be represented as the ring of 3×3 lower-triangular matrices over K.

EXAMPLE 2.7. In the case of the quiver $\widetilde{A_1}$ $1 \underset{\beta}{\overset{\alpha}{\rightrightarrows}} 2$ the hom space from 1 to 2 is 2-dimensional. The path algebra can be represented as $\begin{pmatrix} Ke_1 & 0 \\ K\alpha \oplus K\beta & Ke_2 \end{pmatrix}$.

3. The free abelian category

In what way, then, is a module in the sense defined in Section 2 equivalent to an exact functor on a small abelian category? The equivalence uses Freyd's free abelian category.

THEOREM 3.1. *([**15**, 4.1]) Let $\mathcal{R}$ be any skeletally small preadditive category. There is a full additive embedding j of $\mathcal{R}$ into a skeletally small abelian category* $\mathrm{Ab}(\mathcal{R})$ *defined, up to equivalence, by the property that, given any additive functor M from $\mathcal{R}$ to* **Ab**, *there is a unique-to-equivalence extension through j to an exact functor* $\widetilde{M} : \mathrm{Ab}(\mathcal{R}) \to \mathbf{Ab}$.

$$\begin{array}{ccc} \mathcal{R} & \xrightarrow{j} & \mathrm{Ab}(\mathcal{R}) \\ {\scriptstyle M}\downarrow & \swarrow {\scriptstyle \widetilde{M}} & \\ \mathbf{Ab} & & \end{array}$$

We refer to $\mathrm{Ab}(\mathcal{R})$ as the **free abelian category on** $\mathcal{R}$.

Before continuing, we recall that if $G : \mathcal{A} \to \mathcal{B}$ is an exact functor between abelian categories then the kernel of G, $\mathcal{S} = \{A \in \mathcal{A} : GA = 0\}$, is a **Serre subcategory** of $\mathcal{A}$: that is, if $0 \to A \to B \to C \to 0$ is an exact sequence in $\mathcal{A}$ then $B \in \mathcal{S}$ iff $A, C \in \mathcal{S}$. Conversely, if $\mathcal{S}$ is a Serre subcategory of $\mathcal{A}$ then there is a **quotient category** $\mathcal{A}/\mathcal{S}$, which is abelian, and an exact functor $\mathcal{A} \to \mathcal{A}/\mathcal{S}$

which has kernel $\mathcal{S}$ and which is universal for exact functors from $\mathcal{A}$ with kernel containing $\mathcal{S}$. We give a specific computation of this at Example 3.2.

3.1. The abelian category associated to a module. We can do better, obtaining a unique (to equivalence) abelian category associated to a module M. This is, in a very strong sense, an invariant of the module; compare with the ring associated to a module which, one may argue, is defined at most up to Morita equivalence.

The kernel of $\widetilde{M}$, $\mathcal{S}_M = \{F \in \mathrm{Ab}(\mathcal{R}) : \widetilde{M}F = 0\}$, is a Serre subcategory of $\mathrm{Ab}(\mathcal{R})$ and there is a factorisation of $\widetilde{M}$ as a composition of exact functors through the quotient category $\mathcal{A}(M) = \mathrm{Ab}(\mathcal{R})/\mathcal{S}_M$.

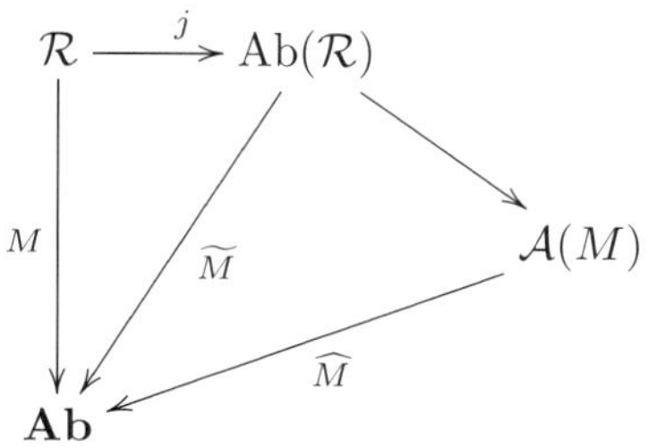

This category $\mathcal{A}(M)$ is the unique-to-equivalence skeletally small abelian category canonically associated to the module M. I will give an interpretation in Sections 4.4 and 9 of $\widetilde{M}$ and $\widehat{M}$ which explains the sense in which the modules seen in Section 1.1 are "essentially the same" (in particular they do have the same associated abelian category $\mathcal{A}(M)$ up to equivalence).

That is how we can produce, from a "module" in the usual sense, an exact functor on a small abelian category. To go in the other direction, we start with a skeletally small abelian category $\mathcal{A}$ and an exact functor $G : \mathcal{A} \to \mathbf{Ab}$. We use the fact (e.g. [**36**, 2.18]) that every such abelian category $\mathcal{A}$ has the form (non-uniquely) $\mathrm{Ab}(\mathcal{R})/\mathcal{S}$ for some skeletally small preadditive $\mathcal{R}$ and some Serre subcategory $\mathcal{S}$ of $\mathrm{Ab}(\mathcal{R})$. Since the localisation functor $\mathrm{Ab}(\mathcal{R}) \to \mathrm{Ab}(\mathcal{R})/\mathcal{S}$ is exact, composition with G gives an exact functor on $\mathrm{Ab}(\mathcal{R})$ and this exact functor restricts to an additive functor, M say, on $\mathcal{R}$ - that is, a module in the usual sense. Furthermore, to that additive functor M on $\mathcal{R}$ we can apply the first process, obtaining $\widehat{M} : \mathcal{A}(M) \to \mathbf{Ab}$ and one may check that this is equivalent to the exact functor, G' say, induced by G on $\mathcal{A}/\ker(G)$. That is, there is an equivalence of categories $\theta : \mathcal{A}(M) \to \mathcal{A}/\ker(G)$ such that $G'\theta$ is equivalent to $\widehat{M}$.

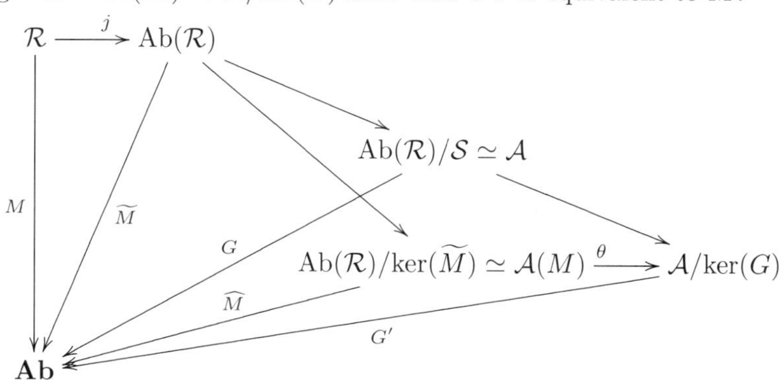

Thus we have the equivalence of the two definitions.

It is rather more interesting to try to compose these processes in the other direction since it raises the question: if we start with an additive functor $M : \mathcal{R} \to$ **Ab**, then can we recover M and $\mathcal{R}$ from $\widehat{M} : \mathcal{A}(M) \to \mathbf{Ab}$? The answer is that we cannot, not even up to Morita equivalence, the point being that, under our proposed definition, a module may be given in many equivalent but very different ways. That is, given modules $M : \mathcal{R} \to \mathbf{Ab}$ and $M_1 : \mathcal{R}_1 \to \mathbf{Ab}$, if there is an equivalence $H : \mathcal{A}(M) \to \mathcal{A}(M_1)$ such that $\widehat{M_1}H$ and $\widehat{M}$ are equivalent functors to **Ab**, then we regard M and M_1 as **the same module**.

3.2. The free abelian category as a functor category. The free abelian category on $\mathcal{R}$, constructed originally in [**15**], has, following [**2**], a nice realisation as the category of finitely presented functors on finitely presented $\mathcal{R}$-modules: $\mathrm{Ab}(\mathcal{R}) = (\mathcal{R}\text{-mod}, \mathbf{Ab})^{\mathrm{fp}}$. The embedding j of $\mathcal{R}$ into $(\mathcal{R}\text{-mod}, \mathbf{Ab})^{\mathrm{fp}}$ is the composition of two Yoneda embeddings, first taking each object p of $\mathcal{R}$ to the representable functor = projective left $\mathcal{R}$-module $(p, -)$ (but thought of as an object of $(\mathcal{R}\text{-mod})^{\mathrm{op}}$) and then taking that object to the corresponding representable functor. The net result is $p \mapsto ((p, -), -)$ with an obvious definition on morphisms. If, for example, $\mathcal{R}$ is just a ring R with one object then this object $*$ is first sent to $(*, -)$, which is just the projective module ${}_RR$, then is sent on to the forgetful functor $({}_RR, -)$ from R-mod to **Ab**.

The extension process from $\mathcal{R}$-modules $M : \mathcal{R} \to \mathbf{Ab}$ to exact functors on $\mathrm{Ab}(\mathcal{R}) = (\mathcal{R}\text{-mod}, \mathbf{Ab})^{\mathrm{fp}}$ also is a two-step process which is described in detail elsewhere (e.g. [**34**, §4]), so I don't repeat it here. I do, however, give details of one, very simple, example next. After that, I describe another way of viewing the extension process, using a different incarnation of the free abelian category.

EXAMPLE 3.2. Take $\mathcal{R}$ to be the path category of the quiver A_2 (Example 2.3 but we now write the objects differently). First consider the functor $\mathcal{R}^{\mathrm{op}} \to (\mathcal{R}\text{-mod})$ (for convenience, I present it this way rather than $\mathcal{R} \to (\mathcal{R}\text{-mod})^{\mathrm{op}}$) given by taking the object i $(i = 1, 2)$ to the representable functor/projective module $(i, -) = P_i$. Explicitly these are given as follows, where we use e_i for the identity map at i and α for the arrow from 1 to 2.

$(1, -)$: $(1,1) = Ke_1$, $(1,2) = K\alpha$, $(1,1) \xrightarrow{(1,\alpha)} (1,2) : e_1 \mapsto \alpha$.

$(2, -)$: $(2,1) = 0$, $(2,2) = Ke_2$

$(\alpha, -) : (2, -) \to (1, -)$: has components $(\alpha, -)_1 = 0 : (2,1) = 0 \to (1,1) = Ke_1$ and $(\alpha, -)_2 : (2,2) = Ke_2 \to (1,2) = K\alpha : e_2 \mapsto \alpha$.

We have the exact sequence

$$0 \to (2, -) \xrightarrow{(\alpha,-)} (1, -) \to (1, -)/\mathrm{im}(\alpha, -) \to 0$$

in $\mathcal{R}$-mod, where the cokernel may be identified as the simple left module S_1, given by $S_1(1) = Ke_1$, $S_1(2) = 0$. We write the sequence in more familiar form: $0 \to P_2 \xrightarrow{i} P_1 \xrightarrow{\pi} S_1 \to 0$.

We know that A_2 has finite representation type and that this is the complete list of indecomposable modules, with every module being a direct sum of copies of these three modules. In general, we would have to consider cokernels of arbitrary maps between direct sums of representable functors in order to find all the indecomposables.

That, the construction of $\mathcal{R}$-mod, was the first stage of the construction of the free abelian category $\mathrm{Ab}(\mathcal{R})$. Now we just repeat the process, taking each finitely presented left $\mathcal{R}$-module M to the representable functor $(M,-):\mathcal{R}\text{-mod}\to\mathbf{Ab}$ and forming cokernels to obtain all the finitely presented functors from $\mathcal{R}$-mod to $\mathbf{Ab}$.

Consider, for instance, the exact sequence $0\to S_1\to P_1\to P_2\to 0$ in $(\mathcal{R}\text{-mod})^{\mathrm{op}}$. This goes to the, non-exact, sequence $0\to(S_1,-)\xrightarrow{(\pi,-)}(P_1,-)\xrightarrow{(i,-)}(P_2,-)\to 0$ in $(\mathcal{R}\text{-mod},\mathbf{Ab})^{\mathrm{fp}}$. We can obtain a couple of indecomposable, indeed simple, functors from this sequence. First, the cokernel $(P_1,-)/\mathrm{im}(\pi,-)$ of $(\pi,-)$ is the functor given by $P_1\mapsto K$, $P_2,S_1\mapsto 0$; second the cokernel of $(i,-)$, which is the functor given on objects by $P_2\mapsto K$, $P_1,S_1\mapsto 0$.

In fact, that's it - there are no more that these five indecomposable finitely presented functors from $\mathcal{R}$-mod to $\mathbf{Ab}$. So, in this particular case, we can completely decribe the free abelian category. We do that in a less bare-hands way, and give a clearer picture of it, at Example 4.7.

3.3. The free abelian category *via* systems of linear equations. There is a very different-seeming construction of the free abelian category, arising from the model theory of modules, see [**23**], [**19**], [**12**] (one can also take a categorical logic approach, see [**7**, §2.2] in particular). In this construction the objects of the free abelian category are pairs of projected systems of linear equations; I explain this.

3.3.1. *pp formulas.* Let R be a ring [I will parenthetically say how to modify everything for general preadditive $\mathcal{R}$]. A **(homogeneous) system of R-linear equations** is an expression of the form

$$\bigwedge_{j=1}^{m}\sum_{i=1}^{n} r_{ij}x_i=0$$

where $x_1,\dots,x_n$ are variables (which can run over the elements of any particular R-module) and the r_{ij} are the elements (or corresponding multiplication maps) of R. (The symbol $\bigwedge$ is read "and", as is $\wedge$, where the use of $\bigwedge/\wedge$ is like that of $\sum/+$.)

[If the preadditive category $\mathcal{R}$ has more than one object, then each variable has a **sort** - the sorts are labelled by the objects p of $\mathcal{R}$ and a variable of sort p runs over elements of the group $M(p)$ in any particular $\mathcal{R}$-module M. So then, for the above expression to make sense, each morphism r_{ij} of $\mathcal{R}$ must be an arrow from p, the sort of x_i, to some $q=q(j)$ - so "0" in the jth equation means the zero element in $M(q)$. For instance, over A_2, $(x_1-\lambda x_2=0)\wedge(\alpha x_1-x_3=0)$, with λ some element of K, is a system of linear equations, where x_1 and x_2 both have sort labelled by the object 1 of $\mathcal{R}$ and x_3 has sort labelled by 2.]

The solution set in any module M of such a system of linear equations is a subgroup of M^n (not necessarily a submodule unless R is commutative). The projection of such a solution set to, say, the first $k\leq n$ coordinates is what is called a **pp-definable subgroup** of M (strictly, a subgroup of M^k pp-definable in M). The terminology "pp", short for "positive primitive", refers to the form of the condition

$$\exists x_{k+1},\dots,x_n\bigwedge_{j=1}^{m}\sum_{i=1}^{n} r_{ij}x_i=0$$

which defines this set. If we think of this as a condition on the variables $x_1, \dots, x_k$ then we refer to the condition itself as a **pp condition** or **pp formula**, using notation such as ϕ or, if we want to show the free (i.e. unquantified) variables, $\phi(x_1, \dots, x_k)$. The solution set in M - the corresponding pp-definable subgroup - that is

$$\{(a_1, \dots, a_k) \in M^k : \exists a_{k+1}, \dots, a_n \in M \sum_{i=1}^{n} r_{ij} a_i = 0 \text{ for each } j = 1, \dots, m\}$$

we write as $\phi(M)$.

[In the many-sorted case, where $\mathcal{R}$ has more than one object, the solution set of such a pp formula in an $\mathcal{R}$-module M will be a subset of a finite product of abelian groups each of the form $M(p)$ for p an object of $\mathcal{R}$. In the A_2 example above, the solution set of the system of two equations is a subgroup of $M(1) \oplus M(1) \oplus M(2)$; if we quantify out the variables x_1 and x_2, to get the pp formula $\exists x_1, x_2 \big((x_1 - \lambda x_2 = 0) \wedge (\alpha x_1 - x_3 = 0)\big)$, then the solution set of this, in any module M, is a subgroup of $M(2)$, namely the image of α.]

If $\psi(\overline{x})$ and $\phi(\overline{x})$ are pp formulas with the same free variables, we write $\psi \leq \phi$ if, for every module M, $\psi(M)$ is contained in $\phi(M)$, that is, if ψ is a stronger condition than ϕ. There is a characterisation (see [**33**, 1.1.13]) of this in terms of the coefficients r_{ij} of the respective systems of equations. If $\psi \leq \phi$ we refer to this as a **pp-pair**, writing ϕ/ψ.

3.3.2. *The category of pp-pairs.* We associate to any module category $\mathcal{R}$-Mod the **category**, $\mathbb{L}(\mathcal{R}\text{-Mod}) = \mathbb{L}^{\text{eq}+}_{\mathcal{R}^{\text{op}}}$, **of pp-pairs**. The objects of this category are the pp-pairs ϕ/ψ. The morphisms are the pp-defined maps; literally they are certain equivalence classes of pp formulas. Precisely, a morphism from $\phi(\overline{x})/\psi(\overline{x})$ to $\phi'(\overline{y})/\psi'(\overline{y})$ is given by a pp formula $\rho(\overline{x}, \overline{y})$ which defines a (necessarily additive) map from $\phi(M)/\psi(M)$ to $\phi'(M)/\psi'(M)$ for every module M. The conditions for a pp formula to define a map (rather than just a relation) are simply expressed: $\phi(\overline{x}) \leq \exists \overline{y}\, \rho(\overline{x}, \overline{y})$, $\big(\rho(\overline{x}, \overline{y}) \wedge \phi(\overline{x})\big) \leq \phi'(\overline{y})$ and $\big(\rho(\overline{x}, \overline{y}) \wedge \psi(\overline{x})\big) \leq \psi'(\overline{y})$ (in every module). Since different pp formulas can define the same function in every module, we take such formulas up to this equivalence as the arrows of $\mathbb{L}^{\text{eq}+}_{\mathcal{R}^{\text{op}}}$. All these conditions, though initially they might seem not checkable since they refer to every module, are equivalent to certain $\mathcal{R}$-bilinear constraints on the coefficients appearing in the formulas involved (use [**33**, 1.1.13]).

This category is, in fact [**19**], abelian and is equivalent to the category $(\mathcal{R}\text{-mod}, \mathbf{Ab})^{\text{fp}}$ ([**12**, 3.2.5]; see, e.g., [**33**, 10.2.30]), hence is an incarnation of $\text{Ab}(\mathcal{R})$. The equivalence with the functor category arises as follows.

One easily checks that homomorphisms preserve solution sets of pp formulas so, for any pp formula ϕ, the assignment $M \mapsto \phi(M)$ defines a functor, which we denote F_ϕ, from $\mathcal{R}$-Mod to $\mathbf{Ab}$. Because (taking the solution set of) a pp formula commutes with direct limits and since every module is a direct limit of finitely presented modules, this functor is determined by its restriction to the subcategory, $\mathcal{R}$-mod, of finitely presented modules. We will use the same notation, F_ϕ, for this restriction. This is a subfunctor of a power of the forgetful functor [in the case that $\mathcal{R}$ has one object; in the general case, a subfunctor of some product of representable-by-objects-of-$\mathcal{R}$ functors]. One may check (using "free realisations" see [**33**, §1.2.2]) that this functor is finitely generated so, since the functor category $(\mathcal{R}\text{-mod}, \mathbf{Ab})^{\text{fp}}$ is locally coherent [**4**], F_ϕ is finitely presented. Furthermore, if $\psi \leq \phi$ then F_ψ is a subfunctor of F_ϕ and we can form the quotient functor $F_{\phi/\psi} = F_\phi/F_\psi$, which takes a module M to $\phi(M)/\psi(M)$. As a quotient of finitely presented functors, this is finitely presented. It may be shown that every finitely presented functor has this form and that, moreover, the natural transformations between finitely presented functors are given by pp formulas (see [**33**, 10.2.30]). This is part of an extensive equivalence between the functorial and model-theoretic views of modules.

4. The context in which a module sits - definable categories, and the functors between these

4.1. Definable categories. When we define a module in the usual way, thereby fixing the ring $\mathcal{R}$, it is natural to regard the module as sitting inside the category $\mathcal{R}$-Mod of all additive functors from $\mathcal{R}$ to $\mathbf{Ab}$. This is a typical Grothendieck abelian category with a generating set of finitely generated projectives. On the other hand, the definition of a module M as an exact functor on an abelian category $\mathcal{A}$ naturally places M within the category, $\mathrm{Ex}(\mathcal{A}, \mathbf{Ab})$, of all such functors. The categories which arise this way are the (**exactly**) **definable** (**additive**) **categories**. These include module categories and all finitely accessible additive categories (see [**1**]) with products but are more general than that - for example, they may have no finitely presented objects apart from 0, see Example 8.5. The fact that there is a canonical choice for $\mathcal{A}$, given M, gives a minimal definable category $\mathrm{Ex}(\mathcal{A}(M), \mathbf{Ab})$ in which M lives; we will see below that there are various senses in which this category is generated by M.

The fact that definable categories are accessible follows from the Downwards Löwenheim-Skolem Theorem (see any text on model theory, e.g. [**20**]). But accessible categories with products need not be definable; for example a Grothendieck category is definable iff it is finitely accessible [**36**, 3.6]. As further examples of (finitely accessible) definable categories, we have the category of comodules over a K-coalgebra and, over many schemes, the category of quasicoherent sheaves.

Definable categories first arose as certain subcategories of module categories: we say that a full subcategory (closed under isomorphism) of a module category $\mathcal{R}$-Mod is a **definable subcategory** if it satisfies the equivalent conditions of the next result, where an embedding $f : M \to N$ between left $\mathcal{R}$-modules is **pure** if for every (finitely presented) right $\mathcal{R}$-module L, the morphism $L \otimes_{\mathcal{R}} M \xrightarrow{1_L \otimes f} L \otimes_{\mathcal{R}} N$ is an embedding (of abelian groups); $\otimes_{\mathcal{R}}$ is illustrated in the example that follows.

THEOREM 4.1. *(see [**33**, 3.4.7]) The following are equivalent for a subcategory $\mathcal{D}$ of $\mathcal{R}$-*Mod*:*
*(i) $\mathcal{D}$ is closed in $\mathcal{R}$-*Mod* under direct products, direct limits and pure submodules;*
*(ii) $\mathcal{D}$ is closed in $\mathcal{R}$-*Mod* under ultraproducts, finite direct sums and pure submodules;*
(iii) there is a set Φ of functors in $(\mathcal{R}\text{-mod}, \mathbf{Ab})^{\mathrm{fp}}$ such that $\mathcal{D} = \{M \in \mathcal{R}\text{-Mod} : FM = 0 \text{ for all } F \in \Phi\}$;
(iii) there is a set Φ of pp-pairs (in the language of $\mathcal{R}$-modules) such that $\mathcal{D} = \{M \in \mathcal{R}\text{-Mod} : \phi(M)/\psi(M) = 0 \text{ for all } \phi/\psi \in \Phi\}$.

EXAMPLE 4.2. I recall how to define tensor product over a small preadditive category, using the path category, $\mathcal{R}$, of A_3 to illustrate this.

A left $\mathcal{R}$-module M, that is a functor from $\mathcal{R}$ to K-Mod, can usefully be identified with its "image", meaning the corresponding representation $M1 \xrightarrow{M\alpha} M2 \xrightarrow{M\beta} M3$ of A_3. A right module L can be identified with a representation, $L1 \xleftarrow{L\alpha} L2 \xleftarrow{L\beta} L3$, of the opposite quiver.

The first part of the definition of $(-) \otimes_{\mathcal{R}} M$ is that its value on a representable right module $(-, i)$ is Mi; for instance if $i = 1$ then $(-, 1)$ is being identified with the representation $K \leftarrow 0 \leftarrow 0$ so $(-, 1) \otimes M = M1$. Taking $i = 2$, hence the representation $K \xleftarrow{1} K \leftarrow 0$, we have $(-, 2) \otimes M = M2$ and, similarly, $(-, 3) \otimes M = M3$.

The rest of the definition of $- \otimes M$ is that it is right exact, which is enough because we've just said how it is defined on the indecomposable projectives. For instance, if L is $0 \leftarrow K \xleftarrow{1} K$, so has projective presentation $(-, 1) \xrightarrow{(-,\alpha\beta)} (-, 3) \to L \to 0$, then, tensoring with M, we obtain the exact sequence $M1 \xrightarrow{M(\alpha\beta)} M3 \to ML \to 0$. So, if M is the representation $K \xrightarrow{1} K \xrightarrow{1} K$ then $L \otimes M = 0$ and, if M is any other indecomposable representation, then $L \otimes M = M3$.

THEOREM 4.3. *(see [**33**, 18.1.6]) A category is exactly definable iff it is equivalent to a definable subcategory of some module category $\mathcal{R}$-Mod. We say simply that such a category is* **definable**.

If $M \in \mathcal{R}$-Mod then write $\langle M \rangle$ for the definable subcategory of $\mathcal{R}$-Mod **generated by** M (in the sense of being the smallest definable subcategory which contains M). Every definable category has a generator in this sense, indeed an **elementary cogenerator**, meaning an object N such that every $M \in \mathcal{D}$ purely embeds in a direct product of copies of N. We say that an object N is **pure-injective** if it is injective over pure embeddings.

THEOREM 4.4. *(see [**33**, §3.4,5.3.52]) Every definable category has the form $\langle M \rangle$ for some module M. Since every definable category is closed under pure-injective hulls, we may take M to be pure-injective. Indeed every definable category has an elementary cogenerator, which may be taken to be a direct product of indecomposable pure-injectives, alternatively the pure-injective hull of a direct sum of indecomposable pure-injectives.*

EXAMPLE 4.5. Within the category $\mathbb{Z}$-Mod $=$ **Ab** of abelian groups, examples of definable subcategories are:
$\langle \mathbb{Z} \rangle$, the torsionfree abelian groups;
$\langle \mathbb{Q}/\mathbb{Z} \rangle$, the category of divisible abelian groups;
$\langle \mathbb{Q} \rangle$, the intersection of these;
$\langle \overline{\mathbb{Z}_{(p)}} \rangle = \langle \mathbb{Z}_{(p)} \rangle$, the localisation of **Ab** at any nonzero prime p;
$\langle A \rangle$, the category of direct sums of copies of direct summands of A, where A is any finite abelian group.

There are further definable subcategories of **Ab** and these are, as are the definable subcategories of any definable category, in natural bijection with the closed sets of the associated Ziegler spectrum. That is a topological space with points being the isomorphism classes of indecomposable pure-injectives and with topology defined using pp-pairs, as originally introduced by Ziegler [**42**], but which can be defined in a variety of ways, see [**33**, §§5.1, 12.4].

The category of torsion abelian groups is not a definable subcategory of **Ab** though, being a finitely accessible category with products (the finite groups are its finitely presented $\varinjlim$-generators) it is a definable category (which is not 'definably embedded' in **Ab**).

4.2. The intrinsic structure of a definable category. If $\mathcal{D}$ is a definable subcategory of a module category $\mathcal{R}$-Mod then the usual pure-exact structure on the latter restricts to an exact structure on $\mathcal{D}$ which is, in fact, intrinsic to $\mathcal{D}$. That follows directly from the fact that a sequence $0 \to A \to B \to C \to 0$ in $\mathcal{D}$ is pure-exact in the sense of $\mathcal{R}$-Mod iff some ultraproduct of it is split. Since ultraproducts are direct limits of products, and $\mathcal{D}$ is closed in $\mathcal{R}$-Mod under these operations, this definition of purity may be used without any reference to such a containing category $\mathcal{R}$-Mod. Thus there is an intrinsic theory of purity in any definable category.

Such a category $\mathcal{D}$ also has an intrinsic model theory, with language based on the category $\mathcal{A}(M)$ where M is any generator for $\mathcal{D}$ as a definable category. This category does not depend on choice of generator M, being equivalent to the category $(\mathcal{D}, \mathbf{Ab})^{\prod\to}$ of additive functors from $\mathcal{D}$ to **Ab** which commute with direct products and direct limits ([**34**, 12.10]). We write fun($\mathcal{D}$) for this skeletally small abelian category. So, in the case of a module category, we have the following ways of obtaining the free abelian category of a ring.

THEOREM 4.6. *For any skeletally small preadditive category $\mathcal{R}$ the categories* $(\mathcal{R}\text{-mod}, \mathbf{Ab})^{\mathrm{fp}}$, $\mathbb{L}^{\mathrm{eq}+}_{\mathcal{R}^{\mathrm{op}}}$, $\mathrm{Ab}(\mathcal{R})$ *and* $\text{fun-}\mathcal{R}^{\mathrm{op}} = (\mathcal{R}\text{-Mod}, \mathbf{Ab})^{\prod\to}$ *are equivalent.*

The canonical (multi-sorted) language for a definable category $\mathcal{D}$ has one sort for each object of fun($\mathcal{D}$), an abelian group structure on each sort, and a function symbol for each arrow of fun($\mathcal{D}$). This general kind of language is discussed in Section 9; we consider a specific example here.

EXAMPLE 4.7. We recompute the functor category fun($\mathcal{D}$) of the category of modules $\mathcal{D} = KA_2$-Mod over the path category of A_2, using some more general theory than we did with the direct computations at Example 3.2.

We use the fact that, for a ring R of finite representation type, such as KA_2, the functor category is equivalent (see, e.g., [**11**, 4.9.4], [**6**, §VI.5]) to the category of right modules over the **Auslander algebra** of R; this is the endomorphism ring $S = \mathrm{End}(M)$ where M is a direct sum of one copy of each of the indecomposable (finitely generated) R-modules. In our example, and continuing the notation of Example 3.2, the Auslander algebra is $S = \mathrm{End}(P_1 \oplus P_2 \oplus S_1)$. Then, by direct computation (let S act on the right as a 3×3 matrix ring and decompose it into indecomposable projectives = rows), S is the path algebra of the quiver $\bullet_1 \xleftarrow{\pi} \bullet_2 \xleftarrow{i} \bullet_3$ with relation $i\pi = 0$. The Auslander-Reiten quiver of Mod-S is the following, where T_i, Q_i, J_i are respectively the simple module at vertex i, the projective cover of T_i, the injective hull of T_i.

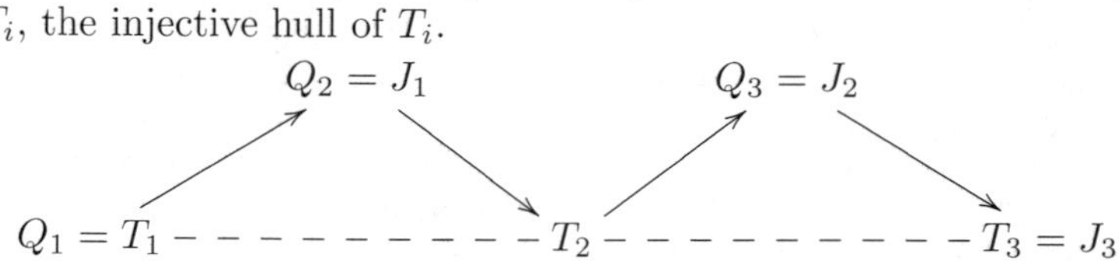

One can check[3] that these are, as functors and as pp-pairs, as follows (where M can be the module above or, since the action on M determines that on any module, arbitrary):
$Q_1 = (S_1, -) = (\alpha x = 0)/(x = 0) : M \mapsto \{m \in e_1 M : \alpha m = 0\}$;
$Q_2 = (P_1, -) = (x = e_1 x)/(x = 0) : M \mapsto e_1 M$;
$T_2 = (P_1, -)/\mathrm{im}(\pi, -) \simeq (P_1, -)/(S_1, -) \simeq (\alpha|x)/(x = 0) : M \mapsto \alpha M$;
$Q_3 = (P_2, -) = (x = e_2 x)/(x = 0) : M \mapsto e_2 M$;
$T_3 = (P_2, -)/\mathrm{im}(i, -) \simeq (P_2, -)/(P_1, -) = (x = e_2 x)/(\alpha|x) : M \mapsto e_2 M/\alpha M$.

In particular S is of finite representation type and so there are just these 5 indecomposable pp-pairs, with every pp sort, that is object of $\mathbb{L}^{\mathrm{eq}+}_{(KA_2)^{\mathrm{op}}}$ being a direct sum of copies of these. In general, even if R is of finite representation type, the Auslander algebra of R need not be, so this is a very pleasant situation in which the category of pp-sorts can be described completely. A similar computation, for the algebra $K[\epsilon : \epsilon^2 = 0]$, can be found in [**27**, §6.8].

4.3. Localisation. Localisation/relativisation is relevant when we move from a definable category to a definable subcategory. The restriction/relativisation of functors (that is, pp-pairs) to the subcategory exactly corresponds to Serre localisation of the associated functor categories.

Precisely, suppose that $\mathcal{D}$ is a definable subcategory of, say, $\mathcal{R}$-Mod. By $\mathbb{L}^{\mathrm{eq}+}(\mathcal{D})$ we denote the category of pp-pairs for $\mathcal{D}$. The objects of $\mathbb{L}^{\mathrm{eq}+}(\mathcal{D})$ may be taken to be the same as those of $\mathbb{L}^{\mathrm{eq}+}_{\mathcal{R}^{\mathrm{op}}}$ (alternatively, the pp-pairs obtained from the intrinsic model theory of $\mathcal{D}$). The morphisms from sort ϕ/ψ to ϕ'/ψ' are the equivalence classes, meaning equivalence *on modules in* $\mathcal{D}$, of pp formulas which define a relation from the group defined by ϕ/ψ to that defined by ϕ'/ψ' which is total and functional in every $D \in \mathcal{D}$.

EXAMPLE 4.8. Take $\mathcal{D}$ to be the definable subcategory $\mathbb{Z}_{(2)}$-Mod of $\mathbb{Z}$-Mod. Since every prime other than 2 is invertible when acting on the objects of $\mathcal{D}$, the sort $(y = y)/(3|y)$ must be isomorphic to the zero object of $\mathbb{L}^{\mathrm{eq}+}(\mathcal{D})$. Represent the zero object as, say, $(x = x)/(x = x)$. The pp formula $\rho(x, y)$ which is just $y = x$, regarded as defining a relation from $(x = x)/(x = x)$ to $(y = y)/(3|y)$ is certainly not functional on $\mathbb{Z}$-Mod - if M is a $\mathbb{Z}$-module then ρ relates the only element of M/M to every element of $M/3M$ - but it is functional on modules in $\mathcal{D}$ since, for these, $M/3M = 0$. For an example of an arrow of $\mathbb{L}^{\mathrm{eq}+}(\mathcal{D})$ which is not in $\mathbb{L}^{\mathrm{eq}+}(\mathbb{Z}\text{-Mod})$, take multiplication by the inverse of 3 on any sort, say on $(x = x)/(x = 0)$ where the relation defined by the formula $\sigma(x, y)$, being $3y = x$, gives a new arrow.

THEOREM 4.9. *(see* [**33**, 12.3.20]*) If $\mathcal{D}$ is a definable subcategory of $\mathcal{R}$-Mod and $\mathcal{S}_{\mathcal{D}}$ denotes the full subcategory of $\mathbb{L}^{\mathrm{eq}+}_{\mathcal{R}^{\mathrm{op}}}$ on those pp-pairs which are* **closed on** *(i.e. zero when evaluated on) $\mathcal{D}$ then the quotient category $\mathbb{L}^{\mathrm{eq}+}_{\mathcal{R}^{\mathrm{op}}}/\mathcal{S}_{\mathcal{D}}$ is naturally equivalent to $\mathbb{L}^{\mathrm{eq}+}(\mathcal{D})$.*

---

[3]For example, first note that Q_2 is the functor $(P_1, -)$, hence has action as given, and similarly for Q_3. Then, hom the exact sequence $0 \to P_2 \xrightarrow{i} P_1 \xrightarrow{\pi} S_1 \to 0$ into M to get that $Q_1(M) = (S_1, M) = \ker(i, M)$ which, since i is just multiplication by α, gives the pp-description of this functor. Since T_2 is the cokernel of the embedding of Q_1 into Q_2, it can be described as the quotient $M \mapsto e_1 M/\{m \in e_1 M : \alpha m = 0\}$, which is isomorphic (definably, by multiplication by α) to αM. And the description of T_3 is also immediate.

The naturality of the equivalence is with respect to the actions on $\mathcal{D}$: if ϕ/ψ is a pp-pair for $\mathcal{R}$-modules then the action of its image in the quotient category is given by the same pp-pair but now restricted to $\mathcal{D}$.

Every small abelian category arises thus, as the category of pp-pairs for some definable category, see [**36**, 2.18].

We give an example of computing such a localisation.

EXAMPLE 4.10. Take $\mathcal{D}$ to be the definable subcategory generated by the indecomposable representation $K \xrightarrow{1} K$ of KA_2. The kernel of the localisation from $\text{fun}(KA_2\text{-Mod}) = \mathbb{L}^{\text{eq}+}_{(KA_2)^{\text{op}}}$ to $\text{fun}(\langle K \xrightarrow{1} K\rangle)$ is the Serre subcategory generated by those indecomposables which are 0 on this representation, that is (continuing the notation from Example 4.7), by Q_1 and T_3. Therefore, in the quotient category, the epimorphism from Q_2 to T_2 becomes an isomorphism (it had kernel Q_1 which is now zero); also the embedding of T_2 in Q_3 becomes an isomorphism (it had cokernel T_3). So we see that there is just one indecomposable object of $\text{fun}(\langle K \xrightarrow{1} K\rangle)$, as we should expect since, note, $\langle K \xrightarrow{1} K\rangle$ is equivalent to the category K-Mod of K-vector spaces and that, one may check (or see [**33**, 10.2.38]), has K-mod for its functor category.

4.4. Imaginaries. If M is an $\mathcal{R}$-module, we may view the corresponding exact functor $\widehat{M} : \mathcal{A}(M) \to \mathbf{Ab}$ as an enriched version of M. It has been enriched by the addition of many new sorts, namely the various $\phi(M)/\psi(M)$ for $\phi/\psi \in \mathbb{L}^{\text{eq}+}_{\mathcal{R}^{\text{op}}}$ (or in the Serre quotient $\mathcal{A}(M)$ of that category), plus the pp-definable functions between these sorts. The model-theoretic view is that these are new kinds of elements of M, "imaginaries", see Section 9. They "are essentially" elements of M in the same sense that one might say this of n-tuples of elements of M. This has proved to be an extremely useful point of view in model theory in general as well as in particular contexts. In the context of modules/additive structures it fits particularly well with the functor-category approach to representation theory.

For example, from this point of view, if A is a finitely presented module, then the elements of (A, M) are essentially elements of M. Also, if A is FP_2, then the elements of $\text{Ext}^1(A, M)$ are essentially elements of M (see 8.2). From this perspective, in the example in Section 1.1, the elements of the $K\widetilde{A}_1$-representation build from a $K[T]$-module M are essentially elements of M, that $K\widetilde{A}_1$-representation being an imaginary sort of M. Indeed, its underlying group has the form $\phi(M)/\psi(M)$ for some pp-pair ϕ/ψ in the language of $K[T]$-modules, and the action of each element of $K\widetilde{A}_1$ on that group is defined by a pp formula.

Put another way, denoting by $M^{\text{eq}+}$ the multi-sorted structure built in this way from M, this is just a slightly different way of viewing the functor $\widehat{M} : \mathcal{A}(M) \to \mathbf{Ab}$, which is evaluation of sorts at M. Namely, it is the image of that functor.

4.5. Interpretation functors. A functor $F : \mathcal{C} \to \mathcal{D}$ between definable categories is an **interpretation functor** if it commutes with direct products and direct limits (it then follows that F preserves pure-exact sequences and pure-injectivity). So these are the structure-preserving functors between definable categories. They are also the functors which are model-theoretic interpretations given by pp formulas - see Section 9. They precisely correspond, contravariantly, to the exact functors between the corresponding abelian functor categories - see Section 7. The functors we saw in Section 1.1 are easily seen to be interpretation functors.

EXAMPLE 4.11. Given a morphism $\theta : R \to S$ of rings, this induces a restriction-of-scalars functor from Mod-S to Mod-R, and that is clearly an interpretation functor.

EXAMPLE 4.12. If ${}_SB_R$ is an (S, R)-bimodule which, as a left S-module is finitely presented, then the functor $- \otimes_S B_R : S\text{-Mod} \to R\text{-Mod}$ is an interpretation functor: since tensor is a left adjoint, it preserves direct limits, and tensoring with a finitely presented module preserves products [**24**].

In particular, if R and S are finite-dimensional algebras then we say that $F = ({}_SB_R \otimes_S -) : S\text{-Mod} \to R\text{-Mod}$ is a **representation embedding** if ${}_SB$ is finitely generated projective and the restriction of F to S-mod preserves indecomposability and reflects isomorphism (but we are not requiring that it must be full on isomorphisms). (The concept is not restricted to finite-dimensional algebras. For instance, if S is $K[T]$ or $K\langle X, Y\rangle$ then the condition is usually required on finite-dimensional, rather than finitely presented, S-modules.)

What is not clear is whether a representation embedding F is an equivalence, in the category of definable categories and interpretation functors, between its domain and its image. That is, can the original category S-Mod be "recovered definably" from its image? It is an open question whether this is always the case, though it is true if we have a strict representation embedding [**30**] or, more generally, if F is a finitely-controlled representation embedding [**17**, 4.3]. From this one can deduce that the conjecture "wild representation type implies undecidable theory of modules" is true for finitely-controlled-wild (including strictly wild) algebras.

I explain, in Section 9, the sense in which interpretation functors are indeed "interpretations".

5. When is a module over a ring?

We say that a module M is **over a ring** if M can be represented as a functor from a single-object preadditive category to **Ab**, equivalently if there is a ring R with 1 such that $\mathcal{A}(M)$ is a quotient of $\mathrm{Ab}(R)$ by a Serre subcategory.

EXAMPLE 5.1. Let $\mathcal{R}$ be the K-linear path category of the quiver A_∞^∞: $\cdots \to -1 \to 0 \to 1 \to 2 \to \ldots$. So, $(i, j) = K$ if $i \leq j$ and $= 0$ otherwise. The category of $\mathcal{R}$-modules is not equivalent to the category of modules over any ring with 1 so there will be many $\mathcal{R}$-modules which are not over any ring.[4]

We will say that an object A in an abelian category $\mathcal{A}$ is an **abelian generator** of $\mathcal{A}$ if $\mathcal{A}$ is the smallest, possibly non-full, abelian subcategory of $\mathcal{A}$ containing A and all its endomorphisms.

PROPOSITION 5.2. *Let M be a module; then M is over a ring iff there is an object U of $\mathcal{A}(M)$ which is an abelian generator of $\mathcal{A}(M)$.*

PROOF. First we recall why $({}_RR, -)$ is an abelian generator of $\mathrm{Ab}(R)$. Let $A \in R$-mod and choose a projective presentation $R^m \xrightarrow{h} R^n \xrightarrow{p} A \to 0$. This induces an exact sequence $0 \to (A, -) \xrightarrow{(p,-)} (R, -)^n \xrightarrow{(h,-)} (R, -)^m$ so, as the

[4]This can be deduced, for example, from the fact that the Ziegler spectrum of this category is not compact: the open sets determined by the pp-pairs $(e_i x = x)/(x = 0)$ give a cover with no finite subcover but, if R is a ring with 1, then its Ziegler spectrum is compact and so, therefore, is the Ziegler spectrum of any definable subcategory, since it is a closed subset.

kernel of $(h,-)$, the functor $(A,-)$ and its inclusion into $(R,-)^n$ is in the abelian subcategory generated by $(R,-)$. Any morphism $(B,-)\to(A,-)$ has the form $(f,-)$ for some $f:A\to B$ in R-mod. Taking also a projective presentation of B, we can see, using that $(R,-)$ is injective (as well as projective), that $(f,-)$ is the restriction of some morphism between powers of $(R,-)$.

$$\begin{array}{ccccccc}
0 & \longrightarrow & (A,-) & \xrightarrow{(p,-)} & (R,-)^n & \xrightarrow{(h,-)} & (R,-)^m \\
 & & {\scriptstyle (f,-)}\uparrow & & {\scriptstyle (f',-)}\uparrow & & \\
0 & \longrightarrow & (B,-) & \xrightarrow{(q,-)} & (R,-)^k & \longrightarrow & (R,-)^l
\end{array}$$

Consider the epi-mono factorisation of $(f,-)$; we claim that each map is in the abelian subcategory generated by $(R,-)$. The epimorphism is there because it is the cokernel of the kernel of $(p,-)(f,-)=(f',-)(q,-)$ and the latter is a morphism in the abelian category generated by $(R,-)$. It then follows from the cokernel property that the monomorphism also is in the subcategory. So the composition, $(f,-)$ of these maps is in the subcategory, as required. Since every object of $\mathrm{Ab}(R)$ is the cokernel of a map such as $(f,-)$, we have all the objects of $\mathrm{Ab}(R)$ in the abelian subcategory generated by $(R,-)$. Finally, given a morphism $F_g\to F_f$ in $\mathrm{Ab}(R)$ we can lift it to a projective presentation and deduce from what we have already, that this morphism also is in the subcategory generated by $\mathrm{Ab}(R)$.

[The argument is equally valid in the case where general $\mathcal{R}$ replaces R, the conclusion being that $\mathrm{Ab}(\mathcal{R})$ is the smallest abelian subcategory containing all the (projective-injective) functors $(P,-)$ where P ranges over any generating set of the category $\mathcal{R}$-proj of finitely generated projective modules.)]

If $\mathcal{A}(M)$ is a Serre quotient of a category of the form $\mathrm{Ab}(R)$ then, since the forgetful functor $({}_RR,-)$ is an abelian generator of $\mathrm{Ab}(R)$, the same is true of its image in $\mathcal{A}(M)$ (since the inverse image under an exact functor of an abelian subcategory is abelian).

For the converse, suppose that $\mathcal{A}(M)$ has an abelian generator F; set $R=\mathrm{End}(F)$. Let M_F denote the functor from the preadditive category R to $\mathcal{A}(M)$ with image the 1-object full subcategory of $\mathcal{A}(M)$ on F. Then this extends to an exact functor from $\mathrm{Ab}(R)$ to $\mathcal{A}(M)$, the image of which is an abelian subcategory which contains F, hence which is all of $\mathcal{A}(M)$. □

6. Elementary duality

There is a duality which applies to everything that we have discussed so far. To every definable category $\mathcal{D}$ there is a corresponding dual definable category $\mathcal{D}^{\mathrm{d}}$, defined below, which, in the case of a category $\mathcal{R}$-Mod of left modules, is the category Mod-$\mathcal{R}$ of right modules. The associated small abelian functor categories, $\mathrm{fun}(\mathcal{D})$ and $\mathrm{fun}(\mathcal{D}^{\mathrm{d}})$, are opposite and there is a natural bijection between definable subcategories of $\mathcal{D}$ and its dual (this follows most directly from the fact that a Serre subcategory of an abelian category is also a Serre subcategory of the opposite category). The pervasiveness of duality throughout all associated structures is clear from certain equivalences of 2-categories, one of which we discuss in the next section. The existence of this duality is clear on the 2-category, $\mathbb{ABEX}$, of small abelian categories with exact functors, which has a natural involution taking each abelian category $\mathcal{A}$ to its opposite $\mathcal{A}^{\mathrm{op}}$.

If $\mathcal{D} = \mathrm{Ex}(\mathcal{A}, \mathbf{Ab})$ then the (**elementary**) **dual** category of $\mathcal{D}$ is $\mathcal{D}^{\mathrm{d}} = \mathrm{Ex}(\mathcal{A}^{\mathrm{op}}, \mathbf{Ab})$ and we have $\mathbb{L}^{\mathrm{eq}+}(\mathcal{D}^{\mathrm{d}}) = \left(\mathbb{L}^{\mathrm{eq}+}(\mathcal{D})\right)^{\mathrm{op}}$, that is $\mathrm{fun}(\mathcal{D}^{\mathrm{d}}) = (\mathrm{fun}(\mathcal{D}))^{\mathrm{op}}$. In particular, $(R\text{-mod}, \mathbf{Ab})^{\mathrm{fp}} \simeq \left((\text{mod-}R, \mathbf{Ab})^{\mathrm{fp}}\right)^{\mathrm{op}}$ ([**5**, §7], [**18**, 5.6]). In the context of the model theory of modules this duality was found first for pp formulas, and termed elementary duality, in [**28**], then extended to the category of pp-pairs (and thence to theories of modules) in [**19**].

7. The bigger picture

What we have described above fits into a framework given by certain equivalences and anti-equivalences of 2-categories. The 2-categories involved are categories of categories; they have categories for their objects, functors for their arrows and the "2" refers simply to the extra structure of natural transformations between functors. One of these 2-categories is $\mathbb{ABEX}$: the objects of this category are the skeletally small abelian categories and the arrows are the exact functors between these; the natural transformations between these exact functors are the 2-arrows. Another is $\mathbb{DEF}$: its objects are the definable additive categories, its arrows are the interpretation functors and, again, natural transformations provide the "2"-structure. There is a contravariant equivalence between these.

THEOREM 7.1. [**38**, 2.3 and comments following that] *There is an anti-equivalence between* $\mathbb{ABEX}$ *and* $\mathbb{DEF}$*:*

$$\mathbb{ABEX} \xrightarrow{\ \simeq^{\mathrm{op}}\ } \mathbb{DEF}\ .$$

Explicitly:

$$\mathcal{A} = \mathrm{fun}(\mathcal{D}) = (\mathcal{D}, \mathbf{Ab})^{\prod\to} \rightleftarrows \mathcal{D} = \mathrm{Ex}(\mathcal{A}, \mathbf{Ab})$$

with the actions on morphisms being given by composition.

In fact, this fits into a larger picture of 2-categories, see the introduction of [**36**] The following result is one instance of this picture. Here $\mathcal{A}(\mathcal{R})$ is the smallest abelian subcategory of Mod-$\mathcal{R}$ which contains the finitely presented $\mathcal{R}$-modules, see Example 8.4, and $\mathcal{R}$-Flat denotes the category of flat left $\mathcal{R}$-modules.

PROPOSITION 7.2. [**37**, 7.1] *If* $\mathcal{R}$ *is any skeletally small preadditive category then*

$$\mathrm{Ex}(\mathcal{A}(\mathcal{R}), \mathbf{Ab}) \simeq \langle \mathcal{R}\text{-Flat}\rangle.$$

If $\mathcal{R}$ *is right coherent, so* $\mathcal{A}(\mathcal{R}) = \text{mod-}\mathcal{R}$ *and* $\mathcal{R}$-Flat *is a definable subcategory of* $\mathcal{R}$-Mod, *then*

$$\mathrm{Ex}(\text{mod-}\mathcal{R}, \mathbf{Ab}) \simeq \mathcal{R}\text{-Flat}.$$

The (elementary) dual (in the sense of Section 6) to this is the following, where Abs-$\mathcal{R}$ is the category of absolutely pure right $\mathcal{R}$-modules (for more on these see Example 8.4).

PROPOSITION 7.3. [**37**, 7.2] *If* $\mathcal{R}$ *is any skeletally small preadditive category then*

$$\mathrm{Ex}(\mathcal{A}(\mathcal{R})^{\mathrm{op}}, \mathbf{Ab}) \simeq \langle \text{Abs-}\mathcal{R}\rangle.$$

If $\mathcal{R}$ *is right coherent, so* Abs-$\mathcal{R}$ *is a definable subcategory of* Mod-$\mathcal{R}$, *then*

$$\mathrm{Ex}((\text{mod-}\mathcal{R})^{\mathrm{op}}, \mathbf{Ab}) \simeq \text{Abs-}\mathcal{R}.$$

8. Examples

Here are some examples to illustrate the definitions and results we have given.

EXAMPLE 8.1. **Morita equivalence:** is a type of interpretation functor. Consider, for example, the equivalence, for any ring R, between R-modules and modules over the ring $M_n(R)$ of $n \times n$ matrices over R. This takes an R-module M to the direct sum M^n regarded naturally as an $M_n(R)$-module. The corresponding exact functor (in the other direction) between categories of pp-pairs takes the "home sort" of $M_n(R)$ (that is, the usual, defining, sort for modules regarded as 1-sorted structures) to the n-th power of the home sort $(R, -)$ for R. Essentially this is the identity functor since the image of that full and faithful functor is all of $\mathbb{L}_{R^{\mathrm{op}}}^{\mathrm{eq}+}$. That is, this is really just making an alternative choice of home sort: $(R^n, -)$ in place of $(R, -)$ in this example, $(P, -)$ where P is a finitely generated projective generator of R-Mod in general.

EXAMPLE 8.2. **Tilting:** Tilting is a relation between module (and other) categories much weaker than Morita equivalence. A right module T_R is a (classical 1-)**tilting module** if it has projective dimension ≤ 1, if $\mathrm{Ext}_R^1(T, T) = 0$ and if there is an embedding of R into a direct sum of copies of T such that the factor module is a direct summand of a direct sum of copies of T. Set $S = \mathrm{End}(T_R)$; then ${}_ST$ is tilting and $R = \mathrm{End}({}_ST)$. If we impose the mild finiteness condition that T_R and ${}_ST$ are FP_2 (have a projective presentation with the first three terms finitely generated) then, [**32**, 4.5], the following subcategories of Mod-R:
$\mathcal{F}(T) = \{X_R : \mathrm{Hom}_R(T, X) = 0\}$;
$\mathcal{G}(T) = \{X_R : \mathrm{Ext}_R^1(T, X) = 0\}$
and the following subcategories of Mod-S:
$\mathcal{Y}(T) = \{N_S : \mathrm{Tor}_1^R({}_ST, N_S) = 0\}$;
$\mathcal{X}(T) = \{N_S : N_S \otimes T = 0\}$
are all definable subcategories.

Moreover, under this assumption, the standard inverse pairs of tilting equivalences, $\mathrm{Hom}_R({}_ST_R, -) : \mathcal{G}(T) \to \mathcal{Y}(T)$ with inverse $- \otimes_S T_R$, and $\mathrm{Ext}_R^1({}_ST_R, -) : \mathcal{F}(T) \to \mathcal{X}(T)$ with inverse $\mathrm{Tor}_1^R({}_ST_R, -)$, are interpretation functors. (More generally, see [**33**, 10.2.35], if T is an (S, R)-bimodule then, provided T_R is FP_{n+1} (has a projective presentation, the first $n + 2$ terms of which are finitely generated), the functor $\mathrm{Ext}_R^n(T, -) : \text{Mod-}R \to \text{Mod-}S$ is an interpretation functor, and provided ${}_ST$ is FP_{n+1} the functor $\mathrm{Tor}_n^S : \text{Mod-}S \to \text{Mod-}R$ is an interpretation functor.)

EXAMPLE 8.3. **4-subspace representations:** Let $\mathcal{C} = \text{Mod-}K[T]$ and let $\mathcal{D}$ be the category of representations of the quiver $\widetilde{D_4}$ shown.

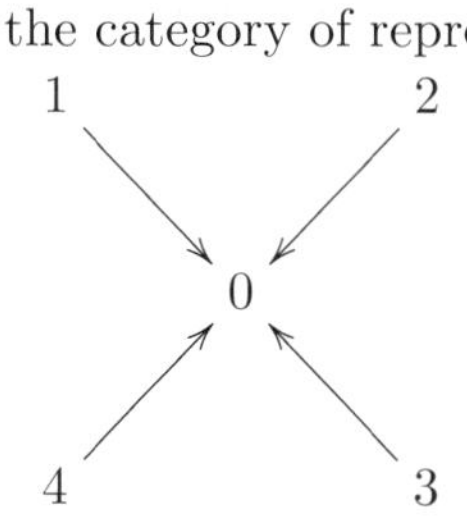

Let $I : \mathcal{C} \to \mathcal{D}$ be defined on a $k[T]$-module M by taking it to the representation

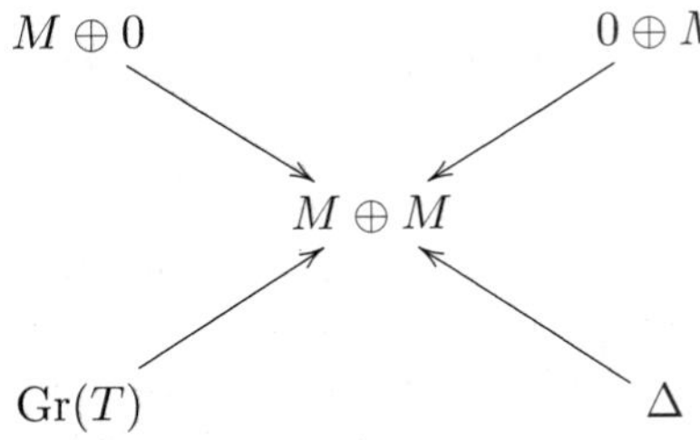

where the maps are inclusions into the direct sum $M \oplus M$, with $\Delta = \{(m, m) : m \in M\}$ the diagonal and $\mathrm{Gr}(T) = \{(m, mT) : m \in M\}$ the graph of multiplication-by-T. The action of I on morphisms is the obvious one and it is clear that this is a functor which commutes with direct products and direct limits, hence an interpretation functor.

The image of I is (e.g. using the rest of this paragraph) a definable subcategory. Given a $\widetilde{D_4}$-representation N in the image of I, we can recover the $k[T]$-module that it came from as follows. For the underlying vectorspace we take $V = e_1 N$ and the action is given, in words, as follows. Let $a \in e_1 N$; there is some (unique, note) $b \in e_2 N$ such that $a + b \in e_4 N$ (here we are identifying the $e_i N$ with their images in $e_0 N$, just to simplify notation). Then there is a unique $c \in e_1 N$ such that $c + b \in e_3 N$; we define $Ta = c$. It is easy to check that N is isomorphic to the image under I of this $K[T]$-module. All this can be expressed using pp formulas. That is, we have just defined an interpretation functor J going back from the image of I to $K[T]$-Mod which is an inverse for I.

This can be found in Baur's paper [**10**, p. 244].

Example 8.4. **Flat modules:** Take R to be a ring (although all this works for $\mathcal{R}$ being a small preadditive category). The definable category, $\langle R\text{-Flat}\rangle$, generated by the left module ${}_R R$ includes all flat modules and equals the category of flat modules iff R is right coherent. Dually, the definable category generated by the injective right R-modules includes all absolutely pure (=fp-injective) modules, where a module is **absolutely pure** if it is a pure submodule of every containing module; it contains exactly these modules iff the ring is right coherent. (Essentially these results go back to Eklof and Sabbagh [**14**], [**41**].) These two categories are dual in the sense of Section 6. Their categories of pp-imaginaries, see 7.2 and 7.3, are obtained as follows.

Consider the functor from $\mathrm{Ab}(R) = (R\text{-mod}, \mathbf{Ab})^{\mathrm{fp}}$ to Mod-R which is evaluation at R: $F \mapsto F({}_R R)$ (the right R-module structure on R induces the right R-module structure on $F({}_R R)$. We define $\mathcal{A}(R)$ to be the image of this functor; it is, see [**38**, §6], the smallest abelian, not necessarily full, subcategory of Mod-R which contains the full subcategory of finitely presented modules. (The objects of $\mathcal{A}(\mathcal{R})$ are, [**38**, 6.4], exactly the kernels of morphisms between finitely presented modules.)

Example 8.5. **Injective modules:** The category of injective right R-modules is a definable subcategory of Mod-R iff R is right noetherian. Taking $R = \mathbb{Z}$ we get the category of direct sums of copies of Prüfer modules $\mathbb{Z}_{p^\infty}$ (p a nonzero prime) and $\mathbb{Q}$. This is a definable category with no finitely presented object apart from 0 (see [**33**, 18.1.1]). By the result stated above, the corresponding functor category = category of pp-pairs is the category of finitely generated abelian groups.

The category of direct sums of copies of Prüfer modules is itself a definable category (see Example 8.7) though not a definable subcategory of Mod-$\mathbb{Z}$ - its Ziegler spectrum is not compact, so it needs a language with infinitely many sorts. We will see in Example 8.7 that its functor category = category of pp-pairs is the category of finite abelian groups.

EXAMPLE 8.6. **Regular rings:** If R is a von Neumann regular ring then, by 7.2 since R is coherent and every module is flat, the category of pp-pairs for R-Mod is mod-R.

EXAMPLE 8.7. **Finite abelian groups:** Given a skeletally small abelian category, we may ask for which definable category is it the functor category? For instance, take the category, fin-$\mathbb{Z}$, of finite abelian groups; can we identify $\mathcal{D} = \mathrm{Ex}(\text{fin-}\mathbb{Z}, \mathbf{Ab})$? We will do this somewhat indirectly.

First, note that fin-$\mathbb{Z}$ is a Serre subcategory of the category, mod-$\mathbb{Z}$, of finitely generated abelian groups. We will use the following result of Krause, where $\mathrm{Pinj}(\mathcal{D})$ denotes the full subcategory on the pure-injective objects of a definable category $\mathcal{D}$.

THEOREM 8.8. *([**21**, 5.1]) Suppose that $\mathcal{A}$ is a small abelian category and that $\mathcal{S}$ is a Serre subcategory. Consider the definable categories $\mathcal{D} = \mathrm{Ex}(\mathcal{A}, \mathbf{Ab})$, $\mathcal{D}'' = \mathrm{Ex}(\mathcal{S}, \mathbf{Ab})$, and $\mathcal{D}' = \mathrm{Ex}(\mathcal{A}/\mathcal{S}, \mathbf{Ab})$ (so $\mathcal{D}'$ is a definable subcategory of $\mathcal{D}$ and $\mathcal{D}''$ is what Krause terms a* **definable quotient category** *of $\mathcal{D}$).*

Then $\mathrm{Pinj}(\mathcal{D}'') \simeq \mathrm{Pinj}(\mathcal{D})/\mathrm{Pinj}(\mathcal{D}')$ where the latter is the stable quotient category of $\mathrm{Pinj}(\mathcal{D})$ with respect to $\mathrm{Pinj}(\mathcal{D}')$ - that is, the objects are those of $\mathrm{Pinj}(\mathcal{D})$ and the morphisms are those of $\mathrm{Pinj}(\mathcal{D})$ modulo those which factor through an object of $\mathrm{Pinj}(\mathcal{D}')$.

In fact, we will apply this to the inclusion $(\text{fin-}\mathbb{Z})^{\mathrm{op}} \to (\text{mod-}\mathbb{Z})^{\mathrm{op}}$ of opposite categories since the computation is a little easier and 8.8 gives us the whole definable category (all the modules in it being injective, in particular pure-injective).

The definable category corresponding to $(\text{mod-}\mathbb{Z})^{\mathrm{op}}$ is, by 7.3, Abs-$\mathbb{Z}$ = Inj-$\mathbb{Z}$ - the category of injective=divisible abelian groups, with Mod-$\mathbb{Q}$ being its definable subcategory annihilated by $(\text{fin-}\mathbb{Z})^{\mathrm{op}}$. So, by 8.8, the category $\mathrm{Ex}((\text{fin-}\mathbb{Z})^{\mathrm{op}}, \mathbf{Ab})$ is the category Inj-$\mathbb{Z}$/Mod-$\mathbb{Q}$ of divisible abelian groups modulo the category of $\mathbb{Q}$-modules. There are no non-zero morphisms from torsion modules to torsionfree ones, so this is just the category of torsion divisible abelian groups - arbitrary direct sums of copies of Prüfer modules.

Going back to the original, opposite, categories, we have the dual definable categories and, using that a definable category is determined by its pure-injectives ([**36**, 3.18]), we conclude that $\mathrm{Ex}(\text{fin-}\mathbb{Z}, \mathbf{Ab})$ is the category of reduced (i.e. without nonzero divisible submodules) torsionfree abelian groups.

EXAMPLE 8.9. **Nori motives:** Let $\mathcal{V}$ be a category of varieties, for example the category $\mathcal{V}_k$ of nonsingular projective varieties and regular maps over an algebraically closed field k. Roughly, the corresponding category of motives is a small abelian category built from $\mathcal{V}$ through which all (co)homology theories, regarded as functors from $\mathcal{V}$ to an abelian category, factor (essentially uniquely). For $\mathcal{V}_k$ there are constructions of "pure motives" whose properties depend on the (open) Grothendieck Standard Conjectures. The pure motives are also expected to be the semisimple objects of the category of "mixed motives" (corresponding to the category of all varieties over k); that category is expected to be recovered, as the heart of a motivic t-structure, from the category of Voevodsky "triangulated motives".

Nori constructed a quiver D_k, the vertices of which are the (X,Y,i) consisting of varieties X,Y with Y a closed subvariety of X, $i \in \mathbb{Z}, i \geq 0$. So this is a large quiver, in particular its path category is a ring in the broad (many-object) sense. The arrows of D_k are defined in such a way that (co)homology theories T are representations of this quiver. Nori then gave a (rather indirect) construction of an associated abelian category of motives for T, in particular taking T to be singular homology, $(X,Y,i) \mapsto \mathrm{H}^i(X(\mathbb{C}),Y(\mathbb{C}))$ to obtain the category of Nori motives.

Barbieri-Viale suggested that there should be a topos-theoretic construction of Nori's category, Caramello saw how to do this by building the syntactic category associated to a regular theory and the details are presented in [**7**]. In fact, the category that Nori constructs can be seen to have the universal property of the category $\mathcal{A}(T)$ where T is the representation of D_k given by singular homology, so this, and the more general T-motives from [**9**], are examples, of a rather different kind, of what we have described, see [**8**].

9. Model theory and interpretations

Many-sorted structures. The idea is that structures may have elements of essentially different kinds or, at least, in completely disjoint parts of the structure. Representations of quivers are natural examples: the vector spaces corresponding to distinct vertices are such separate parts; the term 'sorts' refers to these 'parts'. Even in the case of a module[5] M_R over a ring R, which is naturally 1-sorted, we may consider n-tuples of elements of M as elements of a new sort - that of n-tuples - (one for each n), thus turning M into a multi-sorted structure.

But why stop there? We can add, as new sorts, pp-definable subgroups $\phi(M)$ of M and of its finite powers; we can also add quotients of these by pp-definable equivalence relations. We're restricting to pp formulas in order that all sorts inherit the additive structure, and factoring by pp-definable equivalence relations comes down to forming factor groups $\phi(M)/\psi(M)$.

There are also many definable functions connecting these sorts: coordinate injections and projections in the case of product sorts, the canonical projection from $\phi(M)$ to $\phi(M)/\psi(M)$, but many more as we saw even in the example in Section 1.1.

The resulting (very-)many-sorted structure is denoted by $M^{\mathrm{eq}+}$. Since 'M' is a variable, we should consider the framework behind this. That framework, seen earlier, is that where we take the sorts $\phi(-)/\psi(-)$ to be objects and the pp-definable maps[6] to be arrows. This is an abelian category which, from this point of view, is denoted $\mathbb{L}_R^{\mathrm{eq}+}$ (and which has alternative descriptions, as in 4.6). Then applying a module M as an argument is the functor 'evaluation-at-M' from $\mathbb{L}_R^{\mathrm{eq}+}$ to **Ab** and, as such, it is exact (and was denoted $\widetilde{M}$ in 3.1). Equally, as we have seen, each object in $\mathbb{L}_R^{\mathrm{eq}+}$ can be regarded as a functor from Mod-R to **Ab**

Now, given M, each one of these new sorts $N = \phi(M)/\psi(M)$ of elements is a right module over the ring $S = \mathrm{End}_{\mathbb{L}_R^{\mathrm{eq}+}}(\phi/\psi)$ of pp-definable endomorphisms of the object ϕ/ψ in $\mathbb{L}_R^{\mathrm{eq}+}$. We think of this S-module as "definably contained",

[5]In the earlier parts of this paper, I used mostly left modules but, in this section I will use *right* S-modules, that is S^{op}-modules, to avoid too many ^{op}s appearing elsewhere in the discussion.

[6]That is, pp-definable on *every* R-module, equivalently pp-definable on M if the definable subcategory generated by M is Mod-R; if we meant just pp-definable on M then the resulting category would be a Serre quotient of $\mathbb{L}_R^{\mathrm{eq}+}$, see Section 4.3.

or "interpreted" in M_R. We could also, of course, from the S-module N build the structure $N^{\mathrm{eq}+}$. It might be that, within this structure, we could find a copy of the R-module M by taking a sort of $N^{\mathrm{eq}+}$ with its definable endomorphisms. In that case we would regard M_R and N_S as definably equivalent[7] since each can be constructed/found definably within the other. This is the relation of "being essentially the same" that we obtain for modules if we define them to be exact functors on abelian categories. Since we allow modules over 'multi-sorted rings', that is small preadditive categories, we might use more than one sort when we build new modules inside $^{\mathrm{eq}+}$ of a module. Also, if we just want the weaker relation of being definably contained/interpreted in (rather than definable equivalence) we don't, when interpreting new structures, have to take all the arrows between the chosen sorts in $\mathbb{L}_R^{\mathrm{eq}+}$. That is, we can use any not-necessarily-full subcategory when interpreting new modules in old.

The module M appearing above serves only to fix the definable category $\mathcal{D}$ (whether it be Mod-R or $\langle M \rangle$), so our choice of the sort ϕ/ψ gives us a functor from $\mathcal{D}$ to Mod-S, $S = \mathrm{End}(\phi/\psi)$, which also takes each module $M' \in \mathcal{D}$ to the S-module $\phi(M')/\psi(M')$. This functor will commute with direct limits and direct products. That is, it is an interpretation functor in the sense we defined earlier algebraically (Section 4.5) but which can also be defined in this model-theoretic way. This explains the terminology "interpretation functor".

We can also see why, though the interpretation functor above goes from $\mathcal{D}$ to Mod-S, the exact functor between the associated functor categories goes the other way. For, we found S as (a subring of) $\mathrm{End}_{\mathrm{fun}(\mathcal{D})}(\phi/\psi)$ in fun($\mathcal{D}$). That is, we have an additive functor from the preadditive category S to fun($\mathcal{D}$). By 3.1 that extends to an exact functor from $\mathrm{Ab}(S^{\mathrm{op}}) = \mathrm{fun}(\text{Mod-}S)$ to fun($\mathcal{D}$). The meaning is that, once we have interpreted the action of S on modules $M' \in \mathcal{D}$, the action of every sort of $\mathbb{L}_S^{\mathrm{eq}+}$ is thereby determined.

Of course, what we have just said about interpretation functors for modules over 1-sorted rings applies equally for modules over general small preadditive categories.

That pretty well wraps up what I want to say here. The above discussion of interpretations does, however, depend on the reader having read the rest of the paper so, below, is an independent, and more general, run through the relevant ideas of many-sorted structures, languages and interpretations.

9.1. Languages, structures and interpretations.

The idea. of an interpretation is, roughly, that B is interpretable in A if we can find B definably within A or, to put it another way, if we can see how to manufacture B from A in a definable way. An example from undergraduate mathematics, the formation of the ring $\mathbb{Q}$ from the ring $\mathbb{Z}$, illustrates this quite well: we take the set, $\mathbb{Z}^2$, of pairs of integers, at least those where the second coordinate is non-zero, and define an equivalence relation on those by $(a, b) \sim (c, d)$ if $ad = bc$ (a "definable" relation in the formal sense). We then factor $\mathbb{Z}^2$ by this relation, to get the domain of $\mathbb{Q}$. Then we define the addition and multiplication on $\mathbb{Q}$ in terms of representatives of equivalence classes in the obvious

[7] "Interpretation-equivalent" would fit better with standard terminology in model theory, where it is customary to distinguish definability (taking only definable subsets) and interpretability (also allowing factoring by definable equivalence relations). For our purposes, the distinction seems not worth making.

(again, formally definable) way. What we did there was to produce a new kind of element: starting with the usual elements of $\mathbb{Z}$ - the elements of the "home sort" - we formed the sort of ordered pairs, moved to the definable subset consisting of those where the second coordinate is non-zero, and then factored by a definable equivalence relation to get a new sort which was to be the domain of the new structure. Then we equipped the new sort with a couple of functions (and we could have named the constants 0 and 1 as well), again all done definably in terms of the original structure.

In summary, we can produce new sorts by taking definable subsets of existing ones, by forming finite products of sorts and by factoring sorts by definable equivalence relations. If we take a structure M and close it under these ways of forming new sorts, we get a collection of sorts of elements, usually denoted by M^{eq}. This is not just a collection of sets: there are maps connecting these sorts, for example the projections and injections between product sorts and component sorts, and also the map from a sort to its quotient by a given definable equivalence relation. In general, there are many additional definable maps between sorts and the most complete version of this structure, which we still denote M^{eq}, can naturally be regarded as a category, with the sorts as objects and the definable maps as morphisms. (In that sentence we blurred the distinction between the framework - the category - and the value of this at a particular structure M: for example, the construction above can be applied to any commutative domain. Once we give the definitions, however, the difference can be clarified.)

The details: The meaning of "definable" above is precise: it refers to definability in some formal language appropriate for the structures involved. Suppose that A is a structure of some, possibly many-sorted, **signature**; that means a specified set of sorts (disjoint containers for elements) and a set of function, relation and constants symbols where the sorts (of each domain variable, of the value variable in the case of a function and of each constant symbol) are specified. Suppose also that B is a structure of some signature, in general totally different from that of A. An **interpretation** of B in A consists of the following data: for each sort σ in the signature of B, a sort $\tau(\sigma)$ in the language (built from the signature[8]) of A^{eq}; for each function symbol f in the signature of B and with domain sort $\sigma_1 \times \cdots \times \sigma_n$ and value sort σ, a formula $\phi_f(x'_{\tau(\sigma_1)}, x''_{\tau(\sigma_2)}, \ldots, x^{(n)}_{\tau(\sigma_n)}, x_{\tau(\sigma)})$ in the language for A, where subscripts indicate the sorts of variables; for each relation symbol R in the signature of B and with sort $\sigma_1 \times \cdots \times \sigma_n$, a formula $\phi_R(x'_{\tau(\sigma_1)}, x''_{\tau(\sigma_2)} \ldots, x^{(n)}_{\tau(\sigma_n)})$; for each constant symbol c, of sort σ say, in the signature of B, a formula $\phi_c(x_{\tau(\sigma)})$, the formulas all in the language for A^{eq}.[9]

The conditions on this data are the obvious ones: first that ϕ_f should define a function from the product sort $\tau(\sigma_1) \times \cdots \times \tau(\sigma_n)$ to the sort $\tau(\sigma)$ and that ϕ_c should define a single element; these conditions may be written as sentences in

[8]Given a signature, we can, recursively using the function symbols, build terms from variables and any constants symbols. Then, using equality and any other relations, and then logical connectives and quantifiers, we build formulas of the language based on that signature. The notation for formulas like ϕ_f, ϕ_R shows the free=unquantified=substitutable variables occurring in that formula.

[9]Any formula in this expanded language is equivalent to some formula written using the original signature for A; the new sorts for imaginaries are convenient but not, at this point, essential.

the language for A; write T_τ for this set of sentences. Finally we require that B should be isomorphic to the structure for the signature of B which has, at sort σ, the set $\tau(\sigma)(A)$ and which has, for the interpretations of the function, relation and constants symbols of that signature, the functions, relations and constants defined in A by the corresponding ϕ-formulas. That is, we can definably find, or interpret, B within A.

The extent of an interpretation: This was a description of how a single structure may be interpreted in another, but it is clear that in any structure A' satisfying all the sentences in T_τ the data of the interpretation allows one to interpret in A' some structure B' for the signature of B.

Contravariance: Note that the data of the interpretation goes in one direction (it is a function from something associated to B to something associated to A) whereas the actual interpretation goes in the other direction (from some structures for the signature of A to some structures for that of B).

The additive context/pp-definability: For more detail on the above, one may consult, for instance [**20**, §§5.3, 5.4]. In the additive context, because we want the underlying additive structure to be inherited by definable and interpreted sets, we restrict to pp formulas since these always define subgroups. The restriction also has model-theoretic justification in the light of pp-elimination of quantifiers for modules, e.g. see [**29**, 2.13].[10] The notation $^{\mathrm{eq}+}$ is used in place of ^{eq} to indicate this difference.

Also, the languages that we have used in this paper are functional - no constants other than the (definable) 0 element in each sort and no relation symbols (other than equality). The exclusion of relation symbols can be got around. For instance if we want to look at structures consisting of a K-vector space and a specified subspace, then it would be natural to introduce a 1-ary relation symbol to pick out the subspace. But an alternative is to regard such structures as representations of the quiver A_2 (those where the kernel of the arrow is 0 - these form a definable subcategory of the category of K-representations of A_2). This kind of "coding-up" of relations by introducing new sorts and function symbols can be done in general, see [**23**, p. 703]. Not so for non-zero constants: they cannot be coded up in this way and permitting them *in the formal language* would move us from an additive to an affine world which is not very well-explored and where it is not at all clear that as nice a theory can be developed (e.g. see [**40**]).

The language for $\mathcal{R}$-modules: Suppose that $\mathcal{R}$ is a small preadditive category. We set up a language $\mathcal{L}_\mathcal{R}$ with a sort σ_P for each object P of $\mathcal{R}$ and with, for each morphism $f : P \to Q$, a function symbol, which we denote just by f, from sort σ_P to σ_Q. For each sort we add a binary function symbol (for addition in that sort) and a constant symbol (for the 0 of that sort). We are interested only in additive structures so we impose the axioms which state that the structure in each sort should be that of an abelian group, then we add those which state that each named function f as above should be additive, that is, a morphism of abelian groups from the group of elements of sort σ_P to the group of those of sort σ_Q. We also add axioms expressing the atomic diagram of $\mathcal{R}$; for instance if the

[10]This says roughly that formulas in the language for modules over any ring reduce to simple combinations of what one can say with pp formulas.

composition (or sum) of f and g is h then we add a formula saying so (in terms of their actions). An $\mathcal{L}_{\mathcal{R}}$-structure M which is a model of (that is, satisfies) all these axioms therefore assigns to each object P of $\mathcal{R}$ an abelian group $\sigma_P M$ and to each morphism $f : P \to Q$ of $\mathcal{R}$ a group homomorphism from $\sigma_P M$ to $\sigma_Q M$. In other words, M is a (covariant) additive functor from $\mathcal{R}$ to the category **Ab** of abelian groups. And conversely, every such additive functor is a model of these axioms.

Another language for modules: Suppose that $\mathcal{C}$ is any finitely accessible additive category with products. That is, the finitely presented objects form a skeletally small subcategory $\mathcal{C}^{\mathrm{fp}}$ and every object of $\mathcal{C}$ is a direct limit of finitely presented objects. For example any module category $\mathcal{R}$-Mod is such. We can set up a language, with a sort for each object of (a small but not necessarily skeletal version of) $\mathcal{C}^{\mathrm{fp}}$, a 0 and a symbol for addition on each sort, and a function symbol, from sort Q to sort P for each arrow $f : P \to Q$ in $\mathcal{C}^{\mathrm{fp}}$. The reason for contravariance is that we can now regard each object $C \in \mathcal{C}$ as a structure for this language, setting $\sigma_P C = (P, C)$ - the group of maps from P to C. Since every object of $\mathcal{C}$ is a direct limit of objects of $\mathcal{C}^{\mathrm{fp}}$, every object C is determined by its "elements", that is, by the morphisms from finitely presented objects to it. These morphisms we have split into separate sorts according their domains (for more detail see, [**34**, §18]).

In this language, a module M would be many-sorted, with a sort for each finitely presented module A and the group of elements of M of sort A being (A, M). Of course, the usual language is the restriction of this to the single sort (R, M), "usual" *via* the canonical way of identifying the elements of a module with the morphisms from the free module R to it. The other sorts all lie in $^{\mathrm{eq}+}$ of this single-sorted structure but it can be useful to have them there explicitly as part of a richer language.

Take, for an example of a finitely accessible category with products, the category of torsion abelian groups: the finite abelian groups are the finitely presented objects and product in the category is the the torsion subgroup of the usual product of groups. So the language has a sort for each finite abelian group (taking just the indecomposables would be enough, though then the category of sorts would not be an additive category) and the "elements" of sort, say, $\mathbb{Z}_n$, of a torsion group D form the group $(\mathbb{Z}_n, D)$ of elements of D of order dividing n.

Yet another language for modules: We could, of course, use the full "pp-imaginaries" language, based on $\mathbb{L}_{\mathcal{R}}^{\mathrm{eq}+}$, with a sort for each object of that category and a function symbol for each arrow. This contains the above languages, with each finitely presented module A being replaced by the corresponding representable functor $(A, -)$ (recall, 4.6, that $\mathbb{L}_{\mathcal{R}}^{\mathrm{eq}+}$ is equivalent to the category of finitely presented functors on finitely presented modules). This, furthermore, is the richest possible language which still has the same expressive power as the usual (1-sorted in the case of a ring R) language. "The same expressive power" means that everything which can be said in this richer language can be said (though at greater length) with a formula from the simpler languages. It is richest in the sense that if we were to repeat the $^{\mathrm{eq}+}$ construction, then we would obtain a category equivalent to $\mathbb{L}_{\mathcal{R}}^{\mathrm{eq}+}$, so would not have changed the language (other than adding more copies of things we already had). More discussion of these three kinds of languages is in [**35**].

Definable categories and purity: An alternative, but equivalent, definition of purity to the one given in Section 4.1 is that an embedding $f : M \to N$ of

$\mathcal{R}$-modules is pure if for every pp formula ϕ and tuple $\overline{a}$ from M (matching the free variables of ϕ) we have $\overline{a} \in \phi(M)$ iff $\overline{fa} \in \phi(N)$ (the direction $\Leftarrow$ being the point). This is a weakening of the definition of elementary embedding from model theory, which is this condition for *any* formula (not just pp formulas).

Definable subcategories of categories of modules were first studied in the context of the model theory of modules and this was reflected in the way that they were originally described. Each definable subcategory/subclass of Mod-R can be realised as the class of those modules which are direct summands of models of some complete theory T of modules whose class of models is closed under finite direct sums (hence also under arbitrary direct sums and direct products). This condition on a theory T is denoted $T = T^{\aleph_0}$. By pp-elimination of quantifiers for modules, each such class is described by closure of some set of pp-pairs (condition (iv) in 4.1). My recollection is that the algebraic characterisation came from discussions of Crawley-Boevey, Krause and others; the name is from [**13**].

Interpretation of additive structures, in summary: The process of interpretation can be said both in model-theoretic and algebraic terms. Model-theoretically, we start with the language $\mathcal{L}$ of $\mathcal{R}$-modules or of some more general definable category $\mathcal{C}$. We expand this language to the full pp-imaginary language $\mathcal{L}^{\mathrm{eq}+}(\mathcal{C})$. Then we choose a subset, Σ', of the expanded set of sorts and we choose a subset of the definable functions involving only these sorts and their finite products; that gives us the new, interpreted, language. Each of our original objects $C \in \mathcal{C}$ has a natural expansion, $C^{\mathrm{eq}+}$, to the pp-imaginaries language and this, restricted to the chosen collection of sorts and function symbols, is the structure $D = FC$ interpreted in C. The class of objects obtained this way might not be closed (among "Σ'-modules") under pure subobjects but [**36**, 3.8], after closing in this way, we obtain a definable subcategory $\mathcal{D}$.

Algebraically this is said as follows. We start with a small abelian category $\mathcal{A}$ and we choose a not necessarily full additive subcategory $\mathcal{B}$. The interpretation takes $C \in \mathrm{Ex}(\mathcal{A}, \mathbf{Ab})$ to its restriction to $\mathcal{B}$. We may as well extend C to an exact functor on the abelian subcategory of $\mathcal{A}$ generated by $\mathcal{B}$ (close $\mathcal{B}$ under finite direct sums, kernels and cokernels). So there is no loss in generality in taking $\mathcal{B}$ to be a not necessarily full abelian subcategory of $\mathcal{A}$. From each exact functor C on $\mathcal{A}$ we get an exact functor on this $\mathcal{B}$; we may not get all exact functors on $\mathcal{B}$ this way but, after closing under pure subfunctors, we do. Set $\mathcal{D} = \mathrm{Ex}(\mathcal{B}, \mathbf{Ab})$.

Thus, either way, we get an interpretation functor $F : \mathcal{C} \to \mathcal{D}$ between definable categories, and a corresponding exact functor $\mathbb{L}^{\mathrm{eq}+}(\mathcal{D}) = \mathrm{fun}(\mathcal{D}) \to \mathrm{fun}(\mathcal{C}) = \mathbb{L}^{\mathrm{eq}+}(\mathcal{C})$ between abelian categories.

References

[1] Jiří Adámek and Jiří Rosický, *Locally presentable and accessible categories*, London Mathematical Society Lecture Note Series, vol. 189, Cambridge University Press, Cambridge, 1994. MR1294136

[2] Murray Adelman, *Abelian categories over additive ones*, J. Pure Appl. Algebra **3** (1973), 103–117, DOI 10.1016/0022-4049(73)90026-1. MR0318265

[3] Nathan Broomhead, David Pauksztello, and David Ploog, *Discrete derived categories I: homomorphisms, autoequivalences and t-structures*, Math. Z. **285** (2017), no. 1-2, 39–89, DOI 10.1007/s00209-016-1690-1. MR3598804

[4] Maurice Auslander, *Coherent functors*, Proc. Conf. Categorical Algebra (La Jolla, Calif., 1965), Springer, New York, 1966, pp. 189–231. MR0212070

[5] Maurice Auslander, *Isolated singularities and existence of almost split sequences*, Representation theory, II (Ottawa, Ont., 1984), Lecture Notes in Math., vol. 1178, Springer, Berlin, 1986, pp. 194–242, DOI 10.1007/BFb0075297. MR842486

[6] Maurice Auslander, Idun Reiten, and Sverre O. Smalø, *Representation theory of Artin algebras*, Cambridge Studies in Advanced Mathematics, vol. 36, Cambridge University Press, Cambridge, 1995. MR1314422

[7] L. Barbieri-Viale, O. Caramello and L. Lafforgue: Syntactic categories for Nori motives, arXiv:1506.06113, 2015.

[8] L. Barbieri-Viale and M. Prest, Definable categories and T-motives, Rendiconti Sem. Mat. Univ. Padova, *to appear*.

[9] Luca Barbieri-Viale, $\mathbb{T}$-*motives*, J. Pure Appl. Algebra **221** (2017), no. 7, 1565–1588, DOI 10.1016/j.jpaa.2016.12.017. MR3614967

[10] Walter Baur, *On the elementary theory of quadruples of vector spaces*, Ann. Math. Logic **19** (1980), no. 3, 243–262, DOI 10.1016/0003-4843(80)90011-X. MR595014

[11] D. J. Benson, *Representations and cohomology. I*, Cambridge Studies in Advanced Mathematics, vol. 30, Cambridge University Press, Cambridge, 1991. Basic representation theory of finite groups and associative algebras. MR1110581

[12] K. Burke, Some Model-Theoretic Properties of Functor Categories for Modules, Doctoral Thesis, University of Manchester, 1994.

[13] William Crawley-Boevey, *Infinite-dimensional modules in the representation theory of finite-dimensional algebras*, Algebras and modules, I (Trondheim, 1996), CMS Conf. Proc., vol. 23, Amer. Math. Soc., Providence, RI, 1998, pp. 29–54. MR1648602

[14] Paul Eklof and Gabriel Sabbagh, *Model-completions and modules*, Ann. Math. Logic **2** (1970/1971), no. 3, 251–295, DOI 10.1016/0003-4843(71)90016-7. MR0277372

[15] Peter Freyd, *Representations in abelian categories*, Proc. Conf. Categorical Algebra (La Jolla, Calif., 1965), Springer, New York, 1966, pp. 95–120. MR0209333

[16] Pierre Gabriel, *Des catégories abéliennes* (French), Bull. Soc. Math. France **90** (1962), 323–448. MR0232821

[17] Lorna Gregory and Mike Prest, *Representation embeddings, interpretation functors and controlled wild algebras*, J. Lond. Math. Soc. (2) **94** (2016), no. 3, 747–766, DOI 10.1112/jlms/jdw055. MR3614927

[18] L. Gruson and C. U. Jensen, *Dimensions cohomologiques reliées aux foncteurs* $\varprojlim^{(i)}$ (French), Paul Dubreil and Marie-Paule Malliavin Algebra Seminar, 33rd Year (Paris, 1980), Lecture Notes in Math., vol. 867, Springer, Berlin-New York, 1981, pp. 234–294. MR633523

[19] Ivo Herzog, *Elementary duality of modules*, Trans. Amer. Math. Soc. **340** (1993), no. 1, 37–69, DOI 10.2307/2154545. MR1091706

[20] Wilfrid Hodges, *Model theory*, Encyclopedia of Mathematics and its Applications, vol. 42, Cambridge University Press, Cambridge, 1993. MR1221741

[21] Henning Krause, *Exactly definable categories*, J. Algebra **201** (1998), no. 2, 456–492, DOI 10.1006/jabr.1997.7252. MR1612398

[22] Henning Krause, *Smashing subcategories and the telescope conjecture—an algebraic approach*, Invent. Math. **139** (2000), no. 1, 99–133, DOI 10.1007/s002229900022. MR1728877

[23] T. G. Kucera and M. Prest, *Imaginary modules*, J. Symbolic Logic **57** (1992), no. 2, 698–723, DOI 10.2307/2275302. MR1169204

[24] Helmut Lenzing, *Endlich präsentierbare Moduln* (German), Arch. Math. (Basel) **20** (1969), 262–266, DOI 10.1007/BF01899297. MR0244322

[25] H. Lenzing, *Homological transfer from finitely presented to infinite modules*, Abelian group theory (Honolulu, Hawaii, 1983), Lecture Notes in Math., vol. 1006, Springer, Berlin, 1983, pp. 734–761, DOI 10.1007/BFb0103748. MR722664

[26] Barry Mitchell, *Rings with several objects*, Advances in Math. **8** (1972), 1–161, DOI 10.1016/0001-8708(72)90002-3. MR0294454

[27] S. Perera, Grothendieck Rings of Theories of Modules, Doctoral Thesis, University of Manchester, 2011, available at http://www.maths.manchester.ac.uk/~mprest/publications.

[28] Mike Prest, *Duality and pure-semisimple rings*, J. London Math. Soc. (2) **38** (1988), no. 3, 403–409, DOI 10.1112/jlms/s2-38.3.403. MR972125

[29] Mike Prest, *Model theory and modules*, London Mathematical Society Lecture Note Series, vol. 130, Cambridge University Press, Cambridge, 1988. MR933092

[30] Mike Prest, *Wild representation type and undecidability*, Comm. Algebra **19** (1991), no. 3, 919–929, DOI 10.1080/00927879108824178. MR1102994

[31] Mike Prest, *Remarks on elementary duality*, Ann. Pure Appl. Logic **62** (1993), no. 2, 185–205, DOI 10.1016/0168-0072(93)90174-C. Stability in model theory, III (Trento, 1991). MR1226307

[32] Mike Prest, *Interpreting modules in modules*, Ann. Pure Appl. Logic **88** (1997), no. 2-3, 193–215, DOI 10.1016/S0168-0072(97)00022-5. Joint AILA-KGS Model Theory Meeting (Florence, 1995). MR1600907

[33] Mike Prest, *Purity, spectra and localisation*, Encyclopedia of Mathematics and its Applications, vol. 121, Cambridge University Press, Cambridge, 2009. MR2530988

[34] Mike Prest, *Definable additive categories: purity and model theory*, Mem. Amer. Math. Soc. **210** (2011), no. 987, vi+109, DOI 10.1090/S0065-9266-2010-00593-3. MR2791358

[35] Mike Prest, *Model theory in additive categories*, Models, logics, and higher-dimensional categories, CRM Proc. Lecture Notes, vol. 53, Amer. Math. Soc., Providence, RI, 2011, pp. 231–244. MR2867973

[36] Mike Prest, *Definable additive categories: purity and model theory*, Mem. Amer. Math. Soc. **210** (2011), no. 987, vi+109, DOI 10.1090/S0065-9266-2010-00593-3. MR2791358

[37] Mike Prest, *Definable additive categories: purity and model theory*, Mem. Amer. Math. Soc. **210** (2011), no. 987, vi+109, DOI 10.1090/S0065-9266-2010-00593-3. MR2791358

[38] Mike Prest and Ravi Rajani, *Structure sheaves of definable additive categories*, J. Pure Appl. Algebra **214** (2010), no. 8, 1370–1383, DOI 10.1016/j.jpaa.2009.11.003. MR2593669

[39] Mike Prest, Philipp Rothmaler, and Martin Ziegler, *Absolutely pure and flat modules and "indiscrete" rings*, J. Algebra **174** (1995), no. 2, 349–372, DOI 10.1006/jabr.1995.1131. MR1334216

[40] Ravi Rajani and Mike Prest, *Model-theoretic imaginaries and coherent sheaves*, Appl. Categ. Structures **17** (2009), no. 6, 517–559, DOI 10.1007/s10485-008-9151-6. MR2564121

[41] Gabriel Sabbagh and Paul Eklof, *Definability problems for modules and rings*, J. Symbolic Logic **36** (1971), 623–649, DOI 10.2307/2272466. MR0313050

[42] Martin Ziegler, *Model theory of modules*, Ann. Pure Appl. Logic **26** (1984), no. 2, 149–213, DOI 10.1016/0168-0072(84)90014-9. MR739577

School of Mathematics, Alan Turing Building, University of Manchester, Manchester M13 9PL, United Kingdom

Email address: mprest@manchester.ac.uk

Contemporary Mathematics
Volume **716**, 2018
https://doi.org/10.1090/conm/716/14427

A representation-theoretic approach to recollements of abelian categories

Chrysostomos Psaroudakis

ABSTRACT. Recollement situations were introduced by Beilinson, Bernstein and Deligne in the context of triangulated categories in their seminal paper on perverse sheaves. This is mainly a survey paper on recollements of abelian categories with emphasis on representation-theoretic applications. A large part of the paper is devoted to structural and homological results of recollements of abelian categories. Our aim is to present a unified treatment of the theory of recollements in the context of abelian categories and present several applications in Representation Theory of Artin Algebras. We also include unpublished work with Jorge Vitória on compatible torsion pairs and recollements of hearts.

CONTENTS

2010 *Mathematics Subject Classification.* Primary 16E10, 16E65, 16G, 16G50, 16S50.

Key words and phrases. Recollements of abelian categories, torsion pairs, recollements of hearts, recollements of module categories, extension groups, homological embeddings, eventually homological isomorphisms, stratifying recollements, hereditary algebras, Gorenstein algebras, singularity categories, Cohen-Macaulay modules, (Fg) condition for Hochschild cohomology, representation dimension, finitistic dimension, dominant dimension, recollements of coherent functors.

The author is supported by Deutsche Forschungsgemeinschaft (DFG, grant KO 1281/14 − 1).

1. Introduction

Recollement situations were introduced in the context of triangulated categories by Beilinson, Bernstein and Deligne [**15**] in their axiomatization of Grothendieck's six functors for derived categories of sheaves obtained from stratifications of spaces. Without explicitly writing the definition of a recollement in the context of abelian categories, Beilinson, Bernstein and Deligne [**15**] had already implicitly considered an example of such a situation. More precisely, they had introduced the glueing method of t-structures along a recollement of triangulated categories. The latter situation gives rise to abelian categories by considering the hearts of the t-structures. It turns out that these abelian categories are related via a recollement diagram (see Section 4). In what follows, we first give the definition of a recollement in the context of abelian categories and we provide motivation for our study, we explain what is the goal of this survey and we give a sample of the main results in the paper. Finally we describe the structure of the paper.

Recollements of abelian categories are exact sequences of abelian categories

$$0 \longrightarrow \mathscr{A} \xrightarrow{\ \mathsf{i}\ } \mathscr{B} \xrightarrow{\ \mathsf{e}\ } \mathscr{C} \longrightarrow 0$$

where the inclusion functor i represents $\mathscr{A}$ as a Serre subcategory of $\mathscr{B}$ (i.e. $\mathscr{A}$ is closed under taking subobjects, quotients and extensions) and the functor e represents $\mathscr{C}$ as the quotient category $\mathscr{B}/\mathscr{A}$ (Gabriel quotient, see Section 2), enjoying the additional property that the inclusion functor i admits a left adjoint q and a right adjoint p, and the quotient functor e admits a left adjoint l and a right adjoint r. Throughout the paper such a recollement situation of abelian categories is denoted by the following diagram:

$$\xymatrix{\mathscr{A} \ar[r]^{\mathsf{i}} & \mathscr{B} \ar@/_1.5pc/[l]_{\mathsf{q}} \ar@/^1.5pc/[l]^{\mathsf{p}} \ar[r]^{\mathsf{e}} & \mathscr{C} \ar@/_1.5pc/[l]_{\mathsf{l}} \ar@/^1.5pc/[l]^{\mathsf{r}}} \qquad \mathsf{R_{ab}}(\mathscr{A},\mathscr{B},\mathscr{C})$$

One of the first and fundamental examples of a recollement situation of abelian categories appeared in the construction of perverse sheaves by MacPherson and Vilonen [**87**] (see Example 3.11). Motivated by this work, Franjou and Pirashvili [**37**] investigated when a recollement of abelian categories is equivalent to the one arising by the construction of MacPherson and Vilonen. In particular, they studied properties of recollements of abelian categories and they gave a necessary and sufficient condition for a morphism between recollements of abelian categories to be an equivalence.

In the context of Representation Theory, Cline, Parshall and Scott indicated the connections between recollements of derived categories, highest weight categories, and quasi-hereditary algebras, see [**25–27, 105**]. Typically recollement situations arise from pairs (R, e), where R is a ring and e is an idempotent element in R. In particular, any such pair induces a recollement situation between the module categories over the rings R, R/ReR and eRe (see Example 3.1). It will become clear in the sequel that this can be considered as the universal example of recollements of module categories. Moreover, Cline, Parshall and Scott [**27**] showed that a recollement of module categories as above induces a recollement at the level of derived module categories when the ideal ReR is stratifying. It is well known that the module category of a quasi-hereditary algebra is a highest weight category [**25**]. Moreover, it is also known that the crucial issue to understand a highest weight category is the general problem of lifting a recollement of abelian categories to a recollement at the level of derived categories (i.e. stratifying ideals in the case of module recollements). In this direction, Krause characterized in [**76**] highest weight categories via recollements of abelian categories which lift to derived categories.

In Algebraic Topology recollements of abelian categories appeared in the work of Kuhn [**78**–**80**] on the category of generic representations over a finite field $\mathbb{F}_q$. The latter category is used to study modules over the Steenrod algebra. We also refer to the work of Pirashvili [**106**] on polynomial functors. On the other hand, recollements of abelian categories appear also very naturally in Algebraic Geometry. More precisely, let X be a topological space and U an open subspace of X. Denote by F the complement of U in X. Let $\mathcal{O}_X$ be a sheaf of commutative rings on X and denote by $\mathcal{O}_U$, resp. $\mathcal{O}_F$, the restricted sheaves of rings on U, resp. F. Denote by $\mathsf{Mod}\text{-}\mathcal{O}_X$ the abelian category of sheaves of $\mathcal{O}_X$-modules and similarly we have the abelian categories $\mathsf{Mod}\text{-}\mathcal{O}_F$ and $\mathsf{Mod}\text{-}\mathcal{O}_U$. Then the latter categories are linked via a recollement situation $\mathsf{R}_{\mathsf{ab}}(\mathsf{Mod}\text{-}\mathcal{O}_F, \mathsf{Mod}\text{-}\mathcal{O}_X, \mathsf{Mod}\text{-}\mathcal{O}_U)$. It should be noted that various derived versions of the latter recollement appear in Algebraic Geometry. We refer to the book [**67**] for more details, see also Example 3.13 for a proof of the recollement situation $\mathsf{R}_{\mathsf{ab}}(\mathsf{Mod}\text{-}\mathcal{O}_F, \mathsf{Mod}\text{-}\mathcal{O}_X, \mathsf{Mod}\text{-}\mathcal{O}_U)$.

Classically, in Ring Theory, torsion, torsion-free theories in a module category over a ring R correspond bijectively with idempotent ideals in R. This is the well known bijection of Jans [**63**]. The latter is also in one to one correspondence with recollements of module categories. We generalise this classical result to recollements of abelian categories (see Section 6). On the other hand, and going back to Representation Theory, already Happel [**59**], and recently Angeleri Hügel, Koenig, Liu and Yang in a series of papers, studied connections between recollements of triangulated categories with tilting theory, homological conjectures and stratifications of derived categories of rings, see [**2**–**4**]. It should be noted that a good source of examples of recollements of derived module categories are the liftings of recollements of module categories at the derived level. Recently, recollements of derived module categories were investigated further in [**5**, **6**] using the theory of ladders of recollements of triangulated categories. Ladders of derived categories of rings have attracted a lot of attention due to their connection with several homological and K-theoretic invariants. Another important aspect of ladders of derived categories of rings is the relation to the problem of lifting or restricting recollements to different levels of derived categories, see [**5**, Theorem III]. Moreover, by [**5**, Section 5] the height of the ladder measures how far an algebra is from being derived simple [**122**], i.e. its derived category is not the middle term of a non-trivial recollement with outer terms derived categories of algebras. In this connection, it was shown in [**45**] (using the underlying recollement of module categories) that the derived category of the preprojective algebra of Dynkin type $\mathbb{A}_n$ admits a ladder.

It should be clear from the discussion so far that recollements of abelian or module categories and their derived counterparts appear quite naturally in various settings and are omnipresent in Representation Theory. We mentioned already that typically recollement situations arises from pairs (R, e), where R is a ring and e is an idempotent element of R. More generally, the endomorphism ring of a direct sum of two objects in an additive category gives rise to a recollement situation. In Sections 3 and 4 we present a series of examples of recollements of abelian categories.

Given a recollement $\mathsf{R_{ab}}(\mathscr{A},\mathscr{B},\mathscr{C})$ of abelian categories the *slogan* that we want to promote in this paper is twofold and can be summarised as follows:
(α) Encode information for $\mathscr{B}$ from the (possibly) simpler outer terms $\mathscr{A}$ and $\mathscr{C}$.
(β) Compare (homological) properties for $\mathscr{B}$ and $\mathscr{C}$ using the whole recollement situation $\mathsf{R_{ab}}(\mathscr{A},\mathscr{B},\mathscr{C})$, i.e. also the left side of $\mathsf{R_{ab}}(\mathscr{A},\mathscr{B},\mathscr{C})$.

We have provided motivation and background material on recollements of abelian categories. Our aim in this paper is to present a unified treatment of the theory of recollements of abelian categories based on the above slogans giving emphasis on representation-theoretic applications. This survey paper is mainly based on the papers [**108, 110, 111, 113**]. The main part of this work summarizes, from the above mentioned papers, several structural and homological results on recollements in the context of abelian categories that are used essentially in applications. We also include unpublished work with Jorge Vitória on compatible torsion pairs (Section 9) and on recollements of hearts (Section 10). For the reader's convenience below we collect a list of the main features of this paper:

- We present a series of examples of recollements in the context of abelian categories (including recollements of hearts).
- We show that a recollement whose terms are module categories is equivalent to one induced by an idempotent element [**110**]. We also provide an example of a recollement of $\mathsf{Mod}\text{-}R$ which is not equivalent to a recollement by module categories.
- We study recollements of abelian categories via torsion, torsion-free theories (TTF-triples for short). In particular, we show that there is a bijective correspondence between recollements of abelian categories and certain TTF-triples [**110**].
- We introduce compatible TTF-triples in a recollement $\mathsf{R_{ab}}(\mathscr{A},\mathscr{B},\mathscr{C})$ and show, using Happel-Reiten-Smalø tilt, how to produce new recollements of abelian categories from a given one.
- Given a recollement $\mathsf{R_{ab}}(\mathscr{A},\mathscr{B},\mathscr{C})$ we study how the extension groups along the exact functors $\mathsf{i}\colon \mathscr{A} \longrightarrow \mathscr{B}$ (left side) and $\mathsf{e}\colon \mathscr{B} \longrightarrow \mathscr{C}$ (right side) are related. For the left side, after introducing the notion of a homological embedding of an exact functor [**108**], we characterize it in terms of full embeddings at the level of derived categories [**111**]. For the right side, we characterize when the quotient functor e in a recollement $\mathsf{R_{ab}}(\mathscr{A},\mathscr{B},\mathscr{C})$ is an eventually homological isomorphism.
- We characterize when a recollement of module categories lifts to a recollement at the level of unbounded derived categories. This is due to Cline–Parshall–Scott and we provide a proof using the techniques in [**111**].
- We characterize when a recollement $\mathsf{R_{tr}}(\mathsf{D}(B),\mathsf{D}(A),\mathsf{D}(C))$ of derived module categories is stratifying and we show that this is the case when A is a finite dimensional hereditary algebra [**111**]. We also provide a counterexample due to Angeleri Hügel–Koenig–Liu–Yang [**6**] which shows that not every recollement of derived module categories is stratifying.
- Given an Artin algebra Λ with an idempotent element e of Λ, we present a context where we can compare the algebras Λ and $e\Lambda e$ with respect to Gorensteinness, singularity categories and the finite generation condition

(Fg) for the Hochschild cohomology [**113**]. Moreover, in this context we study the finitistic dimension, the representation dimension and the dominant dimension [**36, 54, 108**].

The structure of the paper section by section is as follows. In Section 2, we collect preliminary notions and results on recollements of abelian categories that are used throughout the paper. We analyse further the definition of a recollement situation in the context of abelian categories by recalling basic facts from the theory of localization of abelian categories due to Gabriel, see subsection 2.1. In the second subsection we define recollements of abelian categories and we provide several remarks on the definition. Moreover, we recall Kuhn's bijection on simple objects in a recollement diagram (see Proposition 2.9). The last subsection deals with the necessary preliminaries on torsion theories in abelian categories that are needed in the sequel. Finally, in Proposition 2.14 we extend the classical bijection between torsion pairs and idempotent radical functors to torsion, torsion-free triples.

In Section 3, we provide a variety of examples of recollements of abelian categories. We present several recollement situations appearing in the general context of Algebra and we close this section with two classical examples, one appearing in Algebraic Topology (Kuhn's Generic Representation Theory, see Example 3.14) and the other one in Algebraic Geometry (sheaves of commutative rings over schemes, see Example 3.13). We also include the trivial example of a splitting recollement which is used later in Section 10. In particular, we characterize in Proposition 3.9 when a recollement of abelian categories splits. The algebraic part of this section (i.e. recollements of module categories) is investigated further in Section 5.

In Section 4, we study recollements of hearts. These are recollements of abelian categories that are obtained from recollements of triangulated categories by considering the asssociated hearts of the glued t-structures, a very useful technique due to Beilinson, Bernstein and Deligne [**15**]. We start by recalling basic facts on t-structures and in Proposition 4.2 we provide a detailed proof that the hearts of the glueing give rise to a recollement situation. This recollement is used in Section 10.

Section 5 is devoted to recollements of module categories. More precisely, we prove in Theorem 5.7 that any recollement of $\mathsf{Mod}\text{-}R$, where R is any associative unital ring, whose terms are module categories is equivalent to a recollement induced by an idempotent element. This solves a conjecture made by Kuhn formulated in the setting of finite dimensional algebras. Along the way we discuss equivalences of recollements of abelian categories, see Lemma 5.2. This completes the material of the first subsection. In the second one, we present an example of a recollement of $\mathsf{Mod}\text{-}R$ which is not equivalent to a recollement by module categories (see Proposition 5.11).

In Section 6, we study recollements of abelian categories in connection with torsion theories. In particular, we show in Theorem 6.5 that there is a bijection between recollements of abelian categories $\mathsf{R}_{\mathsf{ab}}(\mathscr{A}, \mathscr{B}, \mathscr{C})$ and certain torsion, torsion-free triples in $\mathscr{B}$. As a byproduct of our investigations, we obtain in Corollary 6.8 a correspondence between equivalence classes of recollements of $\mathsf{Mod}\text{-}R$, two-sided idempotent ideals of R and torsion, torsion-free triples in $\mathsf{Mod}\text{-}R$. The latter two bijections recover the well known correspondence due to Jans [**63**].

Section 7 is devoted to a study of the homological theory of recollements of abelian categories. In particular, we present several results on the relation between the extension groups along the middle exact functors. This section is divided into two subsections. In the first subsection, we recall the notion of a homological embedding of an exact functor and in Theorem 7.4 we characterize when the exact functor $\mathsf{i}\colon \mathscr{A} \longrightarrow \mathscr{B}$ is a k-homological embedding, i.e. there are isomorphisms $\mathsf{Ext}^n_{\mathscr{A}}(A, A') \cong \mathsf{Ext}^n_{\mathscr{B}}(\mathsf{i}(A), \mathsf{i}(A'))$ for all A, A' in $\mathscr{A}$ and $0 \leq n \leq k$. Also, in Theorem 7.6 we show under certain conditions that an exact functor between abelian cateogries is a homological embedding if and only if the induced derived functor between the unbounded derived categories is fully faithful. We continue with Theorem 7.7, where we give necessary and sufficient conditions for the map $\mathsf{Ext}^n_{\mathscr{B}}(B, B') \longrightarrow \mathsf{Ext}^n_{\mathscr{B}}(\mathsf{e}(B), \mathsf{e}(B'))$ to be an isomorphism for all $0 \leq n \leq k$. A variation of the latter result for higher degrees is given in Theorem 7.11. In this connection, the notion of an eventually homological isomorphism is discussed and in Theorem 7.12 we characterize when the quotient functor in a recollement of abelian categories is an eventually homological isomorphism.

In Section 8, we reprove a result of Cline, Parshall and Scott [**27**] on lifting recollements of module categories to the level of unbounded derived categories. In particular, we show in Theorem 8.3 that a recollement of module categories lifts to derived level if and only if the restriction functor is a homological embedding. Finally, we discuss a result of Kalck and Yang [**66**] on lifting recollements. More precisely, we show in Theorem 8.4 that we can always lift a recollement of module categories to derived categories over differential graded algebras.

In Section 9, given a recollement $\mathsf{R}_{\mathsf{ab}}(\mathscr{A}, \mathscr{B}, \mathscr{C})$ we investigate when a torsion pair in $\mathscr{B}$ restricts to torsion pairs in $\mathscr{A}$ and $\mathscr{C}$. In particular, we show in Proposition 9.1 that a torsion pair in $\mathscr{B}$ always restricts to a torsion pair in $\mathscr{A}$, but for $\mathscr{C}$ we need some natural conditions. In that case, we call such torsion pairs compatible with respect to $\mathsf{R}_{\mathsf{ab}}(\mathscr{A}, \mathscr{B}, \mathscr{C})$. Moreover, in Corollary 9.3, we show that the natural torsion pairs that we always have in a recollement $\mathsf{R}_{\mathsf{ab}}(\mathscr{A}, \mathscr{B}, \mathscr{C})$ are compatible.

In Section 10, we investigate recollements of hearts in connection to compatible torsion pairs and the HRS-tilt (Happel–Reiten–Smalø tilt, see Theorem 9.4). We show in Proposition 10.2 that given a recollement of hearts and a compatible torsion pair we can produce another recollement of hearts. This is applied to the natural torsion pairs that we have in a recollement of abelian categories, see Corollary 10.3. As a consequence of the above results, and using different methods, we obtain in Corollary 10.4 a result by Happel, Reiten and Smalø [**60**] on the HRS-tilt of a torsion pair associated to the module category of a triangular matrix ring. However, the main result of this section is Theorem 10.1 which shows how from a recollement of hearts we can produce infinitely many new recollements of abelian categories. It is motivated from the results before as well as from the glueing method of Beilinson, Bernstein and Deligne discussed in Section 4.

In Section 11, we deal with the question whether every recollement of derived categories of rings is equivalent to a derived version of a recollement of module categories (i.e. every recollement of derived categories is stratifying). This section is divided into three subsections. In the first subsection, we provide in Theorem 11.2 a criterion for such an equivalence of recollements in terms of the glueing of tilting t-structures. In the second subsection, we show in Theorem 11.7 that any recollement of $\mathsf{D}(A)$, where A is a finite dimensional hereditary algebra, is stratifying. In the last subsection, we present an example due to Angeleri Hügel, Koenig, Liu and Yang [**6**] which shows that not any recollement of derived module categories is stratifying.

In Sections 12 and 13, which form a unit, we study applications of recollements of abelian categories in Representation Theory. More precisely, in Section 12 we compare Gorensteinness (Theorem 12.3), singularity categories and Cohen-Macaulay modules (Theorem 12.10 and Corollary 12.13), and the (Fg) condition for Hochschild cohomology (Theorem 12.8) for an algebra Λ and $e\Lambda e$, where e is an idempotent element of Λ. In Section 13, we study the problem of how various homological dimensions, including finitistic, representation, and dominant dimension, of the algebras involved in a recollement of module categories are related. Under certain natural conditions, we present several results on the finitistic dimension, see Theorem 13.1, Theorem 13.2, Theorem 13.3 and Theorem 13.4. Also, in Theorem 13.7 we relate the representation dimension of an Artin algebra Λ with the representation dimension of $e\Lambda e$. Moreover, in Theorem 13.8 we present an interesting relation between representation and finitistic dimension. In particular, we provide situations where the finitistic dimension of an Artin algebra is finite. Finally, we discuss in subsection 13.3 a homological relation between Λ and $e\Lambda e$ with respect to dominant dimension. This is due to Fang and Koenig [**36**].

In Section 14, given a recollement of triangulated categories we show that we can derive a recollement of abelian categories by considering the category of coherent functors, in the sense of Auslander [**8**], of the involved triangulated categories.

Conventions and Notation. In this paper, our subcategories are assumed to be closed under isomorphisms and direct summands. For a ring R we denote by $\mathsf{Mod}\text{-}R$ the category of left R-modules. Usually Λ (or A) stands for an Artin algebra and we denote by $\mathsf{mod}\text{-}\Lambda$ the category of finitely generated left Λ-modules. When we deal with finite dimensional algebras over a field, for simplicity the field is assumed to be algebraically closed. For all unexplained notions and results concerning the representation theory of Artin algebras we refer to [**14**]. Also, we refer to [**94**] for basics on the theory of triangulated categories. For an additive functor $F\colon \mathscr{A} \longrightarrow \mathscr{B}$ between additive categories, we denote by $\mathsf{Im}\, F = \{B \in \mathscr{B} \mid B \cong F(A) \text{ for some } A \in \mathscr{A}\}$ the essential image of F and by $\mathsf{Ker}\, F = \{A \in \mathscr{A} \mid F(A) = 0\}$ the kernel of F.

2. Localizations, Recollements and Torsion Pairs

In this section we recall notions and basic results on recollements of abelian categories which will be used throughout the paper. This section is split into three subsections. The first one deals with localization of abelian categories which is the heart of the definition of a recollement situation. In the second subsection we discuss recollements in the context of abelian categories and we provide several results and remarks in order to understand the definition. The last subsection is devoted to torsion pairs in abelian categories. This important notion will be used in the sequel and its link to recollements will become clear later.

2.1. Localization of Abelian Categories. Let $\mathscr{B}$ be an abelian category. A full subcategory $\mathscr{A}$ of $\mathscr{B}$ is called a **Serre subcategory** provided that it is closed under subobjects, quotients and extensions. Equivalently, $\mathscr{A}$ is a Serre subcategory of $\mathscr{B}$ if for every exact sequence

$$0 \longrightarrow X \longrightarrow Y \longrightarrow Z \longrightarrow 0$$

in $\mathscr{B}$, we have that Y lies in $\mathscr{A}$ if and only if X and Z lie in $\mathscr{A}$. A basic source of examples of Serre subcategories arise as kernels of exact functors between abelian categories. In particular, if $\mathsf{e}\colon \mathscr{B} \longrightarrow \mathscr{C}$ is an exact functor between abelian categories then the kernel $\mathsf{Ker}\,\mathsf{e}$ is a Serre subcategory of $\mathscr{B}$.

Given a Serre subcategory $\mathscr{A}$ of an abelian category $\mathscr{B}$, Gabriel constructed in [**40**] (see also [**41**]) the **quotient category** $\mathscr{B}/\mathscr{A}$ which is defined as follows. The objects in $\mathscr{B}/\mathscr{A}$ are the objects in $\mathscr{B}$. Given two objects M and N in $\mathscr{B}$, then the hom space in $\mathscr{B}/\mathscr{A}$ is defined as the following direct limit:

$$\mathsf{Hom}_{\mathscr{B}/\mathscr{A}}(M,N) = \varinjlim_{\substack{N' \leq N\colon N' \in \mathscr{A} \\ M' \leq M\colon M/M' \in \mathscr{A}}} \mathsf{Hom}_{\mathscr{B}}(M', N/N') \tag{2.1}$$

Note that the pairs of subobjects (M', N') such that both N' and M/M' belong to $\mathscr{A}$ form a directed set and therefore we obtain a direct system of abelian groups $\mathsf{Hom}_{\mathscr{B}}(M', N/N')$. We leave to the reader to check that $\mathscr{B}/\mathscr{A}$ is indeed a category. Moreover, there is a **quotient functor** $\mathsf{e}\colon \mathscr{B} \longrightarrow \mathscr{B}/\mathscr{A}$ which is exact and essentially surjective (is the identity on objects). Then, by Gabriel [**40**] the quotient category $\mathscr{B}/\mathscr{A}$ is abelian and the kernel of the quotient functor $\mathsf{e}\colon \mathscr{B} \longrightarrow \mathscr{A}$ is $\mathsf{Ker}\,\mathsf{e} = \mathscr{A}$. In other words, there is a short exact sequence of abelian categories:

$$0 \longrightarrow \mathscr{A} \xrightarrow{\ \mathsf{i}\ } \mathscr{B} \xrightarrow{\ \mathsf{e}\ } \mathscr{B}/\mathscr{A} \longrightarrow 0$$

where i denotes the inclusion functor. Moreover, we have the following universal property for $\mathscr{B}/\mathscr{A}$: Given any exact functor $\mathsf{e}'\colon \mathscr{B} \longrightarrow \mathscr{D}$ between abelian categories such that $\mathscr{A}$ is contained in $\mathsf{Ker}\,\mathsf{e}'$, there is a unique exact functor $\mathsf{e}''\colon \mathscr{B}/\mathscr{A} \longrightarrow \mathscr{D}$ such that $\mathsf{e}'' \circ \mathsf{e} = \mathsf{e}'$.

Consider now the class of all morphisms $f\colon M \longrightarrow N$ in $\mathscr{B}$ such that $\mathsf{Ker}\,f$ and $\mathsf{Coker}\,f$ belong to $\mathscr{A}$. Denote this class by $\mathsf{S}_{\mathscr{A}}$. Associated with this data, there is a **localization category** $\mathscr{B}[\mathsf{S}_{\mathscr{A}}^{-1}]$ of $\mathscr{B}$ with respect to $\mathsf{S}_{\mathscr{A}}$. It is characterized by the existence of a universal functor $\mathsf{Q}\colon \mathscr{B} \longrightarrow \mathscr{B}[\mathsf{S}_{\mathscr{A}}^{-1}]$ such that for any f in $\mathsf{S}_{\mathscr{A}}$ the map $\mathsf{Q}(f)$ is invertible, and for any functor $\mathsf{Q}'\colon \mathscr{B} \longrightarrow \mathscr{D}$ such that $\mathsf{Q}'(f)$ is invertible for any f in $\mathsf{S}_{\mathscr{A}}$ there is a unique functor $\mathsf{Q}''\colon \mathscr{B}[\mathsf{S}_{\mathscr{A}}^{-1}] \longrightarrow \mathscr{D}$ such that $\mathsf{Q}'' \circ \mathsf{Q} = \mathsf{Q}'$. Then, there is an equivalence of categories:

$$\mathscr{B}/\mathscr{A} \cong \mathscr{B}[\mathsf{S}_{\mathscr{A}}^{-1}]$$

Note that the abelian category $\mathscr{B}/\mathscr{A}$ is also known as the **Gabriel quotient** of $\mathscr{B}$ with respect to $\mathscr{A}$. For more details we refer to [**40, 41**].

Let $\mathscr{A}$ and $\mathscr{B}$ as above. The Serre subcategory $\mathscr{A}$ is called **localising** (respectively, **colocalising**) if the quotient functor $\mathsf{e}\colon \mathscr{B} \longrightarrow \mathscr{B}/\mathscr{A}$ admits a right (respectively, left) adjoint. Moreover, $\mathscr{A}$ is called **bilocalising** if it is both localising and colocalising. In this case, we have the following diagram of abelian categories and additive functors such that $(\mathsf{l}, \mathsf{e}, \mathsf{r})$ is an adjoint triple (this diagram is the predecessor of recollements of abelian categories, see subsection 2.2):

$$\mathscr{A} \xrightarrow{\ \mathsf{i}\ } \mathscr{B} \xrightarrow{\ \mathsf{e}\ } \mathscr{B}/\mathscr{A}, \qquad \mathsf{l}, \mathsf{r}\colon \mathscr{B}/\mathscr{A} \longrightarrow \mathscr{B} \tag{2.2}$$

These properties are related to the structure of the **perpendicular categories** of $\mathscr{A}$, i.e. subcategories orthogonal to $\mathscr{A}$ with respect to the pairings $\mathsf{Hom}_{\mathscr{B}}(-,-)$ and

$\mathsf{Ext}^1_{\mathscr{B}}(-,-)$ (in the sense of Yoneda). More precisely, the perpendicular categories are defined in this context as follows:

$$\begin{aligned} {}^{\perp}\mathscr{A} &= \{B \in \mathscr{B} \mid \mathsf{Hom}_{\mathscr{B}}(B,A) = 0 = \mathsf{Ext}^1_{\mathscr{B}}(B,A), \forall A \in \mathscr{A}\} \\ \mathscr{A}^{\perp} &= \{B \in \mathscr{B} \mid \mathsf{Hom}_{\mathscr{B}}(A,B) = 0 = \mathsf{Ext}^1_{\mathscr{B}}(A,B), \forall A \in \mathscr{A}\} \end{aligned}$$

We recall the following important result of Geigle and Lenzing.

THEOREM 2.1 ([**47**, Lemma 2.1, Proposition 2.2]). *The following statements hold for a Serre subcategory $\mathscr{A}$ of $\mathscr{B}$.*

(i) *The quotient functor $\mathsf{e}\colon \mathscr{B} \longrightarrow \mathscr{B}/\mathscr{A}$ induces fully faithful functors*
$$\mathscr{A}^{\perp} \longrightarrow \mathscr{B}/\mathscr{A} \quad \text{and} \quad {}^{\perp}\mathscr{A} \longrightarrow \mathscr{B}/\mathscr{A}.$$
(ii) *The functor $\mathsf{e}\colon \mathscr{A}^{\perp} \longrightarrow \mathscr{B}/\mathscr{A}$ is an equivalence if and only if $\mathscr{A}$ is localising, in which case a quasi-inverse for e is its right adjoint r.*
(iii) *The functor $\mathsf{e}\colon {}^{\perp}\mathscr{A} \longrightarrow \mathscr{B}/\mathscr{A}$ is an equivalence if and only if $\mathscr{A}$ is colocalising, in which case a quasi-inverse for e is its left adjoint l.*

The next result provides necessary and sufficient conditions for an exact functor between abelian categories to be a quotient functor.

THEOREM 2.2 ([**40**, **41**]). *Let $\mathsf{e}\colon \mathscr{B} \longrightarrow \mathscr{C}$ be an exact functor between abelian categories. Assume that there is an adjoint pair (e,r). The following are equivalent:*

(i) *The functor $\mathsf{r}\colon \mathscr{C} \longrightarrow \mathscr{B}$ is fully faithful.*
(ii) *The functor $\mathsf{r}\colon \mathscr{C} \longrightarrow (\mathsf{Ker}\,\mathsf{e})^{\perp}$ is an equivalence of categories.*
(iii) *The functor e induces an equivalence of categories:$\mathscr{B}/\mathsf{Ker}\,\mathsf{e} \xrightarrow{\simeq} \mathscr{C}$.*

In that case, we have:$(\mathsf{Ker}\,\mathsf{e})^{\perp} = \mathsf{Im}\,\mathsf{r}$ *and* ${}^{\perp}(\mathsf{Im}\,\mathsf{r}) = \mathsf{Ker}\,\mathsf{e}$.

We leave to the reader to check the details of this subsection. The standard references are the thesis of Gabriel [**40**] and the book of Gabriel and Zisman [**41**]. We also refer to the book of Faith [**35**]. Note that in [**41**] the authors don't speak about perpendicular categories, as they work in a non-additive setting in general. They rather speak of what is called local objects with respect to a class of morphisms, see also the paper by Krause on localization theory [**75**, Section 2].

2.2. Recollements of Abelian Categories. This section is devoted to the notion of a recollement of abelian categories. Recollements of triangulated categories were introduced by Beilinson-Bernstein-Deligne [**15**]. However, the definition of a recollement situation in the context of abelian categories was formulated later by Franjou-Pirashvili [**37**] and Kuhn [**79**].

DEFINITION 2.3. A **recollement situation** between abelian categories $\mathscr{A},\mathscr{B}$ and $\mathscr{C}$ is a diagram

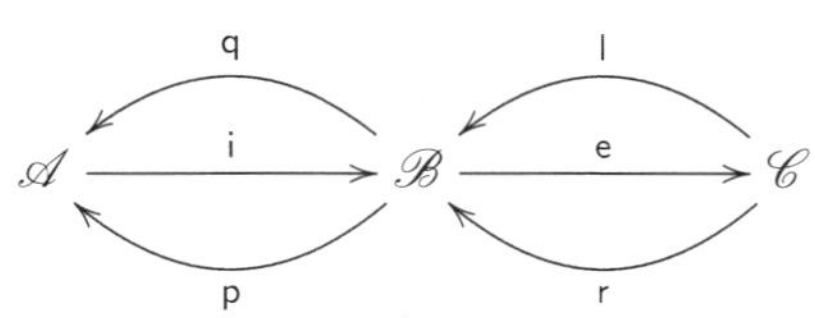

$\mathsf{R}_{\mathsf{ab}}(\mathscr{A},\mathscr{B},\mathscr{C})$

henceforth denoted by $(\mathscr{A},\mathscr{B},\mathscr{C})$, satisfying the following conditions:

1. $(\mathsf{l},\mathsf{e},\mathsf{r})$ is an adjoint triple.
2. $(\mathsf{q},\mathsf{i},\mathsf{p})$ is an adjoint triple.

3. The functors i, l, and r are fully faithful.
4. $\mathsf{Im}\,\mathsf{i} = \mathsf{Ker}\,\mathsf{e}$.

Let $\mathsf{R_{ab}}(\mathscr{A},\mathscr{B},\mathscr{C})$ be a recollement of abelian categories. We denote by $\mu\colon \mathsf{le} \longrightarrow \mathrm{Id}_{\mathscr{B}}$, resp. $\kappa\colon \mathsf{ip} \longrightarrow \mathrm{Id}_{\mathscr{B}}$, the counit of the adjoint pair (l,e), resp. (i,p), and by $\lambda\colon \mathrm{Id}_{\mathscr{B}} \longrightarrow \mathsf{iq}$, resp. $\nu\colon \mathrm{Id}_{\mathscr{B}} \longrightarrow \mathsf{re}$, the unit of the adjoint pair (q,i), resp. (e,r).

In the following remarks we isolate some easily established properties of a recollement situation which will be used in the sequel.

REMARK 2.4. Consider a recollement $\mathsf{R_{ab}}(\mathscr{A},\mathscr{B},\mathscr{C})$.

(i) The functors $\mathsf{e}\colon \mathscr{B} \longrightarrow \mathscr{C}$ and $\mathsf{i}\colon \mathscr{A} \longrightarrow \mathscr{B}$ are exact. This is clear since both functors are left and right adjoints.
(ii) The composition of functors $\mathsf{e}\circ\mathsf{i} = \mathsf{q}\circ\mathsf{l} = \mathsf{p}\circ\mathsf{r} = 0$. Clearly, by the fourth property of Definition 2.3 we have $\mathsf{e}\circ\mathsf{i} = 0$. We show that this yields that $\mathsf{q}\circ\mathsf{l} = 0$. Let A be an object in $\mathscr{A}$. We have the following isomorphisms:

$$\mathsf{Hom}_{\mathscr{A}}(\mathsf{ql}(C),A) \cong \mathsf{Hom}_{\mathscr{B}}(\mathsf{l}(C),\mathsf{i}(A)) \cong \mathsf{Hom}_{\mathscr{C}}(C,\mathsf{ei}(A)) = 0$$

This implies that $\mathsf{ql}(C) = 0$ for all C in $\mathscr{C}$. Similarly we get that $\mathsf{p}\circ\mathsf{r} = 0$.
(iii) The counit $\mathsf{qi} \xrightarrow{\cong} \mathrm{Id}_{\mathscr{A}}$ of the adjoint pair (q,i) is invertible and the unit $\mathrm{Id}_{\mathscr{A}} \xrightarrow{\cong} \mathsf{pi}$ of the adjoint pair (i,p) is invertible. This is standard in Category Theory, see for instance the books [**86**, **88**].
(iv) The counit $\mathsf{er} \xrightarrow{\cong} \mathrm{Id}_{\mathscr{C}}$ of the adjoint pair (e,r) is invertible and the unit $\mathrm{Id}_{\mathscr{C}} \xrightarrow{\cong} \mathsf{el}$ of the adjoint pair (l,e) is invertible.
(v) The functor $\mathsf{e}\colon \mathscr{B} \longrightarrow \mathscr{C}$ is essentially surjective. This is clear by (iv).
(vi) The functor i induces an equivalence between $\mathscr{A}$ and the Serre subcategory $\mathsf{Ker}\,\mathsf{e} = \mathsf{Im}\,\mathsf{i}$ of $\mathscr{B}$. In the sequel we shall view the above equivalence as an identification.
(vii) Since the exact functor $\mathsf{e}\colon \mathscr{B} \longrightarrow \mathscr{C}$ admits a fully faithful left adjoint and a fully faithful right adjoint, it follows from Theorem 2.2 that $\mathscr{A}$ is a localizing and colocalizing subcategory of $\mathscr{B}$ and there is an equivalence of abelian categories $\mathscr{B}/\mathscr{A} \simeq \mathscr{C}$. This means that the abelian category $\mathscr{C}$ is the Gabriel quotient of $\mathscr{B}$ with respect to $\mathscr{A}$. In particular, any recollement of abelian categories $\mathsf{R_{ab}}(\mathscr{A},\mathscr{B},\mathscr{C})$ induces a short exact sequence of abelian categories $0 \longrightarrow \mathscr{A} \longrightarrow \mathscr{B} \longrightarrow \mathscr{C} \longrightarrow 0$ (see subsection 2.1).

REMARK 2.5. Let $\mathsf{e}\colon \mathscr{B} \longrightarrow \mathscr{C}$ be an exact functor between abelian categories which is part of an adjoint triple $(\mathsf{l},\mathsf{e},\mathsf{r})$:

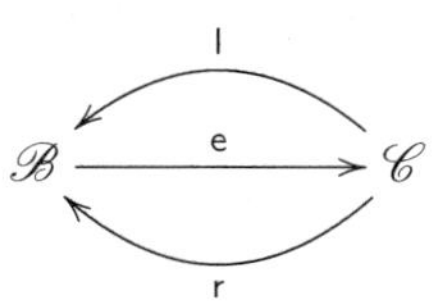

where l or r is fully faithful. We leave to the reader as an exercise to check that the functor l is fully faithful if and only if the functor r is fully faithful. Denote as before by $\mu\colon \mathsf{le} \longrightarrow \mathrm{Id}_{\mathscr{B}}$ the counit of the adjoint pair (l,e) and $\nu\colon \mathrm{Id}_{\mathscr{B}} \longrightarrow \mathsf{re}$ the unit of the adjoint pair (e,r). We define the functor $\mathsf{q}\colon \mathscr{B} \longrightarrow \mathsf{Ker}\,\mathsf{e}$ by $\mathsf{q}(B) = \mathsf{Coker}\,\mu_B$ on objects B in $\mathscr{B}$ and given a morphism $f\colon B \longrightarrow B'$ in $\mathscr{B}$ then $\mathsf{q}(f)$ is the induced morphism $\mathsf{Coker}\,\mu_B \longrightarrow \mathsf{Coker}\,\mu_{B'}$ which makes the next diagram commutative:

$$\begin{array}{ccccccc} \mathsf{le}(B) & \xrightarrow{\mu_B} & B & \longrightarrow & \mathsf{Coker}\,\mu_B & \longrightarrow & 0 \\ \downarrow{\scriptstyle \mathsf{le}(f)} & & \downarrow{\scriptstyle f} & & \downarrow{\scriptstyle \mathsf{q}(f)} & & \\ \mathsf{le}(B') & \xrightarrow{\mu_{B'}} & B' & \longrightarrow & \mathsf{Coker}\,\mu_{B'} & \longrightarrow & 0 \end{array}$$

It is easy to see that q is indeed an additive functor. If we apply the functor e to the sequence $\mathsf{le}(B) \longrightarrow B \longrightarrow \mathsf{Coker}\,\mu_B \longrightarrow 0$, we get by Remark 2.4 (iv) that the map $\mathsf{e}(\mu_B)$ is an isomorphism and thus the object $\mathsf{Coker}\,\mu_B$ lies in $\mathsf{Ker}\,\mathsf{e}$. Similarly, the assignment $B \mapsto \mathsf{p}(B) := \mathsf{Ker}\,\nu_B$ induces a functor $\mathsf{p}\colon \mathscr{B} \longrightarrow \mathsf{Ker}\,\mathsf{e}$. Then it is straightforward to check that $(\mathsf{q}, \mathsf{i}, \mathsf{p})$ is an adjoint triple, where $\mathsf{i}\colon \mathsf{Ker}\,\mathsf{e} \longrightarrow \mathscr{B}$ is the inclusion functor. We infer that the following diagram:

$$\begin{array}{ccccc} & \xleftarrow{\mathsf{q}} & & \xleftarrow{\mathsf{l}} & \\ \mathsf{Ker}\,\mathsf{e} & \xrightarrow{\mathsf{i}} & \mathscr{B} & \xrightarrow{\mathsf{e}} & \mathscr{C} \\ & \xleftarrow[\mathsf{p}]{} & & \xleftarrow[\mathsf{r}]{} & \end{array}$$

is a recollement of abelian categories.

REMARK 2.6. Let $\mathscr{B}$ be an abelian category and $\mathscr{A}$ a Serre subcategory of $\mathscr{B}$. Then from subsection 2.1 we have the exact sequence of abelian categories

$$0 \longrightarrow \mathscr{A} \xrightarrow{\mathsf{i}} \mathscr{B} \xrightarrow{\mathsf{e}} \mathscr{B}/\mathscr{A} \longrightarrow 0$$

It is natural to consider under what conditions on $\mathscr{A}$ and $\mathscr{B}$ we can get diagram (2.2), i.e. a left and a right adjoint of the quotient functor e. If this is the case, then from Remark 2.5 we obtain the recollement $\mathsf{R}_{\mathsf{ab}}(\mathscr{A}, \mathscr{B}, \mathscr{B}/\mathscr{A})$.

For instance, assume that $\mathscr{B}$ is a Grothendieck category and that $\mathscr{A}$ has coproducts, i.e. $\mathscr{A}$ is cocomplete (since $\mathscr{A}$ is abelian). Then we get from [**40**], see also [**35**, Theorem 15.11], that the subcategory $\mathscr{A}$ is localizing and therefore from the proof of Remark 2.5 we obtain the next half recollement:

$$\begin{array}{ccccc} \mathscr{A} & \xrightarrow{\mathsf{i}} & \mathscr{B} & \xrightarrow{\mathsf{e}} & \mathscr{B}/\mathscr{A} \\ & \xleftarrow[\mathsf{p}]{} & & \xleftarrow[\mathsf{r}]{} & \end{array}$$

This means that $0 \longrightarrow \mathscr{A} \longrightarrow \mathscr{B} \longrightarrow \mathscr{B}/\mathscr{A} \longrightarrow 0$ is an exact sequence of abelian categories, (i, p), (e, r) are adjoint pairs and r is fully faithful (see subsection 2.1).

Assume now that $\mathscr{B}$ is Grothendieck category and that $\mathscr{A}$ is a colocalizing subcategory of $\mathscr{B}$, i.e. the quotient functor e has a left adjoint. Then, it follows from [**91**, Proposition 2.2] that $\mathscr{A}$ is also localizing, i.e. e has a right adjoint. Hence, from Remark 2.5 we have the recollement of abelian categories $\mathsf{R}_{\mathsf{ab}}(\mathscr{A}, \mathscr{B}, \mathscr{B}/\mathscr{A})$. We remark that if $\mathscr{B}$ is Grothendieck, then so are $\mathscr{A}$ and $\mathscr{C}$. However, the converse is not in general true. We refer to the paper of Parra-Vitória [**104**] for a thorough study of properties of abelian categories in a recollement situation.

There are many cases where we deal with a subcategory $\mathscr{A}$ in $\mathscr{B}$ such that the inclusion functor admits a left or a right adjoint. Then it is also useful to know when we get an induced adjoint functor between $\mathscr{B}$ and $\mathscr{B}/\mathscr{A}$. In a sense, we consider the converse of Remark 2.5. For more discussions on this issue we refer the reader to Remark 6.4.

REMARK 2.7. Let $\mathsf{R_{ab}}(\mathscr{A},\mathscr{B},\mathscr{C})$ be a recollement of abelian categories and suppose that $\mathscr{B}$ has enough projective and injective objects. Then the functor $\mathsf{q}\colon \mathscr{B} \longrightarrow \mathscr{A}$ preserves projective objects since we have the adjoint pair (q,i) and the functor $\mathsf{i}\colon \mathscr{A} \longrightarrow \mathscr{B}$ is exact. Similarly, the functor $\mathsf{p}\colon \mathscr{B} \longrightarrow \mathscr{A}$ preserves injective objects.

Let A be an object of $\mathscr{A}$. Then for the object $\mathsf{i}(A)$ in $\mathscr{B}$ there exists an epimorphism $P \longrightarrow \mathsf{i}(A)$ with P in $\mathsf{Proj}\,\mathscr{B}$. From Remark 2.4 (iii), if we apply the functor q we obtain the epimorphism $\mathsf{q}(P) \longrightarrow A$ where $\mathsf{q}(P)$ lies in $\mathsf{Proj}\,\mathscr{A}$. This shows that the category $\mathscr{A}$ has enough projective objects and moreover $\mathsf{Proj}\,\mathscr{A} = \mathsf{add}\,\mathsf{q}(\mathsf{Proj}\,\mathscr{B})$. Similarly we derive that $\mathscr{A}$ has enough injective objects and $\mathsf{Inj}\,\mathscr{A} = \mathsf{add}\,\mathsf{p}(\mathsf{Inj}\,\mathscr{B})$.

Suppose that $\mathscr{C}$ has enough projective and injective objects. Then the functor $\mathsf{l}\colon \mathscr{C} \longrightarrow \mathscr{B}$ preserves projectives since we have the adjoint pair (l,e) and the functor $\mathsf{e}\colon \mathscr{B} \longrightarrow \mathscr{C}$ is exact. Similarly, the functor $\mathsf{r}\colon \mathscr{C} \longrightarrow \mathscr{B}$ preserves injective objects.

We continue with the following result which we will use throughout the paper, see [**37**] and [**108**].

PROPOSITION 2.8. *Let $\mathsf{R_{ab}}(\mathscr{A},\mathscr{B},\mathscr{C})$ be a recollement of abelian categories.*

(i) *There are equivalences of categories:*

$$\mathsf{p}\colon \{B \in \mathscr{B} \mid \mathsf{ip}(B) \cong B\} \xrightarrow{\simeq} \mathscr{A} \xleftarrow{\simeq} \{B \in \mathscr{B} \mid \mathsf{iq}(B) \cong B\}\colon \mathsf{q}$$

In particular, an object $B \in \mathscr{B}$ belongs to $\mathsf{i}(\mathscr{A})$ if and only if $B \cong \mathsf{iq}(B)$ if and only if $B \cong \mathsf{ip}(B)$.

(ii) *For every object B in $\mathscr{B}$, we have the exact sequences:*

$$0 \longrightarrow \mathsf{Ker}\,\mu_B \longrightarrow \mathsf{le}(B) \xrightarrow{\mu_B} B \xrightarrow{\lambda_B} \mathsf{iq}(B) \longrightarrow 0 \tag{2.3}$$

$$0 \longrightarrow \mathsf{ip}(B) \xrightarrow{\kappa_B} B \xrightarrow{\nu_B} \mathsf{re}(B) \longrightarrow \mathsf{Coker}\,\nu_B \longrightarrow 0 \tag{2.4}$$

where the objects $\mathsf{Ker}\,\mu_B$ and $\mathsf{Coker}\,\nu_B$ lie in $\mathsf{i}(\mathscr{A})$.

(iii) *For every C in $\mathscr{C}$, there is an exact sequence:*

$$0 \longrightarrow \mathsf{ipl}(C) \longrightarrow \mathsf{l}(C) \xrightarrow{N_C} \mathsf{r}(C) \longrightarrow \mathsf{iqr}(C) \longrightarrow 0 \tag{2.5}$$

PROOF. (i) From Remark 2.4 (iii) we have the natural isomorphisms $\mathsf{qi} \simeq \mathrm{Id}_{\mathscr{A}}$ and $\mathrm{Id}_{\mathscr{A}} \simeq \mathsf{pi}$. Then the statement follows immediately.

(ii) Let B be an object of $\mathscr{B}$. Then, from the counit $\mu_B\colon \mathsf{le}(B) \longrightarrow B$ of the adjoint pair (l,e), we have the exact sequence

$$0 \longrightarrow \mathsf{Ker}\,\mu_B \longrightarrow \mathsf{le}(B) \xrightarrow{\mu_B} B \longrightarrow \mathsf{Coker}\,\mu_B \longrightarrow 0$$

in $\mathscr{B}$. Applying the exact functor $\mathsf{e}\colon \mathscr{B} \longrightarrow \mathscr{C}$ we get from Remark 2.4 (iv) that the map $\mathsf{e}(\mu_B)$ is an isomorphism. This implies that $\mathsf{e}(\mathsf{Ker}\,\mu_B) = 0 = \mathsf{e}(\mathsf{Coker}\,\mu_B)$, i.e. $\mathsf{Ker}\,\mu_B$ and $\mathsf{Coker}\,\mu_B$ are objects of $\mathsf{i}(\mathscr{A})$. Since $\mathsf{Coker}\,\mu_B = \mathsf{i}(A)$ for some $A \in \mathscr{A}$ it follows from (i) that $\mathsf{iq}(\mathsf{Coker}\,\mu_B) \cong \mathsf{Coker}\,\mu_B$. Since the composition of functors $\mathsf{i} \circ \mathsf{q}$ is right exact and $\mathsf{q} \circ \mathsf{l} = 0$ by Remark 2.4 (ii), then the desired exact sequence

(2.3) follows follows from the following commutative diagram:

$$\begin{array}{ccccccc}
\mathsf{le}(B) & \xrightarrow{\mu_B} & B & \longrightarrow & \mathsf{Coker}\,\mu_B & \longrightarrow & 0 \\
\downarrow & & \downarrow{\scriptstyle \lambda_B} & & \simeq \downarrow{\scriptstyle \lambda_{\mathsf{Coker}\,\mu_B}} & & \\
0 & \longrightarrow & \mathsf{iq}(B) & \xrightarrow{\simeq} & \mathsf{iq}(\mathsf{Coker}\,\mu_B) & \longrightarrow & 0
\end{array}$$

Similarly we obtain the exact sequence (2.4).

(iii) Let C be an object in $\mathscr{C}$. Then from (2.3) we have the exact sequence

$$\mathsf{l}(C) \longrightarrow \mathsf{r}(C) \longrightarrow \mathsf{iqr}(C) \longrightarrow 0$$

where the map $\mathsf{l}(C) \longrightarrow \mathsf{r}(C)$ is the composition of the isomorphism $\mathsf{l}(C) \cong \mathsf{ler}(C)$ with the counit map $\mu_{\mathsf{r}(C)}$. Similarly, from (2.4) we obtain the exact sequence

$$0 \longrightarrow \mathsf{ipl}(C) \longrightarrow \mathsf{l}(C) \longrightarrow \mathsf{r}(C)$$

where $\mathsf{l}(C) \longrightarrow \mathsf{r}(C)$ is the composition of the unit map $\nu_{\mathsf{l}(C)} \colon \mathsf{l}(C) \longrightarrow \mathsf{rel}(C)$ with the isomorphism $\mathsf{rel}(C) \cong \mathsf{r}(C)$ (see Remark 2.4). Putting together the above two exact sequences we obtain the exact sequence (2.5). □

Given a recollement $\mathsf{R_{ab}}(\mathscr{A}, \mathscr{B}, \mathscr{C})$ and thinking that this situation is a specially well-behaved decomposition of the underlying category, it is natural to consider if there is a way to get the simples in $\mathscr{B}$ from simple objects of $\mathscr{A}$ and $\mathscr{C}$. In this direction, Kuhn [**78**] showed that this is always the case for the simple objects in a recollement of abelian categories. Recall that an object X in an abelian category $\mathscr{A}$ is simple, if X has no proper nonzero subobject.

For this purpose we need one more functor that exists in $\mathsf{R_{ab}}(\mathscr{A}, \mathscr{B}, \mathscr{C})$. Consider the exact sequence (2.5) in Proposition 2.8. The map N is called the norm [**37**]. Associated with the norm there is a functor $\mathsf{c} \colon \mathscr{C} \longrightarrow \mathscr{B}$, given by $\mathsf{c}(C) = \mathsf{Im}\, N_C$ on objects C in $\mathscr{C}$, called the **intermediate extension functor**. Kuhn defined an extension of an object C in $\mathscr{C}$ to be an object B in $\mathscr{B}$ with an isomorphism $\mathsf{e}(B) \cong C$. In this respect, we see immediately that $\mathsf{c}(C)$ is an extension of C in $\mathscr{C}$ (just apply the functor e to the norm map and use Remark 2.4). The key observation is that the intermediated extension functor preserves simple objects. In this connection, we have the following result.

PROPOSITION 2.9 ([**78**, Proposition 4.7]). *Let $\mathsf{R_{ab}}(\mathscr{A}, \mathscr{B}, \mathscr{C})$ be a recollement. There is a bijection between sets of isomorphism classes of simple objects:*

$$\{\textit{simples in } \mathscr{A}\} \coprod \{\textit{simples in } \mathscr{C}\} \longleftrightarrow \{\textit{simples in } \mathscr{B}\}$$

given by sending a simple A in $\mathscr{A}$ to $\mathsf{i}(A)$ and a simple C in $\mathscr{C}$ to $\mathsf{c}(C)$.

PROOF. We only sketch how the bijection works. First, the functor $\mathsf{i} \colon \mathscr{A} \longrightarrow \mathscr{B}$ induces a bijection between the simple objects in $\mathscr{A}$ and the simple objects in $\mathscr{B}$ which belong to $\mathsf{Ker}\,\mathsf{e}$. Then, it remains to show that the quotient functor $\mathsf{e} \colon \mathscr{B} \longrightarrow \mathscr{C}$ and the intermediate extension functor $\mathsf{c} \colon \mathscr{C} \longrightarrow \mathscr{B}$ induce mutually inverse maps between the simple objects of $\mathscr{B}$ which are not annihilated by e and the simple objects of $\mathscr{C}$. More precisely, it suffices to show that if B is a simple object in $\mathscr{B}$ such that $\mathsf{e}(B)$ is not zero then $\mathsf{e}(B)$ is simple in $\mathscr{C}$ and $\mathsf{ce}(B) \cong B$. We refer the reader to the initial proof of Kuhn for the details. □

For more on the intermediate extension functor we refer to [**78**, Section 4]. We also remark that properties of the intermediate extension functor were used by Crawley-Boevey and Sauter in [**29**] to construct desingularizations of quiver Grassmannians, orbit closures in representation varieties and their desingularizations.

2.3. Torsion Pairs. In this subsection we discuss some aspects of torsion, torsion-free triples, TTF-triples for short, in abelian categories. Throughout this section, $\mathscr{A}$ denotes an abelian category and all subcategories considered are strict[1] (i.e. closed under isomorphisms). We start by defining torsion pairs. For more details on torsion theories in the context of abelian categories we refer to [**30**, **120**].

DEFINITION 2.10. A **torsion pair** in $\mathscr{A}$ is a pair $(\mathcal{X}, \mathcal{Y})$ of full subcategories satisfying the following conditions:

(i) $\mathsf{Hom}_{\mathscr{A}}(\mathcal{X}, \mathcal{Y}) = 0$, i.e. $\mathsf{Hom}_{\mathscr{A}}(X, Y) = 0 \ \forall X \in \mathcal{X}, \ \forall Y \in \mathcal{Y}$;
(ii) For every object $A \in \mathscr{A}$, there are objects X_A in $\mathcal{X}$ and Y^A in $\mathcal{Y}$ and a short exact sequence

$$0 \longrightarrow X_A \longrightarrow A \longrightarrow Y^A \longrightarrow 0$$

Given a torsion pair $(\mathcal{X}, \mathcal{Y})$ in $\mathscr{A}$, we say that $\mathcal{X}$ is a **torsion class** and $\mathcal{Y}$ is a **torsion-free class**. It follows easily from the definition that

$$\begin{aligned} \mathcal{X} &= {}^{\circ}\mathcal{Y} := \{A \in \mathscr{A} \mid \mathsf{Hom}_{\mathscr{A}}(A, Y) = 0, \forall\, Y \in \mathcal{Y}\} \\ \mathcal{Y} &= \mathcal{X}^{\circ} := \{A \in \mathscr{A} \mid \mathsf{Hom}_{\mathscr{A}}(X, A) = 0, \forall\, X \in \mathcal{X}\} \end{aligned}$$

The assignments $A \mapsto \mathsf{R}_{\mathcal{X}}(A) := X_A$ and $A \mapsto \mathsf{L}_{\mathcal{Y}}(A) := Y^A$ yield additive functors $\mathsf{R}_{\mathcal{X}} \colon \mathscr{A} \longrightarrow \mathcal{X}$ and $\mathsf{L}_{\mathcal{Y}} \colon \mathscr{A} \longrightarrow \mathcal{Y}$ respectively. Moreover, we have the adjoint pairs $(\mathsf{i}_{\mathcal{X}}, \mathsf{R}_{\mathcal{X}})$ and $(\mathsf{L}_{\mathcal{Y}}, \mathsf{i}_{\mathcal{Y}})$, where $\mathsf{i}_{\mathcal{X}}$ and $\mathsf{i}_{\mathcal{Y}}$ are the inclusion functors:

(2.6)
$$\mathcal{X} \underset{\mathsf{R}_{\mathcal{X}}}{\overset{\mathsf{i}_{\mathcal{X}}}{\rightleftarrows}} \mathcal{T} \underset{\mathsf{i}_{\mathcal{Y}}}{\overset{\mathsf{L}_{\mathcal{Y}}}{\rightleftarrows}} \mathcal{Y}$$

Hence, $\mathcal{X}$ (respectively, $\mathcal{Y}$) is a coreflective (respectively, reflective) subcategory of $\mathscr{A}$. Moreover, the endofunctors $\mathsf{i}_{\mathcal{X}}\mathsf{R}_{\mathcal{X}}$ and $\mathsf{i}_{\mathcal{Y}}\mathsf{L}_{\mathcal{Y}}$ satisfy the following:

- $\mathsf{i}_{\mathcal{X}}\mathsf{R}_{\mathcal{X}}$ is a **radical functor**. This means that there is natural transformation $\mu \colon \mathsf{i}_{\mathcal{X}}\mathsf{R}_{\mathcal{X}} \longrightarrow \mathrm{Id}_{\mathscr{A}}$ such that μ_A is a monomorphism for all A in $\mathscr{A}$ and $\mathsf{i}_{\mathcal{X}}\mathsf{R}_{\mathcal{X}}(\mathsf{Coker}\,\mu_A) = 0$.
- $\mathsf{i}_{\mathcal{Y}}\mathsf{L}_{\mathcal{Y}}$ is a **coradical functor**. This means that there is a natural transformation $\nu \colon \mathrm{Id}_{\mathscr{A}} \longrightarrow \mathsf{i}_{\mathcal{Y}}\mathsf{L}_{\mathcal{Y}}$ such that μ_A is an epimorphism for all A in $\mathscr{A}$ and $\mathsf{i}_{\mathcal{Y}}\mathsf{L}_{\mathcal{Y}}(\mathsf{Ker}\,\nu_A) = 0$.
- Both $\mathsf{i}_{\mathcal{X}}\mathsf{R}_{\mathcal{X}}$ and $\mathsf{i}_{\mathcal{Y}}\mathsf{L}_{\mathcal{Y}}$ are **idempotent**, i.e. both $\mu_{\mathsf{i}_{\mathcal{X}}\mathsf{R}_{\mathcal{X}}(A)}$ and $\nu_{\mathsf{i}_{\mathcal{Y}}\mathsf{L}_{\mathcal{Y}}(A)}$ are isomorphisms for all A in $\mathscr{A}$.

It should be noted that there are bijections between torsion pairs in $\mathscr{A}$, idempotent radical functors $F \colon \mathscr{A} \longrightarrow \mathscr{A}$ and idempotent coradical functors $G \colon \mathscr{A} \longrightarrow \mathscr{A}$. Thus, the endofunctors $\mathsf{i}_{\mathcal{X}}\mathsf{R}_{\mathcal{X}}$ and $\mathsf{i}_{\mathcal{Y}}\mathsf{L}_{\mathcal{Y}}$ determine the torsion pair uniquely. For more details we refer to [**30**, Theorem 2.8] and [**99**, Theorem 1.2]. Often, torsion and torsion-free classes can be identified by closure properties. Recall that $\mathscr{A}$ is well-powered [**120**] if the class of subobjects of any given object forms a set.

[1] In the literature, the term "replete" is also used.

PROPOSITION 2.11 ([**30**, Theorem 2.3]). *Let $\mathscr{A}$ be a well-powered, complete and cocomplete abelian category. A full subcategory $\mathcal{X}$ is a torsion (respectively, torsion-free) class if and only if it is closed under quotients, extensions and coproducts (respectively, subobjects, extensions and products).*

Recall that a torsion pair $(\mathcal{X}, \mathcal{Y})$ in $\mathscr{A}$ is **hereditary** if $\mathcal{X}$ is closed under subobjects and **cohereditary** if $\mathcal{Y}$ is closed under quotients. We will be interested in classes which are both torsion and torsion-free.

DEFINITION 2.12. A triple $(\mathcal{X}, \mathcal{Y}, \mathcal{Z})$ of full subcategories of $\mathscr{A}$ is called a **torsion torsion-free triple** (**TTF-triple** for short, and $\mathcal{Y}$ is a **TTF-class**) if $(\mathcal{X}, \mathcal{Y})$ and $(\mathcal{Y}, \mathcal{Z})$ are torsion pairs.

Clearly, if $(\mathcal{X}, \mathcal{Y}, \mathcal{Z})$ is a TTF-triple in $\mathscr{A}$, then the torsion pair $(\mathcal{X}, \mathcal{Y})$ is cohereditary and $(\mathcal{Y}, \mathcal{Z})$ is hereditary. By Proposition 2.11, when $\mathscr{A}$ is well-powered, complete and cocomplete, a full subcategory $\mathcal{Y}$ of $\mathscr{A}$ is a TTF-class if and only if it is closed under products, coproducts, extensions, subobjects and quotients.

In Ring Theory, TTF-triples are well understood due to the following result of Jans, which establishes a bijection between TTF-triples in $\mathsf{Mod}\text{-}R$ and idempotent ideals, i.e. two sided ideals I of R such that $I^2 = I$. We state Jans' correspondence below but it will be proved in Section 6, see Corollary 6.8, using our results on TTF-triples of abelian categories. In particular, we will also show that the latter are in bijection with recollements of module categories.

THEOREM 2.13 ([**63**, Corollary 2.2]). *There is a bijection between TTF-triples in $\mathsf{Mod}\text{-}R$ and idempotent ideals of the ring R.*

We close this section with an adaptation of the classical bijection between torsion pairs and idempotent radicals to TTF-triples. This will be used later to prove Theorem 2.13. It is proved in [**110**, Proposition 3.1] but we sketch below the proof.

PROPOSITION 2.14. *Let $\mathscr{A}$ be a well-powered, complete and cocomplete abelian category $\mathscr{A}$. There are bijections between the following classes:*

(i) *TTF-triples in $\mathscr{A}$.*
(ii) *Left exact radical functors $F\colon \mathscr{A} \longrightarrow \mathscr{A}$ preserving products.*
(iii) *Right exact coradical functors $G\colon \mathscr{A} \longrightarrow \mathscr{A}$ preserving coproducts.*

PROOF. (i)$\Longrightarrow$(ii): Let $(\mathcal{X}, \mathcal{Y}, \mathcal{Z})$ be a TTF-triple in $\mathscr{A}$. The TTF-class $\mathcal{Y}$ is a bireflective subcategory in $\mathscr{A}$ and, hence, the inclusion functor $\mathsf{i}_{\mathcal{Y}}\colon \mathcal{Y} \longrightarrow \mathscr{A}$ admits a left adjoint $\mathsf{L}_{\mathcal{Y}}$ and a right adjoint $\mathsf{R}_{\mathcal{Y}}$, see diagram 2.6. Thus, $\mathsf{i}_{\mathcal{Y}}$ is exact and $\mathsf{i}_{\mathcal{Y}}\mathsf{R}_{\mathcal{Y}}$ is a left exact radical functor preserving products. We define a correspondence:

$$\Phi\colon (\mathcal{X}, \mathcal{Y}, \mathcal{Z}) \longmapsto (\mathsf{i}_{\mathcal{Y}}\mathsf{R}_{\mathcal{Y}}\colon \mathscr{A} \longrightarrow \mathscr{A}).$$

(ii)$\Longrightarrow$(i): Given a left exact radical functor $F\colon \mathscr{A} \longrightarrow \mathscr{A}$ preserving products, it is easy to see that F is idempotent and

$$\big(\mathcal{Y}_F := \{A \in \mathscr{A} \mid F(A) = A\}, \mathcal{Y}_F^{\circ} = \mathsf{Ker}\, F\big)$$

is a hereditary torsion pair (see also [**120**, Proposition VI.1.7]). Since F preserves products, $\mathcal{Y}_F$ is closed under products and thus, Proposition 2.11 impies that $\mathcal{Y}_F$ is a TTF-class. Hence, we can associate a TTF-triple to F as follows:

$$\Psi\colon F \longmapsto ({}^{\circ}\mathcal{Y}_F, \mathcal{Y}_F, \mathcal{Y}_F^{\circ})$$

We leave to the reader to check that Φ and Ψ are inverse correspondences. A bijection between (i) and (iii) can be obtained dually. $\square$

We remark that for a functor to be left exact and to preserve products is the same as to commute with limits and dually to be right exact and to preserve coproducts is the same as to commute with colimits (see [**88**]).

3. Examples of Recollements of Abelian Categories

In this section we give a variety of examples of recollements of abelian categories. Our aim is to show that recollement situations of abelian categories are omnipresent in several branches of mathematics (Algebra, Geometry and Topology).

EXAMPLE 3.1. (Idempotents) Let R be a ring and e an idempotent element in R. Then we have the following recollement:

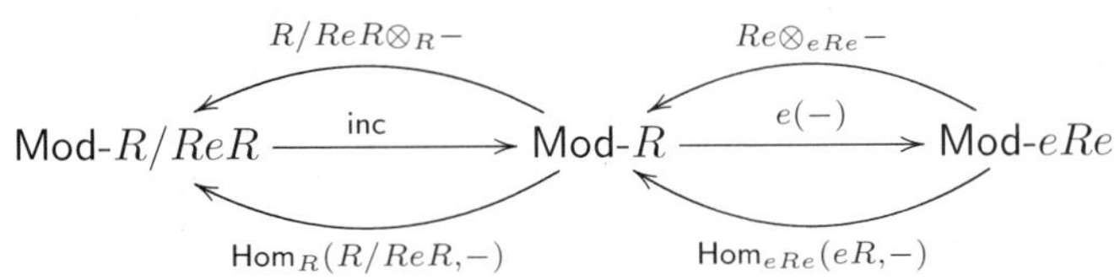

where the category $\mathsf{Mod}\text{-}R/ReR$ of (left) modules over R/ReR is the kernel of the functor $\mathsf{e} = e(-)\colon \mathsf{Mod}\text{-}R \longrightarrow \mathsf{Mod}\text{-}eRe$. Note that if R is semiprimary, then any idempotent ideal I is of the form ReR for an idempotent $e \in R$, see [**33**]. Hence, idempotent ideals in semiprimary rings induce recollements of module categories. The recollement $\mathsf{R}_{\mathsf{ab}}(\mathsf{Mod}\text{-}R/ReR, \mathsf{Mod}\text{-}R, \mathsf{Mod}\text{-}eRe)$ is called **the recollement of** $\mathsf{Mod}\text{-}R$ **induced by the idempotent element** e.

We now turn our attention to recollement arising from a ring with two idempotent elements.

EXAMPLE 3.2. (Morita Contexts) Let R, S be rings, M an S-R-bimodule and N an R-S-bimodule. Let $\phi\colon M \otimes_R N \longrightarrow S$ be an S-S-bimodule homomorphism and let $\psi\colon N \otimes_S M \longrightarrow R$ be an R-R-bimodule homomorphism. Then from the Morita context $\mathcal{M} = (R, N, M, S, \phi, \psi)$ we define the Morita ring:

$$\Lambda_{(\phi,\psi)} = \begin{pmatrix} R & {}_RN_S \\ {}_SM_R & S \end{pmatrix}$$

where the addition of elements of Λ is componentwise and multiplication is given as follows:

$$\begin{pmatrix} r & n \\ m & s \end{pmatrix} \cdot \begin{pmatrix} r' & n' \\ m' & s' \end{pmatrix} = \begin{pmatrix} rr' + \psi(n \otimes m') & rn' + ns' \\ mr' + sm' & ss' + \phi(m \otimes n') \end{pmatrix}$$

For all $m, m' \in M$ and $n, n' \in N$, the following conditions:

$$\phi(m \otimes n)m' = m\psi(n \otimes m') \quad \text{and} \quad n\phi(m \otimes n') = \psi(n \otimes m)n'$$

are needed for the multiplication in $\Lambda_{(\phi,\psi)}$ to be associative.

Recall from [**52**] that the $\Lambda_{(\phi,\psi)}$-modules are tuples (X, Y, f, g) where X is in $\mathsf{Mod}\text{-}R$, Y is in $\mathsf{Mod}\text{-}S$, $f\colon M \otimes_R X \longrightarrow Y$ and $g\colon N \otimes_S Y \longrightarrow X$ such that the following diagrams are commutative:

$$\begin{array}{ccc} N\otimes_S M\otimes_R X & \xrightarrow{1_N\otimes f} & N\otimes_S Y \\ {\scriptstyle \psi\otimes 1_X}\downarrow & & \downarrow{\scriptstyle g} \\ R\otimes_R X & \xrightarrow{\simeq} & X \end{array} \qquad \begin{array}{ccc} M\otimes_R N\otimes_S Y & \xrightarrow{1_M\otimes g} & M\otimes_R X \\ {\scriptstyle \phi\otimes 1_Y}\downarrow & & \downarrow{\scriptstyle f} \\ S\otimes_S Y & \xrightarrow{\simeq} & Y \end{array}$$

If (X, Y, f, g) and (X', Y', f', g') are $\Lambda_{(\phi,\psi)}$-modules, then a morphism $(X, Y, f, g) \longrightarrow (X', Y', f', g')$ in $\mathsf{Mod}\text{-}\Lambda_{(\phi,\psi)}$ is a pair of homomorphisms (a, b) where $a\colon X \longrightarrow X'$ is a morphism in $\mathsf{Mod}\text{-}R$ and $b\colon Y \longrightarrow Y'$ is a morphism in $\mathsf{Mod}\text{-}S$ such that the following diagrams are commutative:

$$\begin{array}{ccc} M\otimes_R X & \xrightarrow{f} & Y \\ {\scriptstyle 1_M\otimes a}\downarrow & & \downarrow{\scriptstyle b} \\ M\otimes_R X' & \xrightarrow{f'} & Y' \end{array} \qquad \begin{array}{ccc} N\otimes_S Y & \xrightarrow{g} & X \\ {\scriptstyle 1_N\otimes b}\downarrow & & \downarrow{\scriptstyle a} \\ N\otimes_S Y' & \xrightarrow{g'} & X' \end{array}$$

Then from Example 3.1 using the idempotent elements $e_1 = \left(\begin{smallmatrix} 1_R & 0 \\ 0 & 0 \end{smallmatrix}\right)$ and $e_2 = \left(\begin{smallmatrix} 0 & 0 \\ 0 & 1_S \end{smallmatrix}\right)$ of $\Lambda_{(\phi,\psi)}$, we derive the following recollements of abelian categories:

$$\mathsf{Mod}\text{-}\Lambda/\Lambda e_1\Lambda \xrightarrow{\mathsf{inc}} \mathsf{Mod}\text{-}\Lambda_{(\phi,\psi)} \xrightarrow{e_1(-)} \mathsf{Mod}\text{-}e_1\Lambda e_1$$
(left adjoints: $\Lambda/\Lambda e_1\Lambda\otimes_\Lambda -$, $\Lambda e_1\otimes_{e_1\Lambda e_1} -$; right adjoints: $\mathsf{Hom}_\Lambda(\Lambda/\Lambda e_1\Lambda, -)$, $\mathsf{Hom}_{e_1\Lambda e_1}(e_1\Lambda, -)$)

$$\mathsf{Mod}\text{-}\Lambda/\Lambda e_2\Lambda \xrightarrow{\mathsf{inc}} \mathsf{Mod}\text{-}\Lambda_{(\phi,\psi)} \xrightarrow{e_2(-)} \mathsf{Mod}\text{-}e_2\Lambda e_2$$
(left adjoints: $\Lambda/\Lambda e_2\Lambda\otimes_\Lambda -$, $\Lambda e_2\otimes_{e_2\Lambda e_2} -$; right adjoints: $\mathsf{Hom}_\Lambda(\Lambda/\Lambda e_2\Lambda, -)$, $\mathsf{Hom}_{e_2\Lambda e_2}(e_2\Lambda, -)$)

where $\mathsf{Mod}\text{-}\Lambda/\Lambda e_1\Lambda \simeq \mathsf{Mod}\text{-}S/\operatorname{Im}\phi$, $\mathsf{Mod}\text{-}e_1\Lambda e_1 \simeq \mathsf{Mod}\text{-}R$, $\mathsf{Mod}\text{-}\Lambda/\Lambda e_2\Lambda \simeq \mathsf{Mod}\text{-}R/\operatorname{Im}\psi$ and $\mathsf{Mod}\text{-}S \simeq \mathsf{Mod}\text{-}e_2\Lambda e_2$. We remark that using the description of the module category $\mathsf{Mod}\text{-}\Lambda_{(\phi,\psi)}$ we can also define in terms of tuples the functors in the above recollement diagrams. We refer to [**53**, Proposition 2.4] for more details. See also Example 3.7 below for an analogue description.

Typically rings of the form $\Lambda_{(\phi,\psi)}$ arise as endomorphism rings $\Lambda = \mathsf{End}_{\mathcal{M}}(X \oplus Y)$, where X and Y are arbitrary objects in an additive category $\mathcal{M}$. Clearly, we have an isomorphism

$$\Lambda \cong \begin{pmatrix} \mathsf{End}_{\mathcal{M}}(X) & \mathsf{Hom}_{\mathcal{M}}(X, Y) \\ \mathsf{Hom}_{\mathcal{M}}(Y, X) & \mathsf{End}_{\mathcal{M}}(Y) \end{pmatrix}$$

Then setting $e_X = \left(\begin{smallmatrix} \mathrm{Id}_X & 0 \\ 0 & 0 \end{smallmatrix}\right)$ and $e_Y = \left(\begin{smallmatrix} 0 & 0 \\ 0 & \mathrm{Id}_Y \end{smallmatrix}\right)$, as above we obtain the following recollements of abelian categories:

$$\mathsf{Mod}\text{-}\mathsf{End}_{\mathcal{M}}(Y)/\operatorname{Im}\phi \longrightarrow \mathsf{Mod}\text{-}\Lambda \longrightarrow \mathsf{Mod}\text{-}\mathsf{End}_{\mathcal{M}}(X)$$

$$\mathsf{Mod}\text{-}\mathsf{End}_{\mathcal{M}}(X)/\operatorname{\mathsf{Im}}\psi \longrightarrow \mathsf{Mod}\text{-}\Lambda \longrightarrow \mathsf{Mod}\text{-}\mathsf{End}_{\mathcal{M}}(Y)$$

For a general study of Morita rings via extensions of abelian categories see [**107**, Chapter 3]. We also refer to [**53**] for a general study of Morita rings in the context of Artin algebras concentrating mainly on representation-theoretic and homological aspects and to [**44**] for Gorenstein homological aspects of Morita rings.

The following example is a special case of the above recollement situation and arises in the Galois theory of noncommutative rings.

EXAMPLE 3.3. (Skew Group Rings and Invariant Subrings) Let R be a ring and G be a finite group of automorphisms of R. Associated with R and G we have the ring of invariants $R^G = \{r \in R \mid g(r) = r \text{ for every } g \in G\}$ and the skew group ring $R * G = \{\sum_{g\in G} r_g g \mid r_g \in R\}$ with multiplication given by the rule $(rg)(sh) = rg^{-1}(s)gh$ for every $r, s \in R$ and $g, h \in G$. Then the ring R is a left and right R^G-module, R is a left and right $R * G$-module and we have bimodule homomorphisms $\phi\colon R \otimes_{R^G} R \longrightarrow R * G$ and $\psi\colon R \otimes_{R*G} R \longrightarrow R^G$, see [**84**]. Then we have the Morita ring

$$\Lambda_{(\phi,\psi)} = \begin{pmatrix} R^G & R \\ R & R*G \end{pmatrix}$$

and by Example 3.2 the following diagrams:

$$\mathsf{Mod}\text{-}R*G/\operatorname{\mathsf{Im}}\phi \longrightarrow \mathsf{Mod}\text{-}\begin{pmatrix} R^G & R \\ R & R*G \end{pmatrix} \longrightarrow \mathsf{Mod}\text{-}R^G$$

$$\mathsf{Mod}\text{-}R^G/\operatorname{\mathsf{Im}}\psi \longrightarrow \mathsf{Mod}\text{-}\begin{pmatrix} R^G & R \\ R & R*G \end{pmatrix} \longrightarrow \mathsf{Mod}\text{-}R*G$$

are recollements of module categories.

We continue with a variation of Example 3.1 in the context of Representation Theory of Artin Algebras.

EXAMPLE 3.4. (Trace Ideals) Let Λ be an Artin algebra and $\mathcal{U}$ a two-sided idempotent ideal in Λ, so $\mathcal{U} = \Lambda e \Lambda$ for some idempotent element of Λ (here we also used the result of Dlab-Ringel [**33**]). Then from [**13**] it follows that $\mathcal{U} = \tau_P(\Lambda)$ for some projective Λ-module P, where $\tau_P(\Lambda)$ is the trace of P in Λ. Recall that $\tau_P(\Lambda)$ is the ideal $\sum \operatorname{\mathsf{Im}} f$ where f ranges over $\mathsf{Hom}_\Lambda(P, \Lambda)$. Also associated with Λ we have the Artin algebras $\Lambda/\mathcal{U}$ and $\Gamma = \mathsf{End}_\Lambda(P)$. Then we have the following recollement of abelian categories:

$$\mathsf{mod}\text{-}\Lambda/\mathcal{U} \xrightarrow{\ \mathsf{inc}\ } \mathsf{mod}\text{-}\Lambda \xrightarrow{\ \mathsf{Hom}_\Lambda(P,-)\ } \mathsf{mod}\text{-}\Gamma$$

(left adjoints: $\Lambda/\mathcal{U}\otimes_\Lambda -$ and $P\otimes_\Gamma -$; right adjoints: $\mathsf{Hom}_\Lambda(\Lambda/\mathcal{U},-)$ and $\mathsf{Hom}_\Gamma(\mathsf{Hom}_\Lambda(P,\Lambda),-)$)

More generally, let R be a ring and P a finitely generated projective R-module. Then as above we have a recollement $(\mathsf{Ker}\,\mathsf{Hom}_R(P,-), \mathsf{Mod}\text{-}R, \mathsf{Mod}\text{-}\Gamma)$, where

$\Gamma = \mathsf{End}_R(P)$. We remark that $\mathsf{Ker}\,\mathsf{Hom}_R(P,-)$ is a module category. Indeed, by a result of Auslander there exists a uniquely determined idempotent ideal $\underline{\alpha}$ of R such that $\mathsf{Ker}\,\mathsf{Hom}_R(P,-) = \mathsf{Mod}\text{-}R/\underline{\alpha}$, see [**10, 11**]. We also provide a proof of this fact, see Proposition 5.6.

The following example shows that the module category of a left Artinian ring can be placed always in the right part of a recollement.

EXAMPLE 3.5. (Artinian Rings) Let Λ be a left Artinian ring. Then by a basic result of Auslander [**9**], there exist a semiprimary ring Γ (i.e. the Jacobson radical $\mathfrak{r}_\Gamma$ is nilpotent and $\Gamma/\mathfrak{r}_\Gamma$ is semisimple) of finite global dimension and an idempotent element e in Γ such that $\Lambda \cong e\Gamma e$. This implies that we have a recollement of module categories $\mathsf{R_{ab}}(\mathsf{Mod}\text{-}\Gamma/\Gamma e\Gamma, \mathsf{Mod}\text{-}\Gamma, \mathsf{Mod}\text{-}\Lambda)$.

The following class of algebras is well known to provide a link between the representation theory of the symmetric and general linear groups .

EXAMPLE 3.6. (Schur Algebras) Let S_r be the symmetric group permuting r elements and let k be a field. Consider the n-dimensional vector space $V = k^n$ where $n \geq r$. The Schur algebra $S_k(n,r)$ is the endomorphism algebra $\mathsf{End}_{kS_r}(V^{\otimes r})$, where $V^{\otimes r}$ is the tensor power of V with r factors. Then there exists an idempotent element e in $S_k(n,r)$ such that $eS_k(n,r)e$ is isomorphic to the group algebra kS_r. This implies that $(\mathsf{mod}\text{-}\mathsf{End}_{kS_r}(V^{\otimes r})/\langle e\rangle, \mathsf{mod}\text{-}\mathsf{End}_{kS_r}(V^{\otimes r}), \mathsf{mod}\text{-}kS_r)$ is a recollement of module categories. In this case, the multiplication functor $e(-)\colon \mathsf{mod}\text{-}\mathsf{End}_{kS_r}(V^{\otimes r}) \longrightarrow \mathsf{mod}\text{-}\mathsf{End}_{kS_r}(V^{\otimes r}$ is called the Schur functor. For more details on Schur algebras we refer to [**55**].

We continue with recollement situations arising from the module categories of preprojective algebras.

EXAMPLE 3.7. (Preprojective Algebras) Let k be an algebraically closed field and Q a finite quiver. Denote by $\overline{Q}$ the double quiver of Q which is obtained from Q by adding for each arrow $a \in Q_1$ an arrow a^* in the opposite direction. Then the preprojective algebra [**48**] is defined as the quotient path algebra

$$\Pi_n(Q) := k\overline{Q}/\langle c\rangle,$$

where $\langle c\rangle$ is the two-sided ideal generated by $c = \Sigma_{a\in Q_1}(a^*a - aa^*)$. Recall that if Q is Dynkin, then $\Pi_n(Q)$ is a finite-dimensional selfinjective algebra.

Let Λ be a finite dimensional k-algebra and denote by $\Pi_n(\Lambda, Q)$ the algebra $\Lambda \otimes_k \Pi_n(Q)$. *We now assume* that the quiver Q is Dynkin of type $\mathbb{A}_n$. This implies that $\Pi_n(\Lambda, Q)$ is a finite dimensional k-algebra and is called the path algebra of $\Pi_n(Q)$ over Λ, see [**85**, subsection 2B]. The module category of $\Pi_n(\Lambda, Q)$ has objects representations of $\Pi_n(Q)$ over Λ, see [**14**], [**85**, Lemma 2.1], [**81**, Proposition 4.12]. More precisely, for $n = 2$ and $n = 3$ we have the following descriptions:

$$\mathsf{Mod}\text{-}\Pi_2(\Lambda, Q) = \{\, X \underset{f}{\overset{g}{\rightleftarrows}} Y \mid g\circ f = 0,\ f\circ g = 0 \text{ and } X, Y, Z \in \mathsf{Mod}\text{-}\Lambda\}$$

$$\mathsf{Mod}\text{-}\Pi_3(\Lambda, Q) = \{\, X \underset{f_1}{\overset{g_1}{\rightleftarrows}} Y \underset{f_2}{\overset{g_2}{\rightleftarrows}} Z \mid g_1\circ f_1 = 0 = f_2\circ g_2,\ f_1\circ g_1 = g_2\circ f_2 \text{ and } X, Y, Z \in \mathsf{Mod}\text{-}\Lambda\}$$

and more generally we have

$$\mathsf{Mod}\text{-}\Pi_n(\Lambda, Q) = \Big\{ \ X_1 \underset{f_1}{\overset{g_1}{\leftrightarrows}} X_2 \underset{f_2}{\overset{g_2}{\leftrightarrows}} X_3 \leftrightarrows \ \cdots \ \underset{f_{n-2}}{\overset{g_{n-2}}{\leftrightarrows}} X_{n-1} \underset{f_{n-1}}{\overset{g_{n-1}}{\leftrightarrows}} X_n \ \Big| $$

$$g_1 \circ f_1 = 0 = f_{n-1} \circ g_{n-1}, \ f_i \circ g_i = g_{i+1} \circ f_{i+1}$$

$$\text{for all } 1 \leq i \leq n-2 \text{ and } X_1, \dots, X_n \in \mathsf{Mod}\text{-}\Lambda \Big\}$$

Restricting to finitely generated Λ-modules, we get the module category $\mathsf{mod}\text{-}\Pi_n(\Lambda, Q)$. An object of $\mathsf{Mod}\text{-}\Pi_n(\Lambda, Q)$ is denoted by $(X_1, \dots, X_n, f_1, g_1, \dots, f_{n-1}, g_{n-1})$ for simplicity. We define the following functors:

(i) The functor $\mathsf{T}_1 \colon \mathsf{Mod}\text{-}\Lambda \longrightarrow \mathsf{Mod}\text{-}\Pi_n(\Lambda, Q)$ is given by

$$X \underset{1_X}{\overset{0}{\leftrightarrows}} X \underset{1_X}{\overset{0}{\leftrightarrows}} X \leftrightarrows \ \cdots \ \underset{1_X}{\overset{0}{\leftrightarrows}} X \underset{1_X}{\overset{0}{\leftrightarrows}} X$$

for a Λ-module X, and given a morphism $a \colon X \longrightarrow X'$ in $\mathsf{Mod}\text{-}\Lambda$ then $\mathsf{T}_1(a) = (a, a, \dots, a)$.

(ii) The functor $\mathsf{T}_2 \colon \mathsf{Mod}\text{-}\Lambda \longrightarrow \mathsf{Mod}\text{-}\Pi_n(\Lambda, Q)$ is given by

$$X \underset{0}{\overset{1_X}{\leftrightarrows}} X \underset{0}{\overset{1_X}{\leftrightarrows}} X \leftrightarrows \ \cdots \ \underset{0}{\overset{1_X}{\leftrightarrows}} X \underset{0}{\overset{1_X}{\leftrightarrows}} X$$

for a Λ-module X, and given a morphism $a \colon X \longrightarrow X'$ in $\mathsf{Mod}\text{-}\Lambda$ then $\mathsf{T}_2(a) = (a, a, \dots, a)$.

(iii) The functor $\mathsf{U}_1 \colon \mathsf{Mod}\text{-}\Pi_n(\Lambda, Q) \longrightarrow \mathsf{Mod}\text{-}\Lambda$ is given by

$$\mathsf{U}_1(X_1, \dots, X_n, f_1, g_1, \dots, f_{n-1}, g_{n-1}) = X_1$$

on objects, and for a morphism $(a_1, \dots, a_n)$ in $\mathsf{Mod}\text{-}\Pi_n(\Lambda, Q)$ we have $\mathsf{U}_1(a_1, \dots, a_n) = a_1$.

(iv) The functor $\mathsf{U}_2 \colon \mathsf{Mod}\text{-}\Pi_n(\Lambda, Q) \longrightarrow \mathsf{Mod}\text{-}\Lambda$ is given by

$$\mathsf{U}_2(X_1, \dots, X_n, f_1, g_1, \dots, f_{n-1}, g_{n-1}) = X_n$$

on objects, and for a morphism $(a_1, \dots, a_n)$ in $\mathsf{Mod}\text{-}\Pi_n(\Lambda, Q)$ we have $\mathsf{U}_2(a_1, \dots, a_n) = a_n$.

(v) The functor $\mathsf{Z}_1 \colon \mathsf{Mod}\text{-}\Pi_{n-1}(\Lambda, Q) \longrightarrow \mathsf{Mod}\text{-}\Pi_n(\Lambda, Q)$ is defined by

$$\mathsf{Z}_1(X_1, \dots, X_{n-1}, f_1, g_1, \dots, f_{n-2}, g_{n-2}) =$$

$$(X_1, \dots, X_{n-1}, 0, f_1, g_1, \dots, f_{n-2}, g_{n-2}, 0, 0)$$

on objects, and for a morphism $(a_1, \dots, a_{n-1})$ in $\mathsf{Mod}\text{-}\Pi_2(\Lambda, Q)$ we have $\mathsf{Z}_2(a_1, \dots, a_{n-1}) = (a_1, \dots, a_{n-1}, 0)$.

(vi) The functor $\mathsf{Z}_2 \colon \mathsf{Mod}\text{-}\Pi_{n-1}(\Lambda, Q) \longrightarrow \mathsf{Mod}\text{-}\Pi_n(\Lambda, Q)$ is defined by

$$\mathsf{Z}_2(X_1, \cdots, X_{n-1}, f_1, g_1, \dots, f_{n-2}, g_{n-2}) =$$

$$(0, X_1, \cdots, X_{n-1}, 0, 0, f_1, g_1, \dots, f_{n-2}, g_{n-2})$$

on objects, and for a morphism $(a_1, \dots, a_{n-1})$ in $\mathsf{Mod}\text{-}\Pi_{n-1}(\Lambda, Q)$ we have $\mathsf{Z}_2(a_1, \dots, a_{n-1}) = (0, a_1, \dots, a_{n-1})$.

Then, it follows from [**45**, Proposition 4.4] that $\mathsf{Mod}\text{-}\Pi_n(\Lambda, Q)$ admits a recollement of module categories as follows:

$$\begin{tikzcd} \mathsf{Mod}\text{-}\Pi_{n-1}(\Lambda,Q) \arrow[r, "\mathsf{Z}_2"] & \mathsf{Mod}\text{-}\Pi_n(\Lambda,Q) \arrow[l, bend right] \arrow[l, bend left] \arrow[r, "\mathsf{U}_1"] & \mathsf{Mod}\text{-}\Lambda \arrow[l, bend right, "\mathsf{T}_1"'] \arrow[l, bend left, "\mathsf{T}_2"] \end{tikzcd}$$

We refer to [**45**] for more details on the above recollement diagram and its connection with the theory of ladders of recollements of derived categories (see Theorem 11.4 for the definition of a ladder).

It should be clear from the examples so far that the main issue for getting all these recollements is the existence of an idempotent element. Hence, it is natural to ask how many from the recollements of module categories are obtained in this way. We answer this problem in Section 5 (in particular, we prove Kuhn's conjecture).

We continue with examples of recollements of abelian categories arising from general abstract constructions.

EXAMPLE 3.8. (**Splitting Recollements**) Let $\mathscr{A}$ and $\mathscr{C}$ be abelian categories and consider the product category $\mathscr{B} := \mathscr{A} \times \mathscr{C}$. Clearly, there is a recollement of abelian categories $\mathsf{R_{ab}}(\mathscr{A}, \mathscr{B}, \mathscr{C})$.

A recollement $\mathsf{R_{ab}}(\mathscr{A}, \mathscr{B}, \mathscr{C})$ is called a **splitting recollement** if it is equivalent to a recollement of the form $\mathsf{R_{ab}}(\mathscr{A}', \mathscr{A}' \times \mathscr{C}', \mathscr{C}')$. For the exact notion of equivalent recollements we refer to Definition 5.1 and the discussion thereafter.

For later use (see Section 10), we characterize below when a recollement of abelian categories $\mathsf{R_{ab}}(\mathscr{A}, \mathscr{B}, \mathscr{C})$ is splitting. We remark that the following result generalizes [**37**, Corollary 8.10] (i.e. we don't assume that the abelian categories in $\mathsf{R_{ab}}(\mathscr{A}, \mathscr{B}, \mathscr{C})$ have enough projectives or enough injectives).

PROPOSITION 3.9. *Let $\mathsf{R_{ab}}(\mathscr{A}, \mathscr{B}, \mathscr{C})$ be a recollement of abelian categories. Then the following statements are equivalent:*

(i) *The recollement $\mathsf{R_{ab}}(\mathscr{A}, \mathscr{B}, \mathscr{C})$ splits.*
(ii) *For every C in $\mathscr{C}$ the norm map $N_C \colon \mathsf{l}(C) \longrightarrow \mathsf{r}(C)$ is an isomorphism.*
(iii) *We have $\mathsf{pl}(C) = 0$ and $\mathsf{qr}(C) = 0$ for every $C \in \mathscr{C}$.*
(iv) *The functors p and q are exact.*

PROOF. (i)$\Longrightarrow$(ii): Consider the recollement $\mathsf{R_{ab}}(\mathscr{A}, \mathscr{A} \times \mathscr{C}, \mathscr{C})$. The left and right adjoint of the (quotient) functor $\mathscr{A} \times \mathscr{C} \longrightarrow \mathscr{C}$ coincide, i.e. they are given by the assignment $C \mapsto (0, C)$. This shows the desired statement.

(ii)$\Longrightarrow$(i): Assume that the norm map $N_C \colon \mathsf{l}(C) \longrightarrow \mathsf{r}(C)$ is an isomorphism for every C in $\mathscr{C}$. Let B be an object of $\mathscr{B}$. From Proposition 2.8 we have the exact sequence $0 \longrightarrow \mathsf{i}(A) \longrightarrow \mathsf{le}(B) \longrightarrow B \longrightarrow \mathsf{iq}(B) \longrightarrow 0$. By assumption we have the isomorphism $\mathsf{le}(B) \cong \mathsf{re}(B)$ and by Remark 2.4 (ii) we have $\mathsf{p} \circ \mathsf{r} = 0$. Using these we obtain the exact sequence

$$0 \longrightarrow \mathsf{le}(B) \xrightarrow{\ \mu_B\ } B \longrightarrow \mathsf{iq}(B) \longrightarrow 0\,. \tag{1}$$

Similarly, from the exact sequence $0 \longrightarrow \mathsf{ip}(B) \longrightarrow B \longrightarrow \mathsf{re}(B) \longrightarrow \mathsf{i}(A') \longrightarrow 0$ (see sequence (2.4)) and since $\mathsf{q} \circ \mathsf{l} = 0$ it follows that the sequence

$$(2) \qquad 0 \longrightarrow \mathsf{ip}(B) \longrightarrow B \xrightarrow{\nu_B} \mathsf{re}(B) \longrightarrow 0$$

is exact. Since $\nu_B \circ (\mu_B \circ N^{-1}_{\mathsf{e}(B)}) = \mathrm{Id}_{\mathsf{re}(B)}$ it follows that the exact sequence (2) splits, and similarly we get that the exact sequence (1) splits as well. We define the functor $\Phi\colon \mathscr{B} \longrightarrow \mathscr{A}\times\mathscr{C}$ by $\Phi(B) = (\mathsf{p}(B), \mathsf{e}(B))$. Then it follows easily that Φ is an equivalence of categories with $\Phi^{-1}\colon \mathscr{A}\times\mathscr{C} \longrightarrow \mathscr{B}$ given by $\Phi^{-1}(A,C) = \mathsf{i}(A)\oplus\mathsf{r}(C)$. Moreover, the following diagram commutes up to natural equivalence:

$$\begin{array}{ccc} \mathscr{B} & \xrightarrow{\mathsf{e}} & \mathscr{C} \\ {\scriptstyle \Phi}\downarrow{\scriptstyle \simeq} & & {\scriptstyle =}\downarrow{\scriptstyle \mathrm{Id}_{\mathscr{C}}} \\ \mathscr{A}\times\mathscr{C} & \longrightarrow & \mathscr{C} \end{array}$$

This implies that the recollement $\mathsf{R}_{\mathsf{ab}}(\mathscr{A},\mathscr{B},\mathscr{C})$ is equivalent to $\mathsf{R}_{\mathsf{ab}}(\mathscr{A},\mathscr{A}\times\mathscr{C},\mathscr{C})$, see Definition 5.1 and Section 5. We infer that the recollement $\mathsf{R}_{\mathsf{ab}}(\mathscr{A},\mathscr{B},\mathscr{C})$ splits.

(ii)$\Longleftrightarrow$(iii): This equivalence follows from Proposition 2.8 (iii) using also that the functor $\mathsf{i}\colon \mathscr{A} \longrightarrow \mathscr{B}$ is fully faithful.

The implication (i)$\Longrightarrow$(iv) is clear. In order to complete the proof it suffices to show that (iv)$\Longrightarrow$(iii).

Before that, we show that p being exact is equivalent to p and r being exact, and the latter is equivalent to $(*)\colon 0 \longrightarrow \mathsf{ip}(B) \longrightarrow B \longrightarrow \mathsf{re}(B) \longrightarrow 0$ being exact for all B in $\mathscr{B}$, see the sequence (2.4). Let $0 \longrightarrow C_1 \xrightarrow{f_1} C_2 \xrightarrow{f_2} C_3 \longrightarrow 0$ be an exact sequence in $\mathscr{C}$. Applying the left exact functor r we get the exact sequence

$$0 \longrightarrow \mathsf{r}(C_1) \xrightarrow{\mathsf{r}(f_1)} \mathsf{r}(C_2) \xrightarrow{\mathsf{r}(f_2)} \mathsf{r}(C_3) \longrightarrow \mathsf{Coker}\,\mathsf{r}(f_2) \longrightarrow 0$$

We claim that $\mathsf{Coker}\,\mathsf{r}(f_2) = 0$. Applying the exact functor p and since $\mathsf{p}\circ\mathsf{r} = 0$ by Remark 2.4 (ii), it follows that $\mathsf{p}(\mathsf{Coker}\,\mathsf{r}(f_2)) = 0$. On the other hand, applying the exact functor e to the above sequence and since there is an isomorphism $\mathsf{e}\circ\mathsf{r} \cong \mathrm{Id}_{\mathscr{C}}$ by Remark 2.4 (iv), we obtain that $\mathsf{Coker}\,\mathsf{r}(f_2) = \mathsf{i}(A)$ for some A in $\mathscr{A}$. Using now Remark 2.4 (iii) we infer that $\mathsf{Coker}\,\mathsf{r}(f_2) = 0$, so the functor r is exact.

Consider the exact sequence (2.4) of Proposition 2.8. Since $\mathsf{e}(\nu_B)$ is an isomorphism, it follows that $\mathsf{Coker}\,\nu_B = \mathsf{i}(A)$ for some A in $\mathscr{A}$. Applying the exact functor p to (2.4), we obtain as before that $\mathsf{Coker}\,\nu_B = 0$. Assume conversely that the sequence $(*)$ is exact for all B in $\mathscr{B}$ and consider a short exact sequence

$0 \longrightarrow B_1 \longrightarrow B_2 \longrightarrow B_3 \longrightarrow 0$ in $\mathscr{B}$. We have the exact commutative diagram:

$$\begin{array}{ccccccccc}
 & & 0 & & 0 & & 0 & & \\
 & & \downarrow & & \downarrow & & \downarrow & & \\
0 & \longrightarrow & \mathsf{ip}(B_1) & \xrightarrow{\kappa_{B_1}} & B_1 & \xrightarrow{\nu_{B_1}} & \mathsf{re}(B_1) & \longrightarrow & 0 \\
 & & \downarrow{\scriptstyle \mathsf{ip}(f_1)} & & \downarrow{\scriptstyle f_1} & & \downarrow{\scriptstyle \mathsf{re}(f_1)} & & \\
0 & \longrightarrow & \mathsf{ip}(B_2) & \xrightarrow{\kappa_{B_2}} & B_2 & \xrightarrow{\nu_{B_2}} & \mathsf{re}(B_2) & \longrightarrow & 0 \\
 & & \downarrow{\scriptstyle \mathsf{ip}(f_2)} & & \downarrow{\scriptstyle f_2} & & \downarrow{\scriptstyle \mathsf{re}(f_2)} & & \\
0 & \longrightarrow & \mathsf{ip}(B_3) & \xrightarrow{\kappa_{B_3}} & B_3 & \xrightarrow{\nu_{B_3}} & \mathsf{re}(B_3) & \longrightarrow & 0 \\
 & & \downarrow & & \downarrow & & \downarrow & & \\
 & & \mathsf{i}(\mathsf{Coker}\,\mathsf{p}(f_2)) & & 0 & & 0 & & \\
 & & \downarrow & & & & & & \\
 & & 0 & & & & & &
\end{array}$$

Note that the map $\mathsf{re}(f_2)$ is an epimorphism since the composition $\nu_{B_3} \circ f_2$ is an epimorphism. Applying the Snake Lemma, we get that $\mathsf{i}(\mathsf{Coker}\,\mathsf{p}(f_2)) = 0$ and therefore $\mathsf{Coker}\,\mathsf{p}(f_2) = 0$ since the functor i is fully faithful. We infer that the functor p is exact. We leave to the reader to prove the dual characterization for the functor q being exact.

(iv)$\Longrightarrow$(iii): For $B = \mathsf{l}(C)$ in $(*)$, we get that the norm map $N_C \colon \mathsf{l}(C) \longrightarrow \mathsf{r}(C)$ is an epimorphism. Then Proposition 2.8 (iii) implies that $\mathsf{qr}(C) = 0$. Similarly, we show that $\mathsf{pl}(C) = 0$ and this completes the proof. □

From recollements where all functors are exact (i.e. splitting recollements), we pass to recollements of comma categories.

EXAMPLE 3.10. (Comma Categories) Let $G \colon \mathscr{B} \longrightarrow \mathscr{A}$ be a right exact functor between abelian categories. The objects of the *comma category* $\mathscr{C} = (G, \mathscr{B}, \mathscr{A})$, also denoted as $\mathscr{C} = (G \downarrow \mathrm{Id})$, are triples (A, B, f) where $f \colon G(B) \longrightarrow A$ is a morphism in $\mathscr{A}$. A morphism $\gamma \colon (A, B, f) \longrightarrow (A', B', f')$ in $\mathscr{C}$ consists of two morphisms $\alpha \colon A \longrightarrow A'$ in $\mathscr{A}$ and $\beta \colon B \longrightarrow B'$ in $\mathscr{B}$ such that $\alpha \circ f = f' \circ G(\beta)$, that is, the following diagram is commutative:

$$\begin{array}{ccc}
G(B) & \xrightarrow{f} & A \\
{\scriptstyle G(\beta)}\downarrow & & \downarrow{\scriptstyle \alpha} \\
G(B') & \xrightarrow{f'} & A'
\end{array}$$

It is well known that the comma category $\mathscr{C}$ is abelian since the functor G is right exact (see [**38**]). Assume that G has a right adjoint $G' \colon \mathscr{A} \longrightarrow \mathscr{B}$ and let $\epsilon \colon GG' \longrightarrow \mathrm{Id}_{\mathscr{A}}$ be the counit and $\eta \colon \mathrm{Id}_{\mathscr{B}} \longrightarrow G'G$ be the unit of the adjoint pair (G, G'). We define the following functors:

(i) The functor $\mathsf{T}_{\mathscr{B}} \colon \mathscr{B} \longrightarrow \mathscr{C}$ is defined by $\mathsf{T}_{\mathscr{B}}(Y) = (G(Y), Y, \mathrm{Id}_{GY})$ on objects Y in $\mathscr{B}$ and given a morphism $\beta \colon Y \longrightarrow Y'$ in $\mathscr{B}$ then $\mathsf{T}_{\mathscr{B}}(\beta) = (G(\beta), \beta)$ is a morphism in $\mathscr{C}$.

(ii) The functor $\mathsf{U}_{\mathscr{B}}\colon \mathscr{C} \longrightarrow \mathscr{B}$ is defined by $\mathsf{U}_{\mathscr{B}}(A,B,f) = B$ on objects (A,B,f) in $\mathscr{C}$ and given a morphism $(\alpha,\beta)\colon (A,B,f) \longrightarrow (A',B',f')$ in $\mathscr{C}$ then $\mathsf{U}_{\mathscr{B}}(\alpha,\beta) = \beta$ is a morphism in $\mathscr{B}$. Similarly we define the functor $\mathsf{U}_{\mathscr{A}}\colon \mathscr{C} \longrightarrow \mathscr{A}$.

(iii) The functor $\mathsf{Z}_{\mathscr{B}}\colon \mathscr{B} \longrightarrow \mathscr{C}$ is defined by $\mathsf{Z}_{\mathscr{B}}(Y) = (0,Y,0)$ on objects Y in $\mathscr{B}$ and given a morphism $\beta\colon Y \longrightarrow Y'$ in $\mathscr{B}$ then $\mathsf{Z}_{\mathscr{B}}(\beta) = (0,\beta)$ is a morphism in $\mathscr{C}$. Similarly we define the functor $\mathsf{Z}_{\mathscr{A}}\colon \mathscr{A} \longrightarrow \mathscr{C}$.

(iv) The functor $\mathsf{q}\colon \mathscr{C} \longrightarrow \mathscr{A}$ is defined by $\mathsf{q}(A,B,f) = \mathsf{Coker}\, f$ on objects (A,B,f) in $\mathscr{C}$ and if $(\alpha,\beta)\colon (A,B,f) \longrightarrow (A',B',f')$ is a morphism in $\mathscr{C}$ then we have the induced morphism $\mathsf{q}(\alpha,\beta)\colon \mathsf{Coker}\, f \longrightarrow \mathsf{Coker}\, f'$.

(v) The functor $\mathsf{H}_{\mathscr{A}}\colon \mathscr{A} \longrightarrow \mathscr{C}$ is defined by $\mathsf{H}_{\mathscr{A}}(X) = (X, G'(X), \epsilon_X)$ on objects X in $\mathscr{A}$ and given a morphism $\alpha\colon X \longrightarrow X'$ in $\mathscr{A}$ then $\mathsf{H}_{\mathscr{A}}(\alpha) = (\alpha, G'(\alpha))$ is a morphism in $\mathscr{C}$.

(vi) The functor $\mathsf{p}\colon \mathscr{C} \longrightarrow \mathscr{B}$ is defined by $\mathsf{p}(A,B,f) = \mathsf{Ker}\,(\eta_B \circ G'(f))$ on objects (A,B,f) in $\mathscr{C}$ and if $(\alpha,\beta)\colon (A,B,f) \longrightarrow (A',B',f')$ is a morphism in $\mathscr{C}$ then we get the morphism $\mathsf{p}(\alpha,\beta)\colon \mathsf{Ker}\,(\eta_B \circ G'(f)) \longrightarrow \mathsf{Ker}\,(\eta_{B'} \circ G'(f'))$.

It is left to the reader to check that the following diagrams:

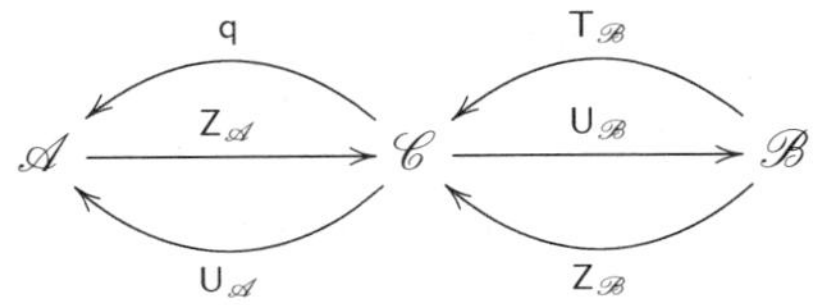

and

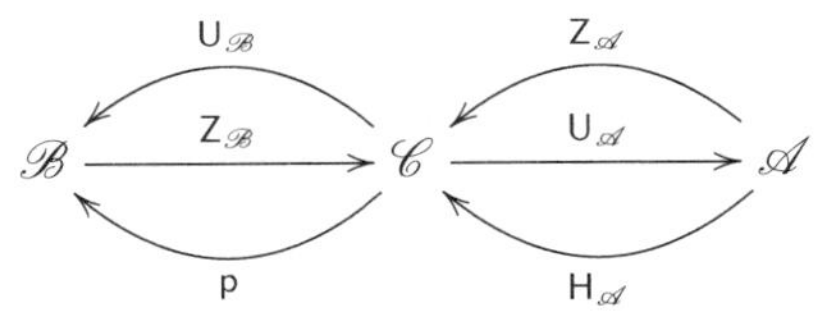

are recollements of abelian categories. Let us give now a specific example of a comma category.

Consider the triangular matrix ring:

$$\Lambda = \begin{pmatrix} R & {}_RN_S \\ 0 & S \end{pmatrix}$$

where R and S are rings and ${}_RN_S$ an R-S-bimodule. It is well known, see for instance [**14**], that the category of left Λ-modules is equivalent to the comma category $\mathscr{C} = (N \otimes_S -, \mathsf{Mod}\text{-}S, \mathsf{Mod}\text{-}R)$. This means precisely that a left Λ-module is given as a triple (X,Y,f) where $f\colon N \otimes_S Y \longrightarrow X$ is a morphism in $\mathsf{Mod}\text{-}R$.

As a special case, consider the triangular matrix ring Λ with all entries a ring R. Then the functors in the recollement situations of $\mathsf{Mod}\text{-}\Lambda$ (which is the morphism category of $\mathsf{Mod}\text{-}R$) are all exact except the functors q and p.

EXAMPLE 3.11. (The MacPherson-Vilonen Category) Let $\mathscr{A}$ and $\mathscr{B}$ be two abelian categories, $G, F\colon \mathscr{A} \longrightarrow \mathscr{B}$ two functors such that F is right exact and G is left exact and let $\xi\colon F \longrightarrow G$ be a natural transformation. Then the MacPherson-Vilonen category [**87**], denoted by $\mathsf{MV}(\xi)$, has objects tuples (A,B,f,g) where A

is in $\mathscr{A}$, B is in $\mathscr{B}$, $f\colon F(A) \longrightarrow B$ and $g\colon B \longrightarrow G(A)$ are morphisms in $\mathscr{B}$ such that the following diagram commutes:

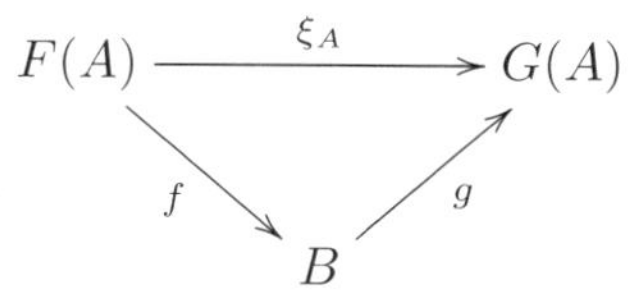

Let (A, B, f, g) and (A', B', f', g') be two objects of $\mathsf{MV}(\xi)$. Then a morphism from $(A, B, f, g) \longrightarrow (A', B', f', g')$ is a pair (α, β), where $\alpha\colon A \longrightarrow A'$ is a morphism in $\mathscr{A}$ and $\beta\colon B \longrightarrow B'$ is a morphism in $\mathscr{B}$, such that the diagram

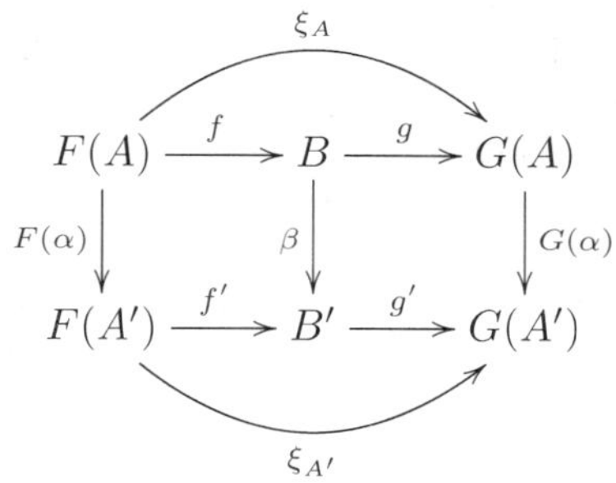

is commutative. Then it follows easily, see [**87**], that the category $\mathsf{MV}(\xi)$ is abelian. We define the following functors:

(i) The functor $\mathsf{l}\colon \mathscr{A} \longrightarrow \mathsf{MV}(\xi)$ is defined by $\mathsf{l}(A) = (A, F(A), \mathrm{Id}_{F(A)}, \xi_A)$ on objects A in $\mathscr{A}$ and given a morphism $a\colon A \longrightarrow A'$ in $\mathscr{A}$ then $\mathsf{l}(a) = (a, F(a))$ is a morphism in $\mathsf{MV}(\xi)$.

(ii) The functor $\mathsf{e}\colon \mathsf{MV}(\xi) \longrightarrow \mathscr{A}$ is defined by $\mathsf{e}(A, B, f, g) = A$ on objects (A, B, f, g) in $\mathsf{MV}(\xi)$ and given a morphism $(\alpha, \beta)\colon (A, B, f, g) \longrightarrow (A', B', f', g')$ in $\mathsf{MV}(\xi)$ then $\mathsf{e}(\alpha, \beta) = \alpha$ is a morphism in $\mathscr{A}$.

(iii) The functor $\mathsf{r}\colon \mathscr{A} \longrightarrow \mathsf{MV}(\xi)$ is defined by $\mathsf{r}(A) = (A, G(A), \xi_A, \mathrm{Id}_{G(A)})$ on objects A in $\mathscr{A}$ and given a morphism $a\colon A \longrightarrow A'$ in $\mathscr{A}$ then $\mathsf{r}(a) = (a, G(a))$ is a morphism in $\mathsf{MV}(\xi)$.

(iv) The functor $\mathsf{i}\colon \mathscr{B} \longrightarrow \mathsf{MV}(\xi)$ is defined by $\mathsf{i}(B) = (0, B, 0, 0)$ on objects B in $\mathscr{B}$ and given a morphism $\beta\colon Y \longrightarrow Y'$ in $\mathscr{B}$ then $\mathsf{i}(\beta) = (0, \beta)$ is a morphism in $\mathsf{MV}(\xi)$.

(v) The functor $\mathsf{q}\colon \mathsf{MV}(\xi) \longrightarrow \mathscr{B}$ is defined by $\mathsf{q}(A, B, f, g) = \mathsf{Coker}\, f$ on objects (A, B, f, g) in $\mathsf{MV}(\xi)$ and if $(\alpha, \beta)\colon (A, B, f, g) \longrightarrow (A', B', f', g')$ is a morphism in $\mathsf{MV}(\xi)$ then $\mathsf{q}(\alpha, \beta)\colon \mathsf{Coker}\, f \longrightarrow \mathsf{Coker}\, f'$ is the induced morphism on cokernels.

(vi) The functor $\mathsf{p}\colon \mathsf{MV}(\xi) \longrightarrow \mathscr{B}$ is defined by $\mathsf{p}(A, B, f, g) = \mathsf{Ker}\, g$ on objects (A, B, f, g) in $\mathsf{MV}(\xi)$ and if $(\alpha, \beta)\colon (A, B, f, g) \longrightarrow (A', B', f', g')$ is a morphism in $\mathsf{MV}(\xi)$ then we get the morphism $\mathsf{p}(\alpha, \beta)\colon \mathsf{Ker}\, g \longrightarrow \mathsf{Ker}\, g'$.

Then it is easy to observe that the following diagram:

$$\mathscr{B} \;\begin{array}{c} \xleftarrow{\;\mathsf{q}\;} \\ \xrightarrow{\;\mathsf{i}\;} \\ \xleftarrow[\;\mathsf{p}\;]{} \end{array}\; \mathsf{MV}(\xi) \;\begin{array}{c} \xleftarrow{\;\mathsf{l}\;} \\ \xrightarrow{\;\mathsf{e}\;} \\ \xleftarrow[\;\mathsf{r}\;]{} \end{array}\; \mathscr{A}$$

is a recollement of abelian categories.

The MacPherson-Vilonen category is used in [**87**] to provide an inductive construction of the category of perverse sheaves $\mathsf{P}(X)$, where X is a stratified topological space or a complex analytic manifold. We refer to [**87**] for more details.

EXAMPLE 3.12. (Functor Categories) Let $\mathscr{C}$ be a skeletally small additive category and let $\mathcal{X}$ be a full subcategory of $\mathscr{C}$. Let $\mathsf{i}\colon \mathcal{X} \longrightarrow \mathscr{C}$ be the inclusion functor. We denote by $\mathsf{Mod}\text{-}\mathscr{C}$ the category of additive functors $\mathscr{C}^{\mathsf{op}} \longrightarrow \mathfrak{Ab}$, where $\mathfrak{Ab}$ denotes the category of abelian groups. The functor category $\mathsf{Mod}\text{-}\mathscr{C}$, also known as the category of $\mathscr{C}$-modules [**10**], is abelian with enough projective objects. Associated with these data, we have the restriction functor $\mathsf{res}\colon \mathsf{Mod}\text{-}\mathscr{C} \longrightarrow \mathsf{Mod}\text{-}\mathcal{X}$ given by $G \mapsto \mathsf{res}\,(G) = G \circ \mathsf{i}$ on objects G in $\mathsf{Mod}\text{-}\mathscr{C}$. It is known that the restriction functor admits a left and right adjoint functor (the left and right Kan extension [**86**]). Let us give an explicit description. For the notion of tensor product of $\mathscr{C}$-modules we refer to [**10, 89**]. For F in $\mathsf{Mod}\text{-}\mathcal{X}$ and C in $\mathscr{C}$ consider the following functors:

$$\begin{aligned} \mathsf{l}(F)(C) &= \mathsf{Hom}_{\mathscr{C}}(C, \mathsf{i}(-)) \otimes_{\mathcal{X}} F \\ \mathsf{r}(F)(C) &= \mathsf{Hom}_{\mathsf{Mod}\text{-}\mathcal{X}}(\mathsf{Hom}_{\mathscr{C}}(\mathsf{i}(-), C), F) \end{aligned}$$

Then, it is not so difficult to check that the triple $(\mathsf{l}, \mathsf{res}, \mathsf{r})$:

$$\mathsf{Mod}\text{-}\mathscr{C} \;\begin{array}{c} \xleftarrow{\;\mathsf{l}\;} \\ \xrightarrow{\;\mathsf{res}\;} \\ \xleftarrow[\;\mathsf{r}\;]{} \end{array}\; \mathsf{Mod}\text{-}\mathcal{X}$$

is an adjoint triple. Moreover, the functor l is fully faithful (equivalently, the functor r is fully faithful) since the functor $\mathsf{i}\colon \mathcal{X} \longrightarrow \mathscr{C}$ is fully faithful, see [**74**, Lemma 4.2]. Then from Remark 2.5 it follows that the diagram

$$\mathsf{Ker}\,\mathsf{res} \;\begin{array}{c} \xleftarrow{\;\mathsf{q}\;} \\ \xrightarrow{\;\mathsf{i}\;} \\ \xleftarrow[\;\mathsf{p}\;]{} \end{array}\; \mathsf{Mod}\text{-}\mathcal{X} \;\begin{array}{c} \xleftarrow{\;\mathsf{l}\;} \\ \xrightarrow{\;\mathsf{res}\;} \\ \xleftarrow[\;\mathsf{r}\;]{} \end{array}\; \mathsf{Mod}\text{-}\mathscr{C} \tag{3.1}$$

is a recollement of abelian categories. It remains to describe the category $\mathsf{Ker}\,\mathsf{res}$.

We denote by $\mathscr{C}/\mathcal{X}$ the stable category whose objects are the objects of $\mathscr{C}$. Given two objects C, C' in $\mathscr{C}$, the morphism space $\mathsf{Hom}_{\mathscr{C}/\mathcal{X}}(C, C')$ is the quotient $\mathsf{Hom}_{\mathscr{C}}(C, C')/\mathsf{I}_{\mathcal{X}}(C, C')$, where $\mathsf{I}_{\mathcal{X}}(C, C')$ is the subgroup of $\mathsf{Hom}_{\mathscr{C}}(C, C')$ consisting of all morphisms which factor through an object of $\mathcal{X}$.

We leave it as an exercise to the reader to show that $\mathsf{Ker}\,\mathsf{res}$ is identified with the full subcategory of $\mathsf{Mod}\text{-}\mathscr{C}$ consisting of all functors $F\colon \mathscr{C}^{\mathsf{op}} \longrightarrow \mathfrak{Ab}$ vanishing on $\mathcal{X}$ and this last category is equivalent to the module category $\mathsf{Mod}\text{-}\mathscr{C}/\mathcal{X}$ over the stable category $\mathscr{C}/\mathcal{X}$. From diagram 3.1 we infer that $\mathsf{R_{ab}}(\mathsf{Mod}\text{-}\mathscr{C}/\mathcal{X}, \mathsf{Mod}\text{-}\mathscr{C}, \mathsf{Mod}\text{-}\mathcal{X})$ is a recollement of abelian categories.

An interesting special case of the above recollement situation is the following. Let Λ be an Artin algebra and let $\mathcal{X}$ be a full subcategory of $\mathsf{mod}\text{-}\Lambda$ containing the projectives. Then we have the recollement $(\mathsf{Mod}\text{-}\underline{\mathcal{X}}, \mathsf{Mod}\text{-}\mathcal{X}, \mathsf{Mod}\text{-}\Lambda)$ where $\underline{\mathcal{X}}$ is the stable category of $\mathcal{X}$ modulo projectives.

We now continue with the following classical recollement situation arising in Algebraic Geometry. For basics on sheaves over topological spaces we refer to the books [**31**, **67**, **68**], and to the Stacks Project [**119**].

EXAMPLE 3.13. (Sheaves over topological spaces) Let X, Y be two topological spaces and $f\colon X \longrightarrow Y$ a continuous map. Let $\mathcal{F}$ be a presheaf of sets on X. Then for every open subset V of Y the pushforward of $\mathcal{F}$ is defined by

$$f_*\mathcal{F}(V) = \mathcal{F}(f^{-1}(V))$$

where $f^{-1}(V)$ is an open subset of X since the map f is continuous. Then we have the pushforward functor $f_*\colon \mathsf{PSh}(X) \longrightarrow \mathsf{PSh}(Y)$, where $\mathsf{PSh}(X)$, resp. $\mathsf{PSh}(Y)$, is the category of presheaves of sets on X, resp. Y. The functor f_* has a left adjoint defined as follows:

$$f_p\colon \mathsf{PSh}(Y) \longrightarrow \mathsf{PSh}(X),\ \mathcal{G} \ \mapsto\ f_p\mathcal{G}(U) = \mathsf{colim}_{f(U)\subseteq V}\mathcal{G}(V)$$

where the colimit runs over the open neighbourhoods V of $f(U)$ in Y. Consider now a sheaf $\mathcal{F}$ of sets on X. We denote by $\mathsf{Sh}(X)$ the category of sheaves of sets on X. Since f is continuous it follows that the pushforward $f_*\mathcal{F}$ of $\mathcal{F}$ is a sheaf of sets on Y. Then we obtain a functor

$$\mathsf{r}\colon \mathsf{Sh}(X) \longrightarrow \mathsf{Sh}(Y),\ \mathcal{F} \ \mapsto\ \mathsf{r}\mathcal{F}(V) = \mathcal{F}(f^{-1}(V))$$

which has a left adjoint. If $\mathcal{G}$ is a sheaf of sets on Y, the left adjoint is defined by

$$\mathsf{e}\colon \mathsf{Sh}(Y) \longrightarrow \mathsf{Sh}(X),\ \mathcal{G} \ \mapsto\ \mathsf{e}\mathcal{G} = (f_p\mathcal{G})^{\mathsf{Sheaf}}$$

where $(f_p\mathcal{G})^{\mathsf{Sheaf}}$ is the sheafification of the presheaf $f_p\mathcal{G}$ and is called the pullback sheaf. In other words, the functor e is the following composition:

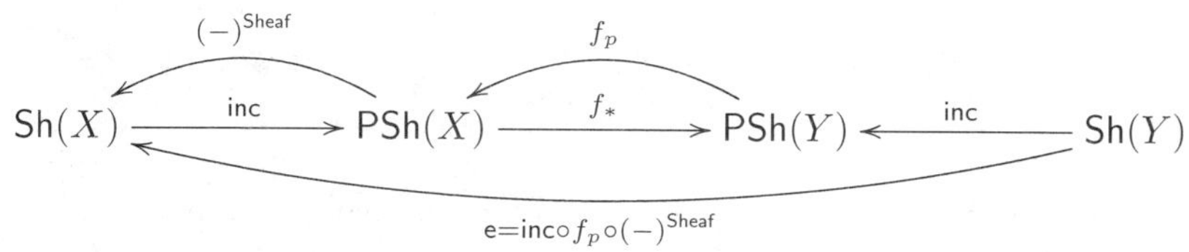

where sheafification $(-)^{\mathsf{Sheaf}}$ is a left adjoint of the inclusion $\mathsf{inc}\colon \mathsf{Sh}(X) \longrightarrow \mathsf{PSh}(X)$ and f_p is the left adjoint of the pushforward f_* on presheaves.

Let X be a topological space as above and consider an open subspace U of X. We denote by F the complement of U in X. For simplicity, we denote again by $\mathsf{Sh}(X)$ the category of sheaves of abelian groups on X. Recall that an abelian sheaf on X is an abelian presheaf on X such that the underlying presheaf of sets is a sheaf. In the setting of abelian sheaves we also have the pullback functor $\mathsf{e}\colon \mathsf{Sh}(X) \longrightarrow \mathsf{Sh}(U)$, the pushforward functor $\mathsf{r}\colon \mathsf{Sh}(U) \longrightarrow \mathsf{Sh}(X)$ and (e, r) is an adjoint pair. However, in this case $(U \subset X)$ the functor e has a simpler description as follows:

$$\mathsf{e}\colon \mathsf{Sh}(X) \longrightarrow \mathsf{Sh}(U), \mathcal{G} \ \mapsto\ \mathsf{e}\mathcal{G}(V) = \mathcal{G}(V)$$

for V an open subset of U, and the sheaf $\mathsf{e}\mathcal{G}$ is called the restriction of $\mathcal{G}$ to U. It turns out that for an open immersion $U \longrightarrow X$ the functor e has a left adjoint defined, first on presheaves, as follows:

$$\mathsf{l}_p \colon \mathsf{PSh}(U) \longrightarrow \mathsf{PSh}(X),\ \mathcal{F} \mapsto \begin{cases} \mathcal{F}(V) & V \subset U \\ 0 & V \not\subset U \end{cases}$$

and then if $\mathcal{F}$ is an abelian sheaf on U we define $\mathsf{l} \colon \mathsf{Sh}(U) \longrightarrow \mathsf{Sh}(X)$ to be $(\mathsf{l}_p\mathcal{F})^{\mathsf{Sheaf}}$, that is, the sheafification of the abelian presheaf $\mathsf{l}_p\mathcal{F}$. Thus, so far we have the following adjoint triple $(\mathsf{l}, \mathsf{e}, \mathsf{r})$:

$$\xymatrix{\mathsf{Sh}(X) \ar[r]^{\mathsf{e}} & \mathsf{Sh}(U) \ar@/_2pc/[l]_{\mathsf{l}} \ar@/^2pc/[l]^{\mathsf{r}}}$$

Since there is a natural isomorphism $\mathsf{el} \simeq \mathrm{Id}_{\mathsf{Sh}(U)}$ it follows that the functor l, and therefore the functor r, is fully faithful. The kernel of the pullback functor e consists of all sheaves $\mathcal{F}$ of abelian groups on X such that for every open subset V of U we have $\mathcal{F}(V) = 0$. Hence $\mathsf{Ker}\,\mathsf{e}$ consists of all sheaves of abelian groups on the complement $F = X \backslash U$ of U. Then from Remark 2.5 we have the following recollement of abelian categories:

$$\xymatrix{\mathsf{Sh}(F) \ar[r]^{\mathsf{i}} & \mathsf{Sh}(X) \ar@/_2pc/[l]_{\mathsf{q}} \ar@/^2pc/[l]^{\mathsf{p}} \ar[r]^{\mathsf{e}} & \mathsf{Sh}(U) \ar@/_2pc/[l]_{\mathsf{l}} \ar@/^2pc/[l]^{\mathsf{r}}}$$

Finally, given a topological space X with an open subset U, consider a sheaf of commutative rings $\mathcal{O}_X$ on X and denote by $\mathcal{O}_U$, resp. $\mathcal{O}_F$, the restricted sheaves of rings on U, resp. F. Then the abelian categories $\mathsf{Mod}\text{-}\mathcal{O}_F$, $\mathsf{Mod}\text{-}\mathcal{O}_X$ and $\mathsf{Mod}\text{-}\mathcal{O}_U$ of sheaves of modules are related via the recollement situation:

$$\xymatrix{\mathsf{Mod}\text{-}\mathcal{O}_F \ar[r]^{\mathsf{i}} & \mathsf{Mod}\text{-}\mathcal{O}_X \ar@/_2pc/[l]_{\mathsf{q}} \ar@/^2pc/[l]^{\mathsf{p}} \ar[r]^{\mathsf{e}} & \mathsf{Mod}\text{-}\mathcal{O}_U \ar@/_2pc/[l]_{\mathsf{l}} \ar@/^2pc/[l]^{\mathsf{r}}}$$

where the above functors are defined similarly as above. We advise the reader to see the book [**31**] for more uses of the above recollement in Algebraic Geometry.

We close this section with the next example of a recollement situation appearing in Algebraic Topology.

EXAMPLE 3.14. (Generic Representations) Let $\mathbb{F}_q$ be a finite field of characteristic p and order $|\mathbb{F}_q| = q = p^s$, where p is a fixed prime. Denote by $M_k(\mathbb{F}_q)$ the multiplicative semigroup of $k \times k$ matrices over $\mathbb{F}_q$ and by $\mathsf{GL}_k(\mathbb{F}_q)$ the general linear group over $\mathbb{F}_q$. Consider the categories of modules $\mathsf{Mod}\text{-}\mathbb{F}_q[\mathsf{GL}_k(\mathbb{F}_q)]$ and $\mathsf{Mod}\text{-}\mathbb{F}_q[M_k(\mathbb{F}_q)]$, where $\mathbb{F}_q[\mathsf{GL}_k(\mathbb{F}_q)]$ is a group algebra and $\mathbb{F}_q[M_k(\mathbb{F}_q)]$ is a semigroup algebra. Consider now the functor category $\mathsf{Fun}(\mathbb{F}_q)$ with objects the functors from finite dimensional $\mathbb{F}_q$-vector spaces to $\mathbb{F}_q$-vector spaces, and with morphisms the natural transformations between such functors. Note that the functor category

$\mathsf{Fun}(\mathbb{F}_q)$ is an abelian category. Let F be an object in $\mathsf{Fun}(\mathbb{F}_q)$. Then for every finite dimensional $\mathbb{F}_q$-vector space V (with $\dim_{\mathbb{F}_q} V = n$) the object $F(V)$ becomes a module over the group algebra $\mathbb{F}_q[\mathsf{GL}_n(V)]$. This observation shows that we can view the objects of $\mathsf{Fun}(\mathbb{F}_q)$ as generic representations of the general linear group over $\mathbb{F}_q$. The category $\mathsf{Fun}(\mathbb{F}_q)$ is called the category of generic representations [**78**]. Then for $R = \mathbb{F}_q[M_k(\mathbb{F}_q)]$ and

$$e = \begin{pmatrix} I_{k-1} & 0 \\ 0 & 0 \end{pmatrix}$$

we obtain from Example 3.1 the following recollement of abelian categories:

$$\xymatrix{ \mathsf{Mod}\text{-}\mathbb{F}_q[\mathsf{GL}_k(\mathbb{F}_q)] \ar[r]^{\mathsf{i}} & \mathsf{Mod}\text{-}\mathbb{F}_q[M_k(\mathbb{F}_q)] \ar@/_2pc/[l]_{\mathsf{q}} \ar@/^2pc/[l]^{\mathsf{p}} \ar[r]^{\mathsf{e}} & \mathsf{Mod}\text{-}\mathbb{F}_q[M_{k-1}(\mathbb{F}_q)] \ar@/_2pc/[l]_{\mathsf{l}} \ar@/^2pc/[l]^{\mathsf{r}} }$$

Note that the ideal ReR consists of all noninvertible matrices and therefore we have $R/ReR = \mathbb{F}_q[\mathsf{GL}_k(\mathbb{F}_q)]$. Moreover, from basic results of Kuhn [**78, 79**] we have the following diagram:

$$\xymatrix{ & \mathsf{Mod}\text{-}\mathbb{F}_q[\mathsf{GL}_3(\mathbb{F}_q)] \ar[d] & \mathsf{Mod}\text{-}\mathbb{F}_q[\mathsf{GL}_2(\mathbb{F}_q)] \ar[d] & & \\ \cdots \ar[r] & \mathsf{Mod}\text{-}\mathbb{F}_q[M_3(\mathbb{F}_q)] \ar[r] \ar[d] \ar@/_1pc/[u] \ar@/^1pc/[u] & \mathsf{Mod}\text{-}\mathbb{F}_q[M_2(\mathbb{F}_q)] \ar[r] \ar[d] \ar@/_1pc/[u] \ar@/^1pc/[u] & \mathsf{Mod}\text{-}\mathbb{F}_q[M_1(\mathbb{F}_q)] \ar[r] & 0 \\ & \mathsf{Mod}\text{-}\mathbb{F}_q[M_2(\mathbb{F}_q)] \ar@/_1pc/[u] \ar@/^1pc/[u] & \mathsf{Mod}\text{-}\mathbb{F}_q[M_1(\mathbb{F}_q)] \ar@/_1pc/[u] \ar@/^1pc/[u] & & }$$

where the vertical diagrams are recollements of module categories and the inverse limit of the horizontal sequence of $\mathsf{Mod}\text{-}\mathbb{F}_q[M_k(\mathbb{F}_q)]$, $k \geq 1$, is the category of generic representations $\mathsf{Fun}(\mathbb{F}_q)$. The functor category $\mathsf{Fun}(\mathbb{F}_q)$ is a fundamental object of study and has strong connections with several different topics, like: Symmetric groups, Schur algebras, Topological Hochschild Homology, Steenrod algebra. We refer to the work of Kuhn [**78–80**] for more details on this topic and the above recollement situations, in particular, their use in generic representation theory.

4. More Examples: Recollements of Hearts

In this section we show that the glueing method on t-structures, due to Beilinson-Bernstein-Deligne [**15**], in a recollement of triangulated categories produces a recollement of abelian categories by considering the associated hearts. The latter situation is what we call recollements of hearts. In other words, we show that the hearts of the glued t-structures are linked via a recollement. This machinery gives us as a method to pass from a recollement of triangulated categories to a recollement of abelian categories. The main aim of this section is to provide a detailed proof of this recollement situation.

Let $\mathsf{R}_{\mathsf{tr}}(\mathcal{U},\mathcal{T},\mathcal{V})$ be a recollement of triangulated categories [**15**]. This means that we have three triangulated categories $\mathcal{U}$, $\mathcal{T}$ and $\mathcal{V}$ and the following triangulated functors between them:

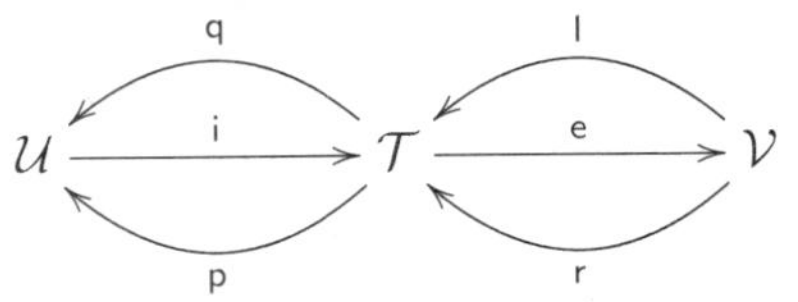

$\mathsf{R}_{\mathsf{tr}}(\mathcal{U},\mathcal{T},\mathcal{V})$

satisfying the analogous axioms as in the abelian case (Definition 2.3), that is, the following hold:

1. $(\mathsf{l},\mathsf{e},\mathsf{r})$ is an adjoint triple.
2. $(\mathsf{q},\mathsf{i},\mathsf{p})$ is an adjoint triple.
3. The functors i, l, and r are fully faithful.
4. $\operatorname{Im}\mathsf{i} = \operatorname{Ker}\mathsf{e}$.

Recall from [**15**] that a **t-structure** in a triangulated category $\mathcal{T}$ with suspension functor [1] is a pair $(\mathcal{T}^{\leq 0},\mathcal{T}^{\geq 0})$ of full subcategories such that setting $\mathcal{T}^{\leq n} = (\mathcal{T}^{\leq 0})[-n]$ and $\mathcal{T}^{\geq n} = (\mathcal{T}^{\geq 0})[-n]$, $\forall n \in \mathbb{Z}$, the following conditions are satisfied:

(i) $\mathsf{Hom}_{\mathcal{T}}(\mathcal{T}^{\leq 0},\mathcal{T}^{\geq 1}) = 0$, i.e. $\mathsf{Hom}_{\mathcal{T}}(X,Y) = 0 \ \ \forall X \in \mathcal{T}^{\leq 0}$ and $\forall Y \in \mathcal{T}^{\geq 1}$.
(ii) $\mathcal{T}^{\leq 0} \subseteq \mathcal{T}^{\leq 1}$ and $\mathcal{T}^{\geq 1} \subseteq \mathcal{T}^{\geq 0}$.
(iii) For every $D \in \mathcal{T}$ there exists a triangle $X \longrightarrow D \longrightarrow Y \longrightarrow X[1]$ such that $X \in \mathcal{T}^{\leq 0}$ and $Y \in \mathcal{T}^{\geq 1}$.

Then it follows from [**15**] that the **heart** $\mathcal{H} = \mathcal{T}^{\leq 0} \cap \mathcal{T}^{\geq 0}$ of a t-structure $(\mathcal{T}^{\leq 0},\mathcal{T}^{\geq 0})$ is an abelian category and we have a full embedding $\epsilon\colon \mathcal{H} \longrightarrow \mathcal{T}$. Note that a sequence $0 \longrightarrow X \longrightarrow Y \longrightarrow Z \longrightarrow 0$ is exact in $\mathcal{H}$ if and only if there is a triangle $X \longrightarrow Y \longrightarrow Z \longrightarrow X[1]$ in $\mathcal{T}$, and moreover for any X, Y in $\mathcal{H}$ there is a functorial isomorphism $\mathsf{Ext}^1_{\mathcal{H}}(X,Y) \cong \mathsf{Hom}_{\mathcal{T}}(X,Y[1])$.

EXAMPLE 4.1. Let $\mathscr{A}$ be an abelian category $\mathscr{A}$. Then the **standard t-structure** $\mathsf{D}_{\mathscr{A}} := (\mathsf{D}^{\leq 0}_{\mathscr{A}},\mathsf{D}^{\geq 0}_{\mathscr{A}})$ in the derived category $\mathsf{D}(\mathscr{A})$ is defined as follows:

$$\mathsf{D}^{\leq 0}_{\mathscr{A}} := \{X \in \mathsf{D}(\mathscr{A}) \mid \mathsf{H}^n_0(X) = 0, \forall n > 0\}$$
$$\mathsf{D}^{\geq 0}_{\mathscr{A}} := \{X \in \mathsf{D}(\mathscr{A}) \mid \mathsf{H}^n_0(X) = 0, \forall n < 0\}$$

We will see below that there is a cohomological functor associated to a t-structure (i.e. a functor sending triangles in $\mathcal{T}$ to long exact sequences in $\mathcal{H}$). In this case, the associated cohomology functors coincide with the cohomologies of the complex (H^n_0, for all n in $\mathbb{Z}$). Moreover, the heart of this t-structure consists of the complexes with cohomologies concentrated in degree zero and, thus, it is equivalent to $\mathscr{A}$.

From now on we fix a recollement of triangulated categories $\mathsf{R}_{\mathsf{tr}}(\mathcal{U},\mathcal{T},\mathcal{V})$. Let $\mathsf{T}_{\mathcal{U}} = (\mathcal{U}^{\leq 0},\mathcal{U}^{\geq 0})$ be a t-structure in $\mathcal{U}$ and $\mathsf{T}_{\mathcal{V}} = (\mathcal{V}^{\leq 0},\mathcal{V}^{\geq 0})$ be a t-structure in $\mathcal{V}$. From [**15**] we have the **glued** t**-structure** $\mathsf{T}_{\mathcal{T}} = (\mathcal{T}^{\leq 0},\mathcal{T}^{\geq 0})$ in $\mathcal{T}$ defined as follows:

$$\mathcal{T}^{\leq 0} = \big\{X \in \mathcal{T} \mid \mathsf{e}(X) \in \mathcal{V}^{\leq 0} \ \text{ and } \ \mathsf{q}(X) \in \mathcal{U}^{\leq 0}\big\}$$

$$\mathcal{T}^{\geq 0} = \big\{X \in \mathcal{T} \mid \mathsf{e}(X) \in \mathcal{V}^{\geq 0} \ \text{ and } \ \mathsf{p}(X) \in \mathcal{U}^{\geq 0}\big\}$$

This is what we usually call: **the glueing method of Beilinson-Bernstein-Deligne**. On the other hand, by [**15**] a t-structure $(\mathcal{T}^{\leq 0},\mathcal{T}^{\geq 0})$ in $\mathcal{T}$ is **glued** with respect to

$\mathsf{R}_{\mathsf{tr}}(\mathcal{U},\mathcal{T},\mathcal{V})$ if and only if $\mathsf{le}(\mathcal{T}^{\leq 0}) \subseteq \mathcal{T}^{\leq 0}$ if and only if $\mathsf{re}(\mathcal{T}^{\geq 0}) \subseteq \mathcal{T}^{\geq 0}$. In this case, the **restrictions** of $(\mathcal{T}^{\leq 0}, \mathcal{T}^{\geq 0})$ to $\mathcal{U}$ and $\mathcal{V}$ satisfy the following relations:

$$\mathcal{U}^{\leq 0} = \mathsf{q}(\mathcal{T}^{\leq 0}),\ \mathcal{U}^{\geq 0} = \mathsf{p}(\mathcal{T}^{\geq 0}) \quad \text{and} \quad \mathcal{V}^{\leq 0} = \mathsf{e}(\mathcal{T}^{\leq 0}),\ \mathcal{V}^{\geq 0} = \mathsf{e}(\mathcal{T}^{\geq 0})$$

Consider now the associated hearts $\mathcal{H}_{\mathcal{U}}$, $\mathcal{H}_{\mathcal{T}}$ and $\mathcal{H}_{\mathcal{V}}$ of the t-structures $\mathsf{T}_{\mathcal{U}}$, $\mathsf{T}_{\mathcal{T}}$ and $\mathsf{T}_{\mathcal{V}}$ respectively. Our goal is to show the following result.

PROPOSITION 4.2. *There is a recollement* $\mathsf{R}_{\mathsf{ab}}(\mathcal{H}_{\mathcal{U}}, \mathcal{H}_{\mathcal{T}}, \mathcal{H}_{\mathcal{V}})$ *of abelian categories:*

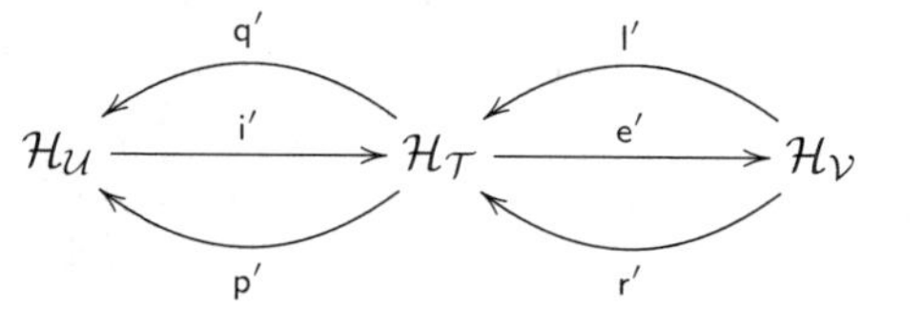

$\mathsf{R}_{\mathsf{ab}}(\mathcal{H}_{\mathcal{U}}, \mathcal{H}_{\mathcal{T}}, \mathcal{H}_{\mathcal{V}})$

The recollement $\mathsf{R}_{\mathsf{ab}}(\mathcal{H}_{\mathcal{U}}, \mathcal{H}_{\mathcal{T}}, \mathcal{H}_{\mathcal{V}})$ is called the **recollement of hearts**. Before we prove the above statement we need to recall some basic facts about t-structures in triangulated categories. For more details we refer to [**15, 49, 68**].

Let $\mathcal{T}$ be a triangulated category with suspension functor [1] and let $(\mathcal{T}^{\leq 0}, \mathcal{T}^{\geq 0})$ be a *t*-structure in $\mathcal{T}$. Associated to the t-structure $(\mathcal{T}^{\leq 0}, \mathcal{T}^{\geq 0})$ are the truncation functors $\tau^{\leq n} \colon \mathcal{T} \longrightarrow \mathcal{T}^{\leq n}$ and $\tau^{\geq n} \colon \mathcal{T} \longrightarrow \mathcal{T}^{\geq n}$. Let T be an object of $\mathcal{T}$. Then there is a triangle $X[n] \longrightarrow T[n] \longrightarrow Y[n] \longrightarrow X[n+1]$ such that $X[n]$ lies in $\mathcal{T}^{\leq 0}$ and $Y[n]$ lies in $\mathcal{T}^{\geq 1}$. Then we obtain the triangle:

$$X \longrightarrow T \longrightarrow Y \longrightarrow X[1]$$

with X in $\mathcal{T}^{\leq n}$ and Y in $\mathcal{T}^{\geq n+1}$. Let $f \colon T \longrightarrow T'$ be a morphism in $\mathcal{T}$ and consider a triangle $X' \longrightarrow T' \longrightarrow Y' \longrightarrow X'[1]$ with X' in $\mathcal{T}^{\leq n}$ and Y' in $\mathcal{T}^{\geq n+1}$. Since $\mathsf{Hom}_{\mathcal{T}}(X, Y') = 0$ we have the following commutative diagram:

$$\begin{array}{ccccccc} X & \longrightarrow & T & \longrightarrow & Y & \longrightarrow & X[1] \\ \downarrow{\scriptstyle g} & & \downarrow{\scriptstyle f} & & \downarrow{\scriptstyle h} & & \downarrow{\scriptstyle g[1]} \\ X' & \longrightarrow & T' & \longrightarrow & Y' & \longrightarrow & X'[1] \end{array}$$

and since $\mathsf{Hom}_{\mathcal{T}}(X, Y'[-1]) = 0$ it follows that the maps g and h are unique. Thus for $T' = T$ and $f = \mathrm{Id}_T$ we infer that the objects X and Y are unique up to a unique isomorphism. Then the truncations functors are defined on objects T in $\mathcal{T}$ by $\tau^{\leq n} \colon \mathcal{T} \longrightarrow \mathcal{T}^{\leq n}$, $\tau^{\leq n}(T) = X$, and $\tau^{\geq n+1} \colon \mathcal{T} \longrightarrow \mathcal{T}^{\geq n+1}$, $\tau^{\geq n+1}(T) = Y$, and given a morphism $f \colon T \longrightarrow T'$ then $\tau^{\leq n}(f) = g$ and $\tau^{\geq n+1}(f) = h$. Hence, for any object T in $\mathcal{T}$ there is a triangle

$$\tau^{\leq n}(T) \longrightarrow T \longrightarrow \tau^{\geq n+1}(T) \longrightarrow \tau^{\leq n}(T)[1]$$

where $\tau^{\leq n}(T)$ lies in $\mathcal{T}^{\leq n}$, $\tau^{\geq n+1}(T)$ lies in $\mathcal{T}^{\geq n+1}$, and the morphism $\tau^{\geq n+1}(T) \longrightarrow \tau^{\leq n}(T)[1]$ is uniquely determined. Note that the truncation functor $\tau^{\leq n}$ is a right adjoint of the inclusion functor $\mathcal{T}^{\leq n} \longrightarrow \mathcal{T}$ and also the truncation functor $\tau^{\geq n+1}$

is a left adjoint of the inclusion functor $\mathcal{T}^{\geq n+1} \longrightarrow \mathcal{T}$ (compare with diagram 2.6):

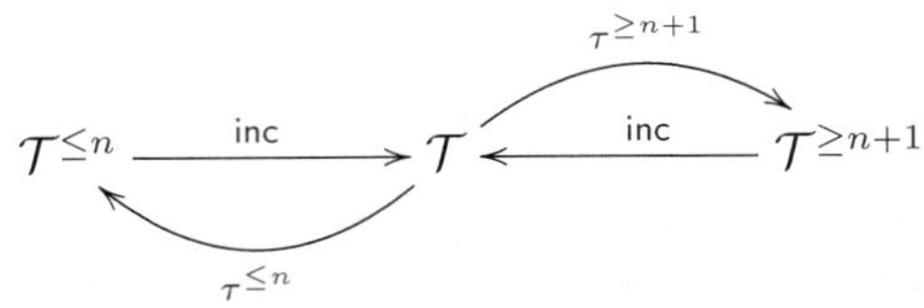

To proceed we need the following functor:

$$\mathsf{H}^0 \colon \mathcal{T} \longrightarrow \mathcal{H} = \mathcal{T}^{\leq 0} \cap \mathcal{T}^{\geq 0}, \ X \mapsto \mathsf{H}^0(X) = \tau^{\geq 0}\tau^{\leq 0}(X) \simeq \tau^{\leq 0}\tau^{\geq 0}(X)$$

It is not difficult to check that (α): the object $\mathsf{H}^0(X) = \tau^{\geq 0}\tau^{\leq 0}(X)$ lies in the heart $\mathcal{H}$, (β): the functor H^0 is cohomological, and (γ): the functors $\tau^{\geq 0}\tau^{\leq 0}$ and $\tau^{\leq 0}\tau^{\geq 0}$ are isomorphic.

Let $\mathcal{T}$ and $\mathcal{V}$ be two triangulated categories with t-structures $(\mathcal{T}^{\leq 0}, \mathcal{T}^{\geq 0})$ and $(\mathcal{V}^{\leq 0}, \mathcal{V}^{\geq 0})$ respectively. Recall that a functor $\mathsf{e} \colon \mathcal{T} \longrightarrow \mathcal{V}$ is called

(i) left t-exact if $\mathsf{e}(\mathcal{T}^{\geq 0}) \subseteq \mathcal{V}^{\geq 0}$,
(ii) right t-exact if $\mathsf{e}(\mathcal{T}^{\leq 0}) \subseteq \mathcal{V}^{\leq 0}$ and
(iii) t-exact if e is both left t-exact and right t-exact.

Assume that the functor e has a right adjoint $\mathsf{r} \colon \mathcal{V} \longrightarrow \mathcal{T}$. Then it is easy to check that the functor e is right t-exact if and only if r is left t-exact. Moreover if these conditions are satisfied we obtain the adjoint pair $(\mathsf{H}^0_{\mathcal{V}} \circ \mathsf{e} \circ \epsilon_{\mathcal{T}}, \mathsf{H}^0_{\mathcal{T}} \circ \mathsf{r} \circ \epsilon_{\mathcal{V}})$ between the hearts $\mathcal{H}_{\mathcal{T}}$ and $\mathcal{H}_{\mathcal{V}}$. We are now ready to prove Proposition 4.2.

Proof of Proposition 4.2. Consider the following diagram:

(4.1)
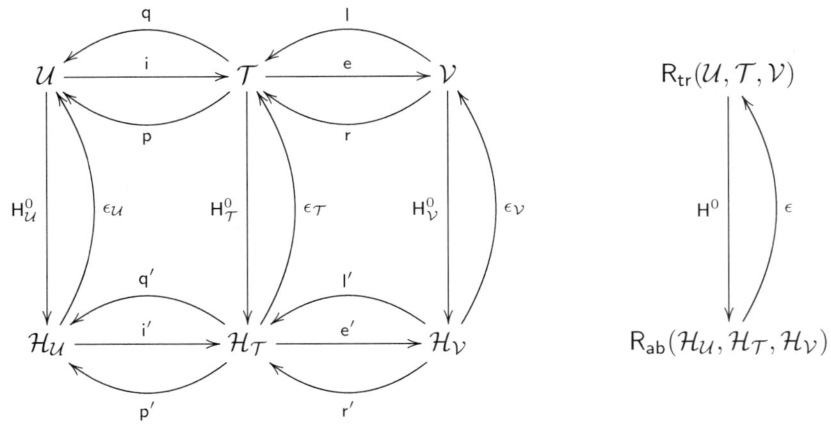

where the functors of $\mathsf{R}_{\mathsf{ab}}(\mathcal{H}_{\mathcal{U}}, \mathcal{H}_{\mathcal{T}}, \mathcal{H}_{\mathcal{V}})$ are defined as follows:

$$\mathsf{q}' = \mathsf{H}^0_{\mathcal{U}} \circ \mathsf{q} \circ \epsilon_{\mathcal{T}} \qquad \mathsf{l}' = \mathsf{H}^0_{\mathcal{T}} \circ \mathsf{l} \circ \epsilon_{\mathcal{V}}$$
$$\mathsf{i}' = \mathsf{H}^0_{\mathcal{T}} \circ \mathsf{i} \circ \epsilon_{\mathcal{U}} \qquad \mathsf{e}' = \mathsf{H}^0_{\mathcal{V}} \circ \mathsf{e} \circ \epsilon_{\mathcal{T}}$$
$$\mathsf{p}' = \mathsf{H}^0_{\mathcal{U}} \circ \mathsf{p} \circ \epsilon_{\mathcal{T}} \qquad \mathsf{r}' = \mathsf{H}^0_{\mathcal{T}} \circ \mathsf{r} \circ \epsilon_{\mathcal{V}}$$

We check the properties of Definition 2.3. We have:

(i) $(\mathsf{l}', \mathsf{e}', \mathsf{r}')$ is an adjoint triple.
From the above and since $(\mathsf{l}, \mathsf{e}, \mathsf{r})$ is an adjoint triple, it remains to show that l is right t-exact, e is left t-exact or right t-exact and r is left t-exact. Let $X \in \mathcal{V}^{\leq 0}$. Then $\mathsf{l}(X) \in \mathcal{T}^{\leq 0}$ since $\mathsf{e}(\mathsf{l}(X)) \simeq X \in \mathcal{V}^{\leq 0}$ and

$\mathsf{q}(\mathsf{l}(X)) = 0 \in \mathcal{U}^{\leq 0}$. Thus l is right t-exact and in the same way we get that e and r are left t-exact.

(ii) $(\mathsf{q}', \mathsf{i}', \mathsf{p}')$ is an adjoint triple.
Similarly as in (i) our statement follows.

(iii) The functors i', l' and r' are fully faithful.
We first show that the functor $\mathsf{l}' \colon \mathcal{H}_{\mathcal{V}} \longrightarrow \mathcal{H}_{\mathcal{T}}$ is fully faithful, equivalently the unit $\mathrm{Id}_{\mathcal{H}_{\mathcal{V}}} \longrightarrow \mathsf{e}'\mathsf{l}'$ is an isomorphism. Let Y be an object of $\mathcal{H}_{\mathcal{V}}$. Then we have

$$\begin{aligned}
\mathsf{e}'\mathsf{l}'(Y) &= \mathsf{H}^0_{\mathcal{V}}\,\mathsf{e}\,\epsilon_{\mathcal{T}}\,\mathsf{H}^0_{\mathcal{T}}\,\mathsf{l}\,\epsilon_{\mathcal{V}}(Y) \\
&= \mathsf{H}^0_{\mathcal{V}}\,\mathsf{e}\,\epsilon_{\mathcal{T}}\,\mathsf{H}^0_{\mathcal{T}}(\mathsf{l}(Y)) \\
&= \mathsf{H}^0_{\mathcal{V}}\,\mathsf{e}\,\epsilon_{\mathcal{T}}(\tau^{\geq 0}(\mathsf{l}(Y))) && \mathsf{l}\colon \text{ right t-exact} \\
&= \mathsf{H}^0_{\mathcal{V}}\,\mathsf{e}\,\tau^{\geq 0}(\mathsf{l}(Y)) \\
&\simeq \mathsf{H}^0_{\mathcal{V}}\,\tau^{\geq 0}\,\mathsf{e}\,\mathsf{l}(Y) && \mathsf{e}\colon \text{ t-exact} \\
&\simeq \mathsf{H}^0_{\mathcal{V}}\,\tau^{\geq 0}(Y) \\
&= Y
\end{aligned}$$

Thus the functor l' is fully faithful. Then since $(\mathsf{l}', \mathsf{e}', \mathsf{r}')$ is an adjoint triple it follows that the functor r' is fully faithful as well. Similarly, using that the functor $\mathsf{i} \colon \mathcal{U} \longrightarrow \mathcal{T}$ is t-exact we derive an isomorphism $\mathsf{q}'\mathsf{i}'(X) \simeq X$ for every X in $\mathcal{H}_{\mathcal{U}}$. Hence the functor $\mathsf{i}' \colon \mathcal{H}_{\mathcal{U}} \longrightarrow \mathcal{H}_{\mathcal{T}}$ is fully faithful.

(iv) $\mathsf{Im}\,\mathsf{i}' = \mathsf{Ker}\,\mathsf{e}'$.
Let $X \in \mathcal{H}_{\mathcal{U}}$. Since the functor i is t-exact it follows that $\mathsf{i}(X)$ lies in $\mathcal{H}_{\mathcal{T}}$. Thus $\mathsf{i}' = \mathsf{i}$ restricting on the heart $\mathcal{H}_{\mathcal{U}}$ and similarly since e is t-exact we have $\mathsf{e}' = \mathsf{e}$ on the heart $\mathcal{H}_{\mathcal{T}}$. Then $\mathsf{e}' \circ \mathsf{i}' = \mathsf{e} \circ \mathsf{i} = 0$.

We infer that $\mathsf{R}_{\mathsf{ab}}(\mathcal{H}_{\mathcal{U}}, \mathcal{H}_{\mathcal{T}}, \mathcal{H}_{\mathcal{V}})$ is a recollement of abelian categories. □

5. Recollements of Module Categories

In Section 3 we studied several examples of recollements of module categories. The common feature of all these examples is that they are constructed using idempotent elements. It is natural to consider how many examples of recollements of module categories we obtain via idempotents. In this connection, Kuhn [**79**] conjectured that any recollement of $\mathsf{Mod}\text{-}\Lambda$ whose terms are module categories, where Λ is a finite dimensional algebra over a field, is equivalent to a recollement induced by an idempotent element (Example 3.1). This section is split into two subsections. In the first one we prove Kuhn's conjecture for general rings and in the second one we provide an example of a recollement of $\mathsf{Mod}\text{-}R$ which is not equivalent to a recollement by module categories. The material of the first subsection is taken from [**110**] which is joint work with Jorge Vitória.

5.1. Kuhn's conjecture. In this subsection our aim is to classify up to equivalence recollements of $\mathsf{Mod}\text{-}R$ whose terms are module categories. For this reason we first need to define the following equivalence relation on the class of recollements of an abelian category $\mathscr{B}$.

DEFINITION 5.1. Two recollements $\mathsf{R_{ab}}(\mathscr{A},\mathscr{B},\mathscr{C})$ and $\mathsf{R_{ab}}(\mathscr{A}',\mathscr{B}',\mathscr{C}')$ are **equivalent** if there are functors $\Phi\colon \mathscr{B} \longrightarrow \mathscr{B}'$ and $\Theta\colon \mathscr{C} \longrightarrow \mathscr{C}'$ which are equivalences of categories such that the diagram below commutes up to natural equivalence:

$$\begin{array}{ccc} \mathscr{B} & \xrightarrow{\mathsf{e}} & \mathscr{C} \\ {\scriptstyle\Phi}\big\downarrow\,\simeq & & \simeq\,\big\downarrow{\scriptstyle\Theta} \\ \mathscr{B}' & \xrightarrow{\mathsf{e}'} & \mathscr{C}' \end{array}$$

i.e. there is a natural equivalence of functors between $\Theta\mathsf{e}$ and $\mathsf{e}'\Phi$.

The following result shows that Definition 5.1 is natural.

LEMMA 5.2. *Let $\mathsf{R_{ab}}(\mathscr{A},\mathscr{B},\mathscr{C})$ and $\mathsf{R_{ab}}(\mathscr{A}',\mathscr{B}',\mathscr{C}')$ be two recollements of abelian categories. The following statements are equivalent:*

(i) *The recollements $\mathsf{R_{ab}}(\mathscr{A},\mathscr{B},\mathscr{C})$ and $\mathsf{R_{ab}}(\mathscr{A}',\mathscr{B}',\mathscr{C}')$ are equivalent.*
(ii) *There are equivalences $\Phi\colon \mathscr{B} \longrightarrow \mathscr{B}'$, $\Psi\colon \mathscr{A} \longrightarrow \mathscr{A}'$ and $\Theta\colon \mathscr{C} \longrightarrow \mathscr{C}'$ such that the six diagrams associated to the six functors of the recollements commute up to natural equivalences (see the diagrams below).*

PROOF. The implication (ii)$\Longrightarrow$(i) is clear.

(i)$\Longrightarrow$(ii): Suppose that there is an equivalence between the recollements of abelian categories $\mathsf{R_{ab}}(\mathscr{A},\mathscr{B},\mathscr{C})$ and $\mathsf{R_{ab}}(\mathscr{A}',\mathscr{B}',\mathscr{C}')$ as in Definition 5.1. Denote by Θ' and Φ' a quasi-inverse of Θ and Φ respectively. We claim that there is a natural isomorphism of functors between $\Phi\mathsf{l}$ and $\mathsf{l}'\Theta$. Since there is a natural isomorphism $\mathsf{e}'\Phi \simeq \Theta\mathsf{e}$, we get a natural isomorphism $\Theta'\mathsf{e}' \simeq \mathsf{e}\Phi'$. Then we have the following natural isomorphisms for any C in $\mathscr{C}$ and B in $\mathscr{B}$:

$$\begin{aligned} \mathsf{Hom}_{\mathscr{B}}(\Phi'\mathsf{l}'\Theta(C),B) &\cong \mathsf{Hom}_{\mathscr{B}'}(\mathsf{l}'\Theta(C),\Phi(B)) \\ &\cong \mathsf{Hom}_{\mathscr{C}'}(\Theta(C),\mathsf{e}'\Phi(B)) \\ &\cong \mathsf{Hom}_{\mathscr{C}}(C,\Theta'\mathsf{e}'\Phi(B)) \\ &\cong \mathsf{Hom}_{\mathscr{C}}(C,\mathsf{e}(B)) \end{aligned}$$

Since any two left adjoints are naturally isomorphic, it follows that $\Phi'\mathsf{l}'\Theta \simeq \mathsf{l}$. Applying the functor Φ we get a natural equivalence of functors:$\mathsf{l}'\Theta \simeq \Phi\mathsf{l}$. Similarly we obtain a natural isomorphism of functors between $\Phi\mathsf{r}$ and $\mathsf{r}'\Theta$. From Remark 2.5, the two recollements are uniquely determined up to equivalence by these functors. Let Ψ be the restriction of Φ to $\mathsf{Ker}\,\mathsf{e}$, which is equivalent to $\mathscr{A}$, and $\mathsf{e}\colon \mathscr{B} \longrightarrow \mathscr{C}$ is the quotient functor of $\mathsf{R_{ab}}(\mathscr{A},\mathscr{B},\mathscr{C})$. Then, the diagram associated with the inclusion functor $\mathsf{i}\colon \mathsf{Ker}\,\mathsf{e} \longrightarrow \mathscr{B}$ clearly commutes and so do the other two, by an adjunction argument analogous to the one above. This completes the proof. $\square$

We summarize below the discussion about equivalent recollements of abelian categories:

$$\mathscr{A} \underset{\mathsf{p}}{\overset{\mathsf{q}}{\leftleftarrows}}\!\xrightarrow{\;\mathsf{i}\;} \mathscr{B} \underset{\mathsf{r}}{\overset{\mathsf{l}}{\leftleftarrows}}\!\xrightarrow{\;\mathsf{e}\;} \mathscr{C} \quad \underset{(\Psi,\Phi,\Theta)}{\xrightarrow{\;\simeq\;}} \quad \mathscr{A}' \underset{\mathsf{p}'}{\overset{\mathsf{q}'}{\leftleftarrows}}\!\xrightarrow{\;\mathsf{i}'\;} \mathscr{B}' \underset{\mathsf{r}'}{\overset{\mathsf{l}'}{\leftleftarrows}}\!\xrightarrow{\;\mathsf{e}'\;} \mathscr{C}'$$

$$\overset{\text{Def.}}{\underset{5.1}{\Longleftrightarrow}} \quad \begin{array}{ccc} \mathscr{B} & \xrightarrow{\mathsf{e}} & \mathscr{C} \\ {\scriptstyle\Phi}\downarrow{\scriptstyle\simeq} & & {\scriptstyle\simeq}\downarrow{\scriptstyle\Theta} \\ \mathscr{B}' & \xrightarrow{\mathsf{e}'} & \mathscr{C}' \end{array}$$

$$\overset{\text{Lemma}}{\underset{5.2}{\Longleftrightarrow}} \quad \begin{array}{ccc} \mathscr{B} & \xleftarrow{\mathsf{l}} & \mathscr{C} \\ {\scriptstyle\Phi}\downarrow{\scriptstyle\simeq} & & {\scriptstyle\simeq}\downarrow{\scriptstyle\Theta} \\ \mathscr{B}' & \xleftarrow{\mathsf{l}'} & \mathscr{C}' \end{array} \qquad \begin{array}{ccc} \mathscr{B} & \xleftarrow{\mathsf{r}} & \mathscr{C} \\ {\scriptstyle\Phi}\downarrow{\scriptstyle\simeq} & & {\scriptstyle\simeq}\downarrow{\scriptstyle\Theta} \\ \mathscr{B}' & \xleftarrow{\mathsf{r}'} & \mathscr{C}' \end{array} \qquad \begin{array}{ccc} \mathscr{B} & \xrightarrow{\mathsf{e}} & \mathscr{C} \\ {\scriptstyle\Phi}\downarrow{\scriptstyle\simeq} & & {\scriptstyle\simeq}\downarrow{\scriptstyle\Theta} \\ \mathscr{B}' & \xrightarrow{\mathsf{e}'} & \mathscr{C}' \end{array}$$

$$\begin{array}{ccc} \mathscr{A} & \xrightarrow{\mathsf{i}} & \mathscr{B} \\ {\scriptstyle\Psi}\downarrow{\scriptstyle\simeq} & & {\scriptstyle\simeq}\downarrow{\scriptstyle\Phi} \\ \mathscr{A}' & \xrightarrow{\mathsf{i}'} & \mathscr{B}' \end{array} \qquad \begin{array}{ccc} \mathscr{A} & \xleftarrow{\mathsf{q}} & \mathscr{B} \\ {\scriptstyle\Psi}\downarrow{\scriptstyle\simeq} & & {\scriptstyle\simeq}\downarrow{\scriptstyle\Theta} \\ \mathscr{A}' & \xleftarrow{\mathsf{q}'} & \mathscr{B}' \end{array} \qquad \begin{array}{ccc} \mathscr{A} & \xleftarrow{\mathsf{p}} & \mathscr{B} \\ {\scriptstyle\Psi}\downarrow{\scriptstyle\simeq} & & {\scriptstyle\simeq}\downarrow{\scriptstyle\Phi} \\ \mathscr{A}' & \xleftarrow{\mathsf{p}'} & \mathscr{B}' \end{array}$$

We remark that equivalences of recollements whose outer equivalence functors (Ψ and Θ in Lemma 5.2) are the identity functors have been studied by Franjou and Pirashvili [**37**]. Equivalences of recollements in the context of triangulated categories also appear in the work of Parshall and Scott [**105**, Theorem 2.5].

We formulate the following definition from [**110**].

DEFINITION 5.3. A recollement of $\mathsf{Mod}\text{-}R$ is called a **recollement by module categories** if it is equivalent to a recollement in which the categories involved are module categories.

In order to study recollements of $\mathsf{Mod}\text{-}R$, it is natural to look at its bireflective subcategories (the inclusion functor admits a left and a right adjoint, i.e. the subcategory is reflective and coreflective respectively) which are classified by epimorphisms in the category of unitary rings. Recall that a ring homomorphism $f\colon R \longrightarrow S$ is an epimorphism if and only if the restriction functor $f_*\colon \mathsf{Mod}\text{-}S \longrightarrow \mathsf{Mod}\text{-}R$ is fully faithful [**120**]. The following result of Gabriel and de la Peña [**43**], see also [**47**], states that all bireflective subcategories of $\mathsf{Mod}\text{-}R$ arise in this way. Recall that two ring epimorphisms $f\colon R \longrightarrow S$ and $g\colon R \longrightarrow S'$ lie in the same epiclass of R, if there is a ring isomorphism $h\colon S \longrightarrow S'$ such that $g = h \circ f$.

THEOREM 5.4 ([**43**, Theorem 1.2]). *There is a bijection between epiclasses of R and bireflective subcategories of $\mathsf{Mod}\text{-}R$, defined by assigning to an epimorphism $f\colon R \longrightarrow S$ the subcategory $\mathcal{X}_S := \mathsf{Im}\, f_*$. Moreover, a full subcategory $\mathcal{X}$ of $\mathsf{Mod}\text{-}R$ is bireflective if and only if it is closed under products, coproducts, kernels and cokernels.*

Some properties of a ring epimorphism $f\colon R \longrightarrow S$ can be seen from the bireflective subcategory $\mathcal{X}_S$. In particular we have the following result.

LEMMA 5.5 ([**116**, Theorem 4.8]). *Let $f\colon R \longrightarrow S$ be a ring epimorphism. Then $\mathcal{X}_S$ is extension-closed if and only if $\mathsf{Tor}_1^R(S,S) = 0$.*

Given a ring epimorphism $f\colon R \longrightarrow S$, consider the unit $\psi_M\colon M \longrightarrow M \otimes_R S$ of the adjunction $(- \otimes_R S, f_*)$ where M is a right R-module. Note that ψ_N is an isomorphism for all N in $\mathcal{X}_S$. In fact, ψ_M is the left $\mathcal{X}_S$-approximation of the right R-module M and the map $f\colon R \longrightarrow S$, regarded as a morphism in $\mathsf{Mod}\text{-}R$, is the left $\mathcal{X}_S$-approximation ψ_R.

The left part of a recollement of $\mathsf{Mod}\text{-}R$ by module categories is a bireflective Serre subcategory. In Section 6 we will study these subcategories further. In the following result we describe these categories in terms of ring epimorphisms. Similar results can be found in [**10**, Section 7] and in [**47**, Proposition 5.3].

PROPOSITION 5.6. *Let $\mathcal{Y}$ be a bireflective Serre subcategory of* $\mathsf{Mod}\text{-}R$. *Then there is an idempotent ideal I of R such that $\mathcal{Y}$ is the essential image of the restriction functor induced by the ring epimorphism $f\colon R \longrightarrow R/I$.*

PROOF. Since $\mathcal{Y}$ is a bireflective subcategory of $\mathsf{Mod}\text{-}R$, by Theorem 5.4 there is a ring epimorphism $f\colon R \longrightarrow S$, for some ring S, such that $f_*(\mathsf{Mod}\text{-}S) = \mathcal{Y}$. We claim that f is surjective. Since $\mathcal{Y}$ is a TTF-class, we have the torsion pair $({}^{\circ}\mathcal{Y}, \mathcal{Y})$ and the composition $\mathsf{i}_{\mathcal{Y}}\mathsf{L}_{\mathcal{Y}}$ ($\mathsf{i}_{\mathcal{Y}}$ being the inclusion $\mathcal{Y} \longrightarrow \mathsf{Mod}\text{-}R$ and $\mathsf{L}_{\mathcal{Y}}$ its left adjoint, see diagram 2.6 in subsection 2.3) is the idempotent coradical functor sending a module M to its torsion-free part. In particular the unit of this adjunction is surjective on every R-module. On the other hand, since $f_*(\mathsf{Mod}\text{-}S) = \mathcal{Y}$, it follows that $\mathsf{i}_{\mathcal{Y}}\mathsf{L}_{\mathcal{Y}}$ is naturally equivalent to $f_*(-\otimes_R S)$. Thus, the map $\psi_M\colon M \longrightarrow M\otimes_R S$ is surjective for every R-module M. In particular, $f = \psi_R$ is surjective. Since $f_*(\mathsf{Mod}\text{-}S)$ is closed under extensions, for $I = \mathsf{Ker}\, f$ we have by Lemma 5.5 that

$$0 = \mathsf{Tor}_1^R(S, S) = \mathsf{Tor}_1^R(R/I, R/I) \cong I/I^2.$$

We infer that $\mathcal{Y} = f_*(\mathsf{Mod}\text{-}R/I)$ with $I^2 = I$. □

Kuhn conjectured the following for recollements of module categories.

CONJECTURE (Kuhn [**79**]). Let Λ be a finite dimensional algebra over a field. Then any recollement of $\mathsf{Mod}\text{-}\Lambda$ by module categories is equivalent to a recollement induced by an idempotent element.

The main result in [**110**] proves this conjecture for any ring R. We give below a sketch of the proof.

THEOREM 5.7 ([**110**, Theorem 5.3, Corollary 5.5]). *A recollement of* $\mathsf{Mod}\text{-}R$ *is a recollement by module categories if and only if it is equivalent to a recollement induced by an idempotent element of a ring S, Morita equivalent to R. If, furthermore, R is semiprimary, then any recollement of* $\mathsf{Mod}\text{-}R$ *is equivalent to a recollement induced by an idempotent element of R.*

PROOF. First, note that by Proposition 5.6 any recollement of $\mathsf{Mod}\text{-}R$ is equivalent to a recollement of the following form:

$$\mathsf{Mod}\text{-}R/I \xrightarrow{\ \mathsf{inc}\ } \mathsf{Mod}\text{-}R \xrightarrow{\ \mathsf{e}\ } \mathscr{C}_I, \tag{5.1}$$

with left adjoints $-\otimes_R R/I\colon \mathsf{Mod}\text{-}R \to \mathsf{Mod}\text{-}R/I$ and $\mathsf{l}\colon \mathscr{C}_I \to \mathsf{Mod}\text{-}R$, and right adjoints $\mathsf{Hom}_R(R/I,-)\colon \mathsf{Mod}\text{-}R \to \mathsf{Mod}\text{-}R/I$ and $\mathsf{r}\colon \mathscr{C}_I \to \mathsf{Mod}\text{-}R$,

where I is some idempotent ideal of R and $\mathscr{C}_I$ denotes the corresponding quotient category. Suppose that the ideal I is generated by an idempotent element e in R. Then by Example 3.1 it follows immediately that $\mathscr{C}_I$ is equivalent to $\mathsf{Mod}\text{-}eRe$.

Assume conversely that (5.1) is equivalent to a recollement by module categories. Since $\mathscr{C}_I$ is equivalent to a module category, there exists a small projective generator P of $\mathscr{C}_I$. We denote by C the ring $\mathsf{End}_{\mathscr{C}_I}(P)$ and by Θ the equivalence $\mathsf{Hom}_{\mathscr{C}_I}(P,-)\colon \mathscr{C}_I \longrightarrow \mathsf{Mod}\text{-}C$. The projective object $\mathsf{l}(P)$ is small since e commutes with coproducts and P is small. Note that in a module category an object is small if and only if it is finitely generated [**120**]. This implies that there is a surjective map $p\colon R^{\oplus n} \longrightarrow \mathsf{l}(P)$, for some n in $\mathbb{N}$, which splits since $\mathsf{l}(P)$ is a projective R-module. Thus, there is an injective map $h\colon \mathsf{l}(P) \longrightarrow R^{\oplus n}$ such that $ph = \mathrm{Id}_{\mathsf{l}(P)}$. Let $S = \mathsf{End}_R(R^{\oplus n})$ and let $\Phi := \mathsf{Hom}_R(R^{\oplus n}, -)$ be the Morita equivalence between $\mathsf{Mod}\text{-}R$ and $\mathsf{Mod}\text{-}S$. Denote by e the element hp. Then it is easy to check that eS is the image of $\Phi(h)\Phi(p)$ and it is isomorphic to $\Phi(\mathsf{l}(P))$ in $\mathsf{Mod}\text{-}S$. Using that l and Φ are fully faithful we obtain a ring isomorphism $C \cong eSe$. Note that we used the ring isomorphism $eSe \longrightarrow \mathsf{End}_S(eS)$ which sends an element in eSe to the endomorphism given by left multiplication with it. Finally, using that Φ is fully faithful and the adjoint pair (l, e) we obtain the natural isomorphism $\Theta\mathsf{e}(M) = \mathsf{Hom}_{\mathscr{C}_I}(P, \mathsf{e}(M)) \cong \mathsf{Hom}_R(\mathsf{l}(P), M) \cong \mathsf{Hom}_S(\Phi(\mathsf{l}(P)), \Phi(M)) = \mathsf{Hom}_S(eS, \Phi(M))$. This shows that the functors $\Theta\colon \mathscr{C}_I \longrightarrow \mathsf{Mod}\text{-}eSe$ and $\Phi\colon \mathsf{Mod}\text{-}R \longrightarrow \mathsf{Mod}\text{-}S$ form an equivalence of recollements from $\mathsf{R_{ab}}(\mathsf{Mod}\text{-}A/I, \mathsf{Mod}\text{-}A, \mathscr{C}_I)$ to $\mathsf{R_{ab}}(\mathsf{Mod}\text{-}S/SeS, \mathsf{Mod}\text{-}S, \mathsf{Mod}\text{-}eSe)$.

Assume that R is semiprimary. Then from [**33**, Statement 6] we know that every idempotent ideal I is of the form ReR for some idempotent element e of R. Hence, if I is the idempotent ideal associated to a recollement of $\mathsf{Mod}\text{-}R$, see (5.1), it follows that the recollement (5.1) is $\mathsf{R_{ab}}(\mathsf{Mod}\text{-}R/ReR, \mathsf{Mod}\text{-}R, \mathsf{Mod}\text{-}eRe)$, i.e. the recollement of $\mathsf{Mod}\text{-}R$ induced by the idempotent element e (Example 3.1). □

Since a finite dimensional algebra Λ over a field is a semiprimary ring (the Jacobson radical $\mathfrak{r}$ of Λ is nilpotent and $\Lambda/\mathfrak{r}$ is semisimple), it follows that any recollement situation of $\mathsf{Mod}\text{-}\Lambda$ arises from one induced by an idempotent element of Λ.

5.2. An Example. We close this section with an example of a recollement of $\mathsf{Mod}\text{-}R$ which is not equivalent to a recollement by module categories, that is, an example which is not induced by an idempotent element (Theorem 5.7).

We start by recalling some basics on von Neuman regular rings.

DEFINITION 5.8. A ring R is called **von Neumann regular** if for every element a in R there exists an element x in R such that $a = axa$.

A von Neuman regular ring is also called absolutely flat. This is due to the characterization that over a von Neumann regular ring every left R-module is flat [**82**, Theorem 4.21]. In what follows we need the following result.

LEMMA 5.9 ([**121**, Proposition 2.1]). *Let R be a von Neumann regular ring and P a projective R-module. Then every finitely generated submodule of P is a direct summand of P, i.e. P is regular.*

From now on and until the end of this section we fix a field k. Set

$$R := \prod_{i=1}^{\infty} k \qquad \text{and} \qquad I := \bigoplus_{i=1}^{\infty} k.$$

We have the following easy observation.

LEMMA 5.10. *Let k be a field and R, I as above. Then R is a unital commutative von Neumann regular ring and I is a two-sided idempotent ideal of R.*

PROOF. Since k is a field it follows immediately that k is a von Neumann regular ring. Then since the multiplication of elements in R is componentwise we infer that R is a unital commutative von Neumann regular ring. It is easy to verify that I is a two-sided idempotent ideal of R. Note that in a von Neumann regular ring every ideal is idempotent. □

Let $\pi\colon R \longrightarrow R/I$ be the canonical surjection and $\pi_*\colon \mathsf{Mod}\text{-}R/I \longrightarrow \mathsf{Mod}\text{-}R$ the restriction functor. As in Theorem 5.7 consider the recollement of $\mathsf{Mod}\text{-}R$ induced by the ideal I:

$$\mathsf{Mod}\text{-}R/I \xrightarrow{\;\pi_*\;} \mathsf{Mod}\text{-}R \xrightarrow{\;\mathsf{e}\;} \mathscr{C}_I \qquad (5.2)$$

with left adjoints $-\otimes_R R/I\colon \mathsf{Mod}\text{-}R \to \mathsf{Mod}\text{-}R/I$ and $\mathsf{l}\colon \mathscr{C}_I \to \mathsf{Mod}\text{-}R$, and right adjoints $\mathsf{Hom}_R(R/I,-)\colon \mathsf{Mod}\text{-}R \to \mathsf{Mod}\text{-}R/I$ and $\mathsf{r}\colon \mathscr{C}_I \to \mathsf{Mod}\text{-}R$.

The statement of this subsection claims the following.

Proposition 5.11. *The recollement of* $\mathsf{Mod}\text{-}R$ *induced by the idempotent ideal* I *is not equivalent to any recollement induced by an idempotent element.*

Proof. We will show that the category $\mathscr{C}_I$ is not equivalent to a category of modules. Then the recollement (5.2) cannot be equivalent to a recollement by module categories and therefore Theorem 5.7 implies that it cannot be induced by an idempotent element. Recall that the above recollement satisfies $\mathsf{Im}\,\mathsf{l} \subseteq \mathsf{Ker}(-\otimes_R R/I)$. Also, for any R-module M, $M \otimes_R R/I$ is isomorphic to M/MI. Thus, we have

$$\mathscr{C}_I \cong \mathsf{Im}\,\mathsf{l} \subseteq \{M \in \mathsf{Mod}\text{-}R \mid MI = M\}$$

Assume that $\mathscr{C}_I$ is a module category and let P be a small projective generator in $\mathscr{C}_I$. Then $Q := \mathsf{l}(P)$ is a small (i.e., finitely generated) projective R-module such that $QI = Q$. We write S_j for the one dimensional simple module corresponding to the copy of the field in position j, with $j \in \mathbb{N}$.

First we claim that there is n in $\mathbb{N}$ such that $\mathsf{Hom}_R(S_n, Q) \neq 0$. Indeed, by the Hom-tensor adjunction we have the isomorphism:

$$\mathsf{Hom}_R(Q \otimes_R I, Q) \cong \mathsf{Hom}(Q, \mathsf{Hom}_R(I, Q)).$$

Since Q is projective and by Remark 2.4 (ii) we have $Q \otimes_R R/I = 0$, we conclude that $Q \otimes_R I \cong Q$ and, thus, $\mathsf{Hom}_R(I, Q) \neq 0$. Since $I = \oplus_{i=1}^{\infty} S_i$, we have the claim.

Secondly, since Q is regular (see Lemma 5.9), every simple S_n embedded in Q is a summand of Q. Therefore, if we define the set

$$J := \{i \in \mathbb{N} \mid \mathsf{Hom}(S_i, Q) \neq 0\},$$

it is then clear that, since each S_i is simple, $\oplus_{i\in J} S_i$ is a summand of Q. Since Q is a finitely generated R-module, this implies that J is a finite set. Observe now that, for all n, S_n lies in $\mathsf{Im}\,\mathsf{l}$ and that S_n is projective since

$$R_R := \prod_{i=1}^{n} S_i \oplus S_n \oplus \prod_{i=n+1}^{\infty} S_i$$

Choose now n in $\mathbb{N} \setminus J$. Clearly $\mathsf{Hom}(Q, S_n) = 0$, since any map would have to be surjective, thus splitting and forcing n to lie in J. This, however, implies that $\mathsf{Hom}_{\mathscr{C}_I}(P, \mathsf{e}(S_n)) = 0$, which is a contradiction under the assumption that P is a projective generator for $\mathscr{C}_I$. □

We remark that more examples of this nature occur in the class of rings considered in the work of Saorin [**115**].

6. Recollements of Abelian Categories and TTF-triples

It is well known that recollements of triangulated categories correspond bijectively to TTF-triples in triangulated categories, see [**15**, Section 1.4.4], [**17**, Remark 2.14] and [**96**, Section 4.2]. Our purpose in this section is to show an analogous result for recollements in the context of abelian categories. In particular, we will establish in Theorem 6.5 a correspondence between recollements of abelian categories up to equivalence and certain TTF-triples. In this connection, we first need to discuss (co)localizations with respect to a torsion theory.

Let $\mathscr{A}$ be an abelian category and $(\mathcal{X}, \mathcal{Y})$ a torsion pair in $\mathscr{A}$. Localisations and colocalisations with respect to $(\mathcal{X}, \mathcal{Y})$ first appeared in [**40**], see also [**99**] and [**120**]. We recall some basic notions from [**99**].

A torsion pair $(\mathcal{X}, \mathcal{Y})$ is **strongly hereditary** (respectively, **strongly cohereditary**), if there is a functor $\mathsf{L}\colon \mathscr{A} \longrightarrow \mathscr{A}$ (respectively, $\mathsf{C}\colon \mathscr{A} \longrightarrow \mathscr{A}$) the **localisation** (respectively, the **colocalisation**) **functor with respect to** $(\mathcal{X}, \mathcal{Y})$, and a natural transformation $\phi\colon \mathrm{Id}_{\mathscr{A}} \longrightarrow \mathsf{L}$ (respectively, $\psi\colon \mathsf{C} \longrightarrow \mathrm{Id}_{\mathscr{A}}$) such that, for all A in $\mathscr{A}$ the following conditions hold:

(i) $\mathsf{Ker}\,\phi_A, \mathsf{Coker}\,\phi_A \in \mathcal{X}$ (respectively, $\mathsf{Ker}\,\psi_A, \mathsf{Coker}\,\psi_A \in \mathcal{Y}$);
(ii) $\mathsf{L}(A) \in \mathcal{Y}$ (respectively, $\mathsf{C}(A) \in \mathcal{X}$);
(iii) $\mathsf{L}(A)$ is $\mathcal{X}$-**divisible** (respectively, $\mathsf{C}(A)$ is $\mathcal{Y}$-**codivisible**), meaning that the functor $\mathsf{Hom}_{\mathscr{A}}(-, \mathsf{L}(A))$ (respectively, $\mathsf{Hom}_{\mathscr{A}}(\mathsf{C}(A), -)$) is exact on exact sequences, $0 \to K \to M \to N \to 0$, with $N \in \mathcal{X}$ (respectively, $K \in \mathcal{Y}$).

The embedding of $\mathsf{Im}\,\mathsf{L}$ in $\mathscr{A}$, known as the **Giraud subcategory** of $\mathscr{A}$ associated with $(\mathcal{X}, \mathcal{Y})$, admits an exact left adjoint such that L is given by the composition of the functors and ϕ is the unit of this adjunction, see [**99**]. Also, $\mathsf{Im}\,\mathsf{L}$ is the full subcategory of $\mathcal{X}$-divisible objects of $\mathcal{Y}$. Dual statements holds for $\mathsf{Im}\,\mathsf{C}$.

DEFINITION 6.1. A TTF-triple $(\mathcal{X}, \mathcal{Y}, \mathcal{Z})$ in $\mathscr{A}$ is called **strong** if $(\mathcal{X}, \mathcal{Y})$ is strongly cohereditary and $(\mathcal{Y}, \mathcal{Z})$ is strongly hereditary.

We recall the following useful result of Tachikawa and Ohtake.

THEOREM 6.2 ([**98**, Theorem 1.6], [**99**, Theorem 1.8-1.8*]). *Let $\mathscr{A}$ be an abelian category. If $\mathscr{A}$ has enough projectives (respectively, injectives), then a torsion pair is cohereditary (respectively, hereditary) if and only if it is strongly cohereditary (respectively, strongly hereditary).*

The next result, which is [**110**, Lemma 3.2], identifies TTF-classes which are localising and colocalising. We provide the proof.

LEMMA 6.3. *Let $(\mathcal{X}, \mathcal{Y}, \mathcal{Z})$ be a TTF-triple in $\mathscr{A}$. The following statements hold.*

(i) *The TTF-class $\mathcal{Y}$ is a localising subcategory of $\mathscr{A}$ if and only if $(\mathcal{Y}, \mathcal{Z})$ is a strongly hereditary torsion pair.*
(ii) *The TTF-class $\mathcal{Y}$ is a colocalising subcategory if and only if $(\mathcal{X}, \mathcal{Y})$ is a strongly cohereditary torsion pair.*

PROOF. (i) ($\Longrightarrow$) Let $\mathsf{r}\colon \mathscr{A}/\mathcal{Y} \longrightarrow \mathscr{A}$ be the right adjoint of the quotient functor $\mathsf{e}\colon \mathscr{A} \longrightarrow \mathscr{A}/\mathcal{Y}$. Given an object A in $\mathscr{A}$, consider the unit map $\nu_A\colon A \longrightarrow \mathsf{re}(A)$. We claim that this is a localization with respect to $(\mathcal{Y}, \mathcal{Z})$. From Theorem 2.1, the object $\mathsf{re}(A)$ lies in $\mathcal{Y}^{\perp}$ and, thus, in $\mathcal{Z} = \mathcal{Y}^{\circ}$. Since the functor e

is exact and $\mathsf{e} \circ \mathsf{r} \simeq \mathrm{Id}_{\mathscr{A}/\mathcal{Y}}$ (Remark 2.4 (iv)), it follows that $\mathsf{e}(\mathsf{Ker}\,\phi_A) = 0 = \mathsf{e}(\mathsf{Coker}\,\phi_A)$. Thus, both the objects $\mathsf{Ker}\,\phi_A$ and $\mathsf{Coker}\,\phi_A$ lie in $\mathcal{Y}$. Finally, since $\mathsf{Ext}^1_{\mathscr{A}}(Y, \mathsf{re}(A)) = 0$ for all Y in $\mathcal{Y}$, we get that the object $\mathsf{re}(A)$ is $\mathcal{Y}$-divisible.

($\Longleftarrow$) Let L and $\phi\colon \mathrm{Id}_{\mathscr{A}} \longrightarrow \mathsf{L}$ be the associated localisation functor and natural transformation, respectively, and $\mathsf{e}\colon \mathscr{A} \longrightarrow \mathscr{A}/\mathcal{Y}$ the quotient functor. Recall that $\mathsf{L} = \mathsf{i} \circ \mathsf{l}$, where $\mathsf{i}\colon \mathcal{G} \longrightarrow \mathscr{A}$ is the inclusion functor of the Giraud subcategory $\mathcal{G} := \mathsf{Im}\,\mathsf{L}$ in $\mathscr{A}$ and l is its left adjoint. We claim that $\mathsf{e} \circ \mathsf{i}\colon \mathcal{G} \longrightarrow \mathscr{A}/\mathcal{Y}$ is an equivalence. Since the composition $\mathsf{e} \circ \phi$ is a natural equivalence between e and eL, it follows that $\mathsf{e} \circ \mathsf{i}$ is essentially surjective. On the other hand, by the description (2.1) of morphisms in $\mathscr{A}/\mathcal{Y}$ and using $\mathcal{Y}$-divisibility we obtain that $\mathsf{e} \circ \mathsf{i}$ is fully faithful. Finally, since both $\mathsf{e} \circ \mathsf{i}$ and l have right adjoints then so does $(\mathsf{e} \circ \mathsf{i}) \circ \mathsf{l} \simeq \mathsf{e} \circ \mathsf{L} \simeq \mathsf{e}$.

(ii) The second statement is proved dually and is left to the reader. □

REMARK 6.4. We state two useful observations about Lemma 6.3.

(i) In Lemma 6.3 we in fact prove that, if $\mathcal{Y}$ is a localising subcategory and r is the right adjoint of the quotient functor $\mathsf{e}\colon \mathscr{A} \longrightarrow \mathscr{A}/\mathcal{Y}$, then the Giraud subcategory $\mathcal{G}$ associated to the strongly hereditary torsion pair $(\mathcal{Y}, \mathcal{Z})$ coincides with $\mathcal{Y}^{\perp}$. Similarly, the Co-Giraud subcategory $\mathcal{H}$ of $\mathscr{A}$ induced by the strongly cohereditary torsion pair $(\mathcal{X}, \mathcal{Y})$ (formed by the $\mathcal{Y}$-codivisible objects of $\mathcal{X}$) coincides with the subcategory ${}^{\perp}\mathcal{Y}$.

(ii) Let $(\mathcal{Y}, \mathcal{Z})$ be a strongly hereditary torsion pair. Then from the proof of Lemma 6.3 and from (the proof of) Remark 2.5 we get the next diagram:

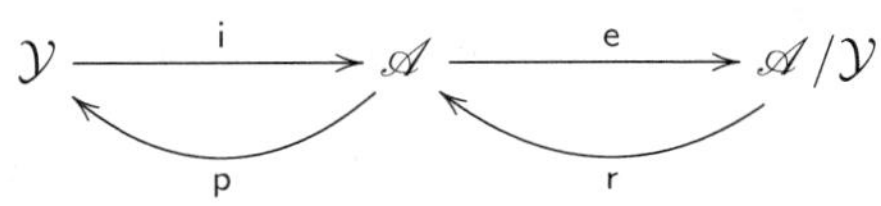

where $0 \longrightarrow \mathcal{Y} \longrightarrow \mathscr{A} \longrightarrow \mathscr{A}/\mathcal{Y} \longrightarrow 0$ is an exact sequence of abelian categories and (i, p), (e, r) are adjoint pairs. Note that $\mathcal{Y}$ is a Serre subcategory of $\mathscr{A}$ since $(\mathcal{Y}, \mathcal{Z})$ is a hereditary torsion pair. In other words, the above diagram is a half recollement, since we also have for free that the functor r is fully faithful (see Theorem 2.2). Compare this with Remark 2.6.

Dually, if $(\mathcal{X}, \mathcal{Y})$ is a strongly cohereditary torsion pair then the diagram:

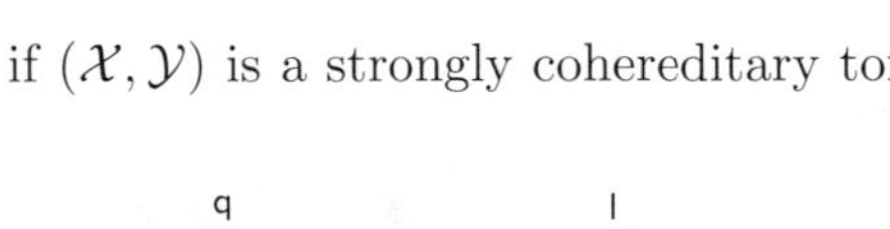

is a half recollement, i.e. the quotient functor e admits a left adjoint.

We are now ready to prove the main result of this section. It is proved in [**110**, Theorem 4.3] but for completeness we sketch below the proof.

THEOREM 6.5. *Let $\mathscr{B}$ be an abelian category. The following are in bijection.*

(i) *Equivalence classes of recollements of abelian categories $\mathsf{R}_{\mathsf{ab}}(\mathscr{A}, \mathscr{B}, \mathscr{C})$;*
(ii) *Strong TTF-triples $(\mathcal{X}, \mathcal{Y}, \mathcal{Z})$ in $\mathscr{B}$;*
(iii) *Bilocalising TTF-classes $\mathcal{Y}$ of $\mathscr{B}$;*
(iv) *Bilocalising Serre subcategories $\mathcal{Y}$ of $\mathscr{B}$.*

PROOF. (i)$\Longrightarrow$(ii): Let $\mathsf{R}_{\mathsf{ab}}(\mathscr{A},\mathscr{B},\mathscr{C})$ be a recollement of $\mathscr{B}$. We first show that $(\mathsf{Ker}\,\mathsf{q},\mathsf{i}(\mathscr{A}),\mathsf{Ker}\,\mathsf{p})$ is a TTF-triple in $\mathscr{B}$. The adjoint triple $(\mathsf{q},\mathsf{i},\mathsf{p})$ ensures that

$$\mathsf{Hom}_{\mathscr{B}}(\mathsf{Ker}\,\mathsf{q},\mathsf{i}(\mathscr{A}))=0=\mathsf{Hom}_{\mathscr{B}}(\mathsf{i}(\mathscr{A}),\mathsf{Ker}\,\mathsf{p})$$

Let B be an object of $\mathscr{B}$. From Proposition 2.8, we have an exact sequence

$$0 \longrightarrow \mathsf{Ker}\,\mu_B \longrightarrow \mathsf{le}(B) \xrightarrow{\mu_B} B \longrightarrow \mathsf{iq}(B) \longrightarrow 0$$

where $\mathsf{Ker}\,\mu_B$ lies in $\mathsf{i}(\mathscr{A})$. Applying the right exact functor q to the sequence and using that the composition $\mathsf{q}\circ\mathsf{l}=0$, see Remark 2.4 (ii), we get that $\mathsf{q}(\mathsf{Im}\,\mu_B)=0$. Thus, we have the exact sequence

$$0 \longrightarrow \mathsf{Im}\,\mu_B \longrightarrow B \longrightarrow \mathsf{iq}(B) \longrightarrow 0$$

with $\mathsf{Im}\,\mu_B$ in $\mathsf{Ker}\,\mathsf{q}$ and $\mathsf{iq}(B)$ in $\mathsf{i}(\mathscr{A})$. This shows that $(\mathsf{Ker}\,\mathsf{q},\mathsf{i}(\mathscr{A}))$ is a torsion pair. Similarly, the exact sequence induced by the unit $\nu_B\colon B \longrightarrow \mathsf{re}(B)$ (see Proposition 2.8) can be used to show that $(\mathsf{i}(\mathscr{A}),\mathsf{Ker}\,\mathsf{p})$ is a torsion pair in $\mathscr{B}$. Since $\mathsf{R}_{\mathsf{ab}}(\mathscr{A},\mathscr{B},\mathscr{C})$ is a recollement, we have that $\mathsf{i}(\mathscr{A})$ is a bilocalising subcategory of $\mathscr{B}$. Hence, from Lemma 6.3 it follows that $(\mathsf{Ker}\,\mathsf{q},\mathsf{i}(\mathscr{A}),\mathsf{Ker}\,\mathsf{p})$ is a strong TTF-triple. Note that this TTF-triple depends only on the equivalence class of the recollement.

(ii)$\Longrightarrow$(i): Let $(\mathcal{X},\mathcal{Y},\mathcal{Z})$ be a strong TTF-triple in $\mathscr{B}$. Since $\mathcal{Y}$ is a TTF-class, it follows that it is bireflective and hence the embedding $\mathsf{i}\colon \mathcal{Y} \longrightarrow \mathscr{B}$ admits a left adjoint q and a right adjoint p. It is also a Serre subcategory and we consider the quotient functor $\mathsf{e}\colon \mathscr{B} \longrightarrow \mathscr{B}/\mathcal{Y}$. Since the triple is strong it follows from Lemma 6.3 that $\mathcal{Y}$ is bilocalising. Thus, e has both left and right adjoints, l and r respectively, which are fully faithful (because the compositions er and el are naturally equivalent to $\mathrm{Id}_{\mathscr{B}/\mathcal{Y}}$, see [**40**]). Hence, the following diagram:

$$\xymatrix{\mathcal{Y} \ar[r]^{\mathsf{i}} & \mathscr{B} \ar@/_1.5pc/[l]_{\mathsf{q}} \ar@/^1.5pc/[l]^{\mathsf{p}} \ar[r]^{\mathsf{e}} & \mathscr{B}/\mathcal{Y} \ar@/_1.5pc/[l]_{\mathsf{l}} \ar@/^1.5pc/[l]^{\mathsf{r}}} \qquad \mathsf{R}_{\mathsf{ab}}(\mathcal{Y},\mathscr{B},\mathscr{B}/\mathcal{Y})$$

is a recollement of abelian categories. Clearly these correspondences are inverse to each other, up to equivalence of recollements.

Since $\mathsf{i}(\mathscr{A})$ is a bilocalising TTF-class as well as a Serre subcategory, the bijection between (i) and (ii) easily implies the bijections between (i), (iii) and (iv). □

Assuming enough projectives and injectives for $\mathscr{B}$, the above bijection becomes more clear.

COROLLARY 6.6. *If $\mathscr{B}$ has enough projectives and injectives, then the equivalence classes of recollements of $\mathscr{B}$ are in bijection with the TTF-triples in $\mathscr{B}$.*

PROOF. Let $(\mathcal{X},\mathcal{Y},\mathcal{Z})$ be a TTF-triple in $\mathscr{B}$. Since $\mathscr{B}$ has enough projectives and injectives it follows from Theorem 6.2 that every TTF-triple in $\mathscr{B}$ is strong. The result then is a consequence of Theorem 6.5. □

As a consequence of Theorem 6.5, we have the following notable equivalences.

COROLLARY 6.7. *Let $\mathsf{R}_{\mathsf{ab}}(\mathscr{A},\mathscr{B},\mathscr{C})$ be a recollement of $\mathscr{B}$, $\mathcal{G}$ be the Giraud subcategory associated to the torsion pair $(\mathsf{i}(\mathscr{A}),\mathsf{Ker}\,\mathsf{p})$ and $\mathcal{H}$ the Co-Giraud* *subcategory associated to the torsion pair* $(\mathsf{Ker}\,\mathsf{q}, \mathsf{i}(\mathscr{A}))$. *Then, the quotient functor* e *induces:*

(i) *an equivalence* $\mathsf{i}(\mathscr{A})^{\perp} = \mathcal{G} = \mathsf{Im}\,\mathsf{r} \xrightarrow{\simeq} \mathscr{B}/\mathsf{i}(\mathscr{A})$,

(ii) *an equivalence* ${}^{\perp}\mathsf{i}(\mathscr{A}) = \mathcal{H} = \mathsf{Im}\,\mathsf{l} \xrightarrow{\simeq} \mathscr{B}/\mathsf{i}(\mathscr{A})$ *and*

(iii) *an equivalence* $\mathsf{Ker}\,\mathsf{q} \cap \mathsf{Ker}\,\mathsf{p} \xrightarrow{\simeq} \mathscr{B}/\mathsf{i}(\mathscr{A})$.

PROOF. Statements (i) and (ii) follow immediately from the fact that the TTF-triple $(\mathsf{Ker}\,\mathsf{q}, \mathsf{i}(\mathscr{A}), \mathsf{Ker}\,\mathsf{p})$ is strong and from Remark 6.4. Statement (iii) is well-known for TTF-triples, see [**50**, Theorem 1.9] or [**17**, Proposition 1.3]. □

We close this section by recovering Jans' bijection between TTF-triples and idempotent ideals (stated already in Theorem 2.13) and classifiy also equivalence classes of recollements of Mod-R.

COROLLARY 6.8. (Jans bijection) *There is a bijection between equivalence classes of recollements of* Mod-R, TTF-*triples in* Mod-R *and idempotent two-sided ideals of* R. *In particular the bijections are given as follows:*

- $\{$*equivalence classes of recollements of* Mod-$R\}$ $\longleftrightarrow$ $\{$*TTF-triples in* Mod-$R\}$

$$\mathscr{A} \xrightarrow{\ \mathsf{i}\ } \text{Mod-}R \xrightarrow{\ \mathsf{e}\ } \mathscr{C} \quad (\text{with adjoints } \mathsf{q}, \mathsf{p} \text{ of } \mathsf{i} \text{ and } \mathsf{l}, \mathsf{r} \text{ of } \mathsf{e}) \qquad \longmapsto \qquad (\mathsf{Ker}\,\mathsf{q}, \mathsf{i}(\mathscr{A}), \mathsf{Ker}\,\mathsf{p})$$

$$(\mathcal{Y}, \text{Mod-}R, (\text{Mod-}R)/\mathcal{Y}) \quad \longleftarrow\!\!\dashv \quad (\mathcal{X}, \mathcal{Y}, \mathcal{Z})$$

- $\{$*idempotent two-sided ideals of* $R\}$ $\longleftrightarrow$ $\{$*TTF-triples in* Mod-$R\}$

$$I \trianglelefteq R,\ I^2 = I \quad \longmapsto \quad ({}^{\circ}\mathcal{Y}_F, \mathcal{Y}_F := \{X \in \text{Mod-}R \mid F(X) = X\}, \mathcal{Y}_F^{\circ})$$

$$\textit{and} \quad F = f_*(- \otimes_R R/I) \colon \text{Mod-}R \longrightarrow \text{Mod-}R$$

$$I \quad \longleftarrow\!\!\dashv \quad (\mathcal{X}, \mathcal{Y}, \mathcal{Z}),\ \mathcal{Y} = f_*(\text{Mod-}R/I)$$

PROOF. The bijection between equivalence classes of recollements and TTF-triples follows from Corollary 6.6, since Mod-R has enough projectives and injectives. The bijection between TTF-triples and idempotent ideals of R can be seen as a consequence of Proposition 2.14. Indeed, the bijection in that proposition assigns to a TTF-triple a functor which, by Proposition 5.6, is precisely $f_*(- \otimes_R R/I)$ for some idempotent ideal I and $f \colon R \longrightarrow R/I$ the canonical projection, thus uniquely determined by the ideal I. Conversely, given an idempotent ideal I and the quotient map $f \colon R \longrightarrow R/I$ it is easy to check that $f_*(- \otimes_R R/I)$ is a right exact idempotent coradical endofunctor of Mod-R preserving coproducts. Indeed, for any module X in Mod-R we have the isomorphism $X \otimes_R R/I \simeq X/XI$. Thus we have the exact sequence $0 \longrightarrow XI \longrightarrow X \longrightarrow X/XI \longrightarrow 0$ in Mod-R and since the ideal I is idempotent it follows that $XI \otimes_R R/I \simeq XI/XI^2 \simeq XI/XI = 0$. Hence the functor $f_*(- \otimes_R R/I)$ is right exact coradical and preserves coproducts since the functors f_* and $- \otimes_R R/I$ are left adjoints. □

7. Homological Theory of Recollements

In this section we study homological properties of the abelian categories involved in a recollement situation $\mathsf{R_{ab}}(\mathscr{A},\mathscr{B},\mathscr{C})$. Since the middle functors in $\mathsf{R_{ab}}(\mathscr{A},\mathscr{B},\mathscr{C})$ are exact it is natural to investigate the extension groups along these functors. More precisely, since the functors $\mathsf{i}\colon \mathscr{A} \longrightarrow \mathscr{B}$ and $\mathsf{e}\colon \mathscr{B} \longrightarrow \mathscr{C}$ are exact, they induce homomorphisms of abelian groups between the extension groups:

$$\mathsf{i}^m\colon \mathsf{Ext}^m_{\mathcal{A}}(A, A') \longrightarrow \mathsf{Ext}^m_{\mathcal{B}}(\mathsf{i}(A), \mathsf{i}(A'))$$

and

$$\mathsf{e}^m\colon \mathsf{Ext}^m_{\mathcal{B}}(B, B') \longrightarrow \mathsf{Ext}^m_{\mathcal{C}}(\mathsf{e}(B), \mathsf{e}(B'))\ .$$

It is therefore natural to formulate the following problem.

PROBLEM 7.1. *Characterize when the above maps are isomorphisms for* $0 \leq m \leq k$.

This section is split into two subsections. In the first subsection we study the left side of the above problem, i.e. the extension groups induced by the functor i. In this connection, we discuss the notion of a homological embedding of an exact functor and we provide a characterization in terms of derived categories. In the second subsection we study the right side of Problem 7.1, i.e. the extension groups along the functor e. Moreover, we discuss when the functor e induces isomorphisms of extension groups in higher degrees.

7.1. The left side and homological embeddings. Motivated by the work of Geigle-Lenzing [**47**] and in connection with the above problem the following notion was introduced in [**108**].

DEFINITION 7.2. An exact functor $\mathsf{i}\colon \mathscr{A} \longrightarrow \mathscr{B}$ between abelian categories is called a k-**homological embedding**, $k \geq 0$, if the map

$$\mathsf{i}^n_{X,Y}\colon \mathsf{Ext}^n_{\mathscr{A}}(X, Y) \longrightarrow \mathsf{Ext}^n_{\mathscr{B}}(\mathsf{i}(X), \mathsf{i}(Y))$$

is invertible, for all X, Y in $\mathscr{A}$ and $0 \leq n \leq k$. The functor i is called a **homological embedding**, if i is a k-homological embedding for all $k \geq 0$.

REMARK 7.3. Due to Oort [**100**] we have the next two observations.

(i) Consider a recollement $(\mathscr{A},\mathscr{B},\mathscr{C})$ of abelian categories. Then the functor $\mathsf{i}\colon \mathscr{A} \longrightarrow \mathscr{B}$ is a 1-homological embedding since there is an isomorphism $\mathsf{i}^n_{X,Y}\colon \mathsf{Ext}^n_{\mathscr{A}}(X, Y) \longrightarrow \mathsf{Ext}^n_{\mathscr{B}}(\mathsf{i}(X), \mathsf{i}(Y))$ for every $X, Y \in \mathscr{A}$ and $n = 0, 1$. This can be easily proved using that $\mathscr{A}$ is a Serre subcategory of $\mathscr{B}$.

(ii) If $\mathsf{i}\colon \mathscr{A} \longrightarrow \mathscr{B}$ is a k-homological embedding, then the map

$$\mathsf{i}^{k+1}_{X,Y}\colon \mathsf{Ext}^{k+1}_{\mathscr{A}}(X, Y) \longrightarrow \mathsf{Ext}^{k+1}_{\mathscr{B}}(\mathsf{i}(X), \mathsf{i}(Y))$$

is a monomorphism for every X and Y in $\mathscr{A}$.

We denote by $\mathsf{R}^n\mathsf{p}$ the n-th right derived functor of $\mathsf{p}\colon \mathscr{B} \longrightarrow \mathscr{A}$ and by $\mathsf{L}_n\mathsf{q}$ the n-th left derived functor of $\mathsf{q}\colon \mathscr{B} \longrightarrow \mathscr{A}$.

The following result provides a solution to the left side of Problem 7.1.

THEOREM 7.4. ([**108**, Theorem 3.9]) *Let* $(\mathscr{A},\mathscr{B},\mathscr{C})$ *be a recollement of abelian categories and assume that* $\mathscr{B}$ *has enough projective and injective objects. Then the* *following statements are equivalent.*

(i) *The functor* $\mathsf{i}\colon \mathscr{A} \longrightarrow \mathscr{B}$ *is a* k*-homological embedding.*
(ii) $\mathsf{Ext}^n_{\mathscr{B}}(\mathsf{i}(P), \mathsf{i}(Y)) = 0$, *for every* $P \in \mathsf{Proj}\,\mathscr{A}$, $Y \in \mathscr{A}$ *and* $1 \leq n \leq k$.
(iii) $\mathsf{Ext}^n_{\mathscr{B}}(\mathsf{i}(X), \mathsf{i}(I)) = 0$, *for every* $I \in \mathsf{Inj}\,\mathscr{A}$, $X \in \mathscr{A}$ *and* $1 \leq n \leq k$.
(iv) $\mathsf{Ext}^n_{\mathscr{B}}(\mathsf{i}(P), \mathsf{i}(I)) = 0$, *for every* $P \in \mathsf{Proj}\,\mathscr{A}$, $I \in \mathsf{Inj}\,\mathscr{A}$ *and* $1 \leq n \leq k$.
(v) *For all* $P \in \mathsf{Proj}\,\mathscr{B}$ *the object* $\mathsf{Im}\,\mu_P$ *has a projective resolution of the form*

$$\mathsf{l}(Q_{k-1}) \longrightarrow \cdots \longrightarrow \mathsf{l}(Q_1) \longrightarrow \mathsf{l}(Q_0) \longrightarrow \mathsf{Im}\,\mu_P \longrightarrow 0$$

where $\mu\colon \mathsf{le} \longrightarrow \mathrm{Id}_{\mathcal{B}}$ *and each* Q_j *lies in* $\mathsf{Proj}\,\mathscr{C}$.
(vi) *For all* $I \in \mathsf{Inj}\,\mathscr{B}$ *the object* $\mathsf{Im}\,\nu_I$ *has an injective coresolution of the form*

$$0 \longrightarrow \mathsf{Im}\,\nu_I \longrightarrow \mathsf{r}(J_0) \longrightarrow \mathsf{r}(J_1) \longrightarrow \cdots \longrightarrow \mathsf{r}(J_{k-1})$$

where $\nu\colon \mathrm{Id}_{\mathscr{B}} \longrightarrow \mathsf{re}$ *and each* J_i *lies in* $\mathsf{Inj}\,\mathscr{C}$.
(vii) $\mathsf{R}^n\mathsf{p}(\mathsf{i}(I)) = 0$ *for every* $I \in \mathsf{Inj}\,\mathscr{A}$ *and* $1 \leq n \leq k$.
(viii) $\mathsf{R}^n\mathsf{p}(\mathsf{i}(A)) = 0$ *for every* $A \in \mathscr{A}$ *and* $1 \leq n \leq k$.
(ix) $\mathsf{L}_n\mathsf{q}(\mathsf{i}(P)) = 0$ *for every* $P \in \mathsf{Proj}\,\mathscr{A}$ *and* $1 \leq n \leq k$.
(x) $\mathsf{L}_n\mathsf{q}(\mathsf{i}(A)) = 0$ *for every* $A \in \mathscr{A}$ *and* $1 \leq n \leq k$.

We continue with an example which shows that there are functors which are k-homological embeddings but fail to be $(k+1)$-homological embeddings.

EXAMPLE 7.5. Let Λ be the path algebra of the quiver

$$1 \longrightarrow 2 \longrightarrow \cdots \longrightarrow k \longrightarrow k+1 \longrightarrow k+2$$

modulo the ideal generated by the paths of length 2. Then from Example 3.4 we have the following recollement of module categories:

$$\mathsf{mod}\text{-}\Lambda/\mathcal{U} \xrightarrow{\ \mathsf{inc}\ } \mathsf{mod}\text{-}\Lambda \xrightarrow{\ \mathsf{Hom}_\Lambda(P,-)\ } \mathsf{mod}\text{-}\Gamma$$

(with left adjoints $\Lambda/\mathcal{U}\otimes_\Lambda -$ and $P\otimes_\Gamma -$ above, and right adjoints $\mathsf{Hom}_\Lambda(\Lambda/\mathcal{U},-)$ and $\mathsf{Hom}_\Gamma(\mathsf{Hom}_\Lambda(P,\Lambda),-)$ below)

where $\Gamma = \mathsf{End}_\Lambda(P_3 \oplus \cdots \oplus P_{k+1})$ and the idempotent ideal $\mathcal{U} = \tau_{P_3\oplus\cdots\oplus P_{k+1}}(\Lambda)$ is the trace of the projective Λ-module $P_3 \oplus \cdots \oplus P_{k+1}$ in Λ.

Auslander, Platzeck and Todorov proved in [**13**] that $\mathcal{U}$ is k-idempotent but not $k+1$-idempotent, i.e. the map $\mathsf{i}^n_{X,Y}\colon \mathsf{Ext}^n_{\Lambda/\mathcal{U}}(X,Y) \longrightarrow \mathsf{Ext}^n_\Lambda(X,Y)$ is invertible, for all X, Y in $\mathsf{mod}\text{-}\Lambda/\mathcal{U}$ and $n = 0, 1, \ldots, k$, but it is not invertible for $n = k+1$. It follows that the functor $\mathsf{inc}\colon \mathsf{mod}\text{-}\Lambda/\mathcal{U} \longrightarrow \mathsf{mod}\text{-}\Lambda$ is a k-homological embedding but not a $(k+1)$-homological embedding. We refer to [**13**] for more details.

In Section 8 we discuss homological ring epimorphisms. From Theorem 8.1, homological ring epimorphisms $f\colon R \longrightarrow S$ give rise to examples of homological embeddings and the derived functor $f_*\colon \mathsf{D}(S) \longrightarrow \mathsf{D}(R)$ is fully faithful. In the next result we provide a characterization for homological embeddings $\mathsf{i}\colon \mathscr{A} \longrightarrow \mathscr{B}$ of abelian categories that generalizes the latter statement. Note that we do not assume $\mathscr{B}$ to be a Serre subcategory of $\mathscr{A}$ (compare [**124**, Theorem 2.1]).

The main tool for proving Theorem 7.6 below is Brown representability. Recall that a triangulated category $\mathcal{T}$ with coproducts **satisfies Brown representability** if every cohomological functor $H\colon \mathcal{T}^{op} \longrightarrow \mathsf{Mod}\text{-}\mathbb{Z}$ which sends coproducts to products is representable, i.e. $H \cong \mathsf{Hom}_{\mathcal{T}}(-, X)$ for some X in $\mathcal{T}$. The notion of left-completeness provides examples of triangulated categories satisfying Brown representability. Recall that the derived category $\mathsf{D}(\mathscr{A})$ of an abelian category $\mathscr{A}$ is **left-complete** if $\mathsf{D}(\mathscr{A})$ has products and any object X is isomorphic to a Milnor limit of its standard truncations $(\tau^{\geq n} X)_{n \leq 0}$ with respect to the canonical maps $\tau^{\geq n-1} X \longrightarrow \tau^{\leq n} X$, for $n \leq 0$. We refer to [**92–95**] for more details on Brown representability and its consequences, and we also refer to [**72**] for the terminology Milnor limit. The second main result in this section is the following.

THEOREM 7.6. ([**111**, Theorem 6.9]) *Let $\mathscr{B}$ be an abelian category such that $\mathsf{D}(\mathscr{B})$ has products and coproducts and let $\mathscr{A}$ be a Grothendieck category such that $\mathsf{D}(\mathscr{A})$ is left-complete. If $\mathsf{i} \colon \mathscr{A} \longrightarrow \mathscr{B}$ is an exact full embedding such that the derived functor $\mathsf{i} \colon \mathsf{D}(\mathscr{A}) \longrightarrow \mathsf{D}(\mathscr{B})$ preserves products and coproducts, then the following statements are equivalent.*

(i) *The functor $\mathsf{i} \colon \mathscr{A} \longrightarrow \mathscr{B}$ is a homological embedding.*
(ii) *The derived functor $\mathsf{i} \colon \mathsf{D}(\mathscr{A}) \longrightarrow \mathsf{D}(\mathscr{B})$ is fully faithful.*

If, in addition, the derived category $\mathsf{D}(\mathscr{B})$ is left-complete, then the statements (i) *and* (ii) *are also equivalent to:*

(iii) *The derived functor i induces a triangle equivalence between $\mathsf{D}(\mathscr{A})$ and the full subcategory $\mathsf{D}_{\mathscr{A}}(\mathscr{B})$ of $\mathsf{D}(\mathscr{B})$, whose objects are complexes of objects in $\mathscr{B}$ with cohomologies in $\mathsf{i}(\mathscr{A})$.*

Clearly, the derived category of a ring is left-complete. The above result is used in Section 8 for lifting recollements of module categories to the derived level.

7.2. The right side and eventually homological isomorphisms. We continue with the right side of Problem 7.1. In this connection we have the next result.

THEOREM 7.7. ([**108**, Propositions 3.3 and 3.4, Theorem 3.10]) *Let $\mathsf{R}_{\mathsf{ab}}(\mathscr{A}, \mathscr{B}, \mathscr{C})$ be a recollement of abelian categories and assume that $\mathscr{B}$ and $\mathscr{C}$ have enough projective and injective objects. Let B be an object in $\mathscr{B}$.*

(i) *The following statements are equivalent:*
  (a) *The map*

$$\mathsf{e}^n_{B,B'} \colon \mathsf{Ext}^n_{\mathscr{B}}(B, B') \xrightarrow{\cong} \mathsf{Ext}^n_{\mathscr{C}}(\mathsf{e}(B), \mathsf{e}(B'))$$

  is an isomorphism for every object B' in $\mathscr{B}$ and $0 \leq n \leq k$.
  (b) *The object B has a projective resolution of the form*

$$\mathsf{l}(Q_{k+1}) \longrightarrow \cdots \longrightarrow \mathsf{l}(Q_1) \longrightarrow \mathsf{l}(Q_0) \longrightarrow B \longrightarrow 0$$

  where each Q_j lies in $\mathsf{Proj}\,\mathscr{C}$.
  (c) *$\mathsf{Ext}^n_{\mathscr{B}}(B, \mathsf{i}(A)) = 0$ for every A in $\mathscr{A}$ and $0 \leq n \leq k+1$.*
  (d) *$\mathsf{Ext}^j_{\mathscr{B}}(B, \mathsf{i}(I)) = 0$ for every I in $\mathsf{Inj}\,\mathscr{A}$ and $0 \leq n \leq k+1$.*

(ii) *The following statements are equivalent:*
  (a) *The map*

$$\mathsf{e}^n_{B',B} \colon \mathsf{Ext}^n_{\mathscr{B}}(B', B) \xrightarrow{\cong} \mathsf{Ext}^n_{\mathscr{C}}(\mathsf{e}(B'), \mathsf{e}(B))$$

  is an isomorphism for every object B' in $\mathscr{B}$ and $0 \leq n \leq k$.

(b) *The object B has an injective coresolution of the form*

$$0 \longrightarrow B \longrightarrow \mathsf{r}(J_0) \longrightarrow \mathsf{r}(J_1) \longrightarrow \cdots \longrightarrow \mathsf{r}(J_{k+1})$$

where each J_i lies in $\mathsf{Inj}\,\mathscr{C}$).

(c) $\mathsf{Ext}^n_{\mathscr{B}}(\mathsf{i}(A), B) = 0$ *for every A in $\mathscr{A}$ and $0 \leq n \leq k+1$.*

(d) $\mathsf{Ext}^n_{\mathscr{B}}(\mathsf{i}(P), B) = 0$ *for every P in $\mathsf{Proj}\,\mathscr{A}$ and $0 \leq n \leq k+1$.*

It turns out that the solution to Problem 7.1 is closely related and dependent on the behaviour of the subcategory of $\mathscr{B}$ consisting of all objects admitting a truncated projective (or injective) resolution by projectives (or injectives) coming from the quotient category $\mathscr{C}$ via the section functor l (or r). This subcategory is also related with the k-generalized perpendicular subcategory of $\mathscr{B}$ associated to $\mathscr{A}$. We refer to [**108**] for more details. We remark that Theorem 7.4 and Theorem 7.7 generalize related results of Auslander, Platzeck and Todorov [**13**], formulated in the setting of finitely generated modules over an Artin algebra equipped with an idempotent ideal (see Example 3.4), and Geigle-Lenzing [**47**], formulated in the context of classical perpendicular categories and homological ring epimorphisms.

As a consequence of Theorem 7.4 and Theorem 7.7 we have the following.

COROLLARY 7.8. *Let $\mathsf{R_{ab}}(\mathscr{A}, \mathscr{B}, \mathscr{C})$ be a recollement and assume that $\mathscr{B}$ and $\mathscr{C}$ have enough projective and injective objects. Then the following are equivalent.*

(i) *The functor $\mathsf{i}\colon \mathscr{A} \longrightarrow \mathscr{B}$ is a $(k+1)$-homological embedding.*

(ii) *For all $P \in \mathsf{Proj}\,\mathscr{B}$ the object $\mathsf{Im}\,\mu_P$ has a projective resolution of the form*

$$\mathsf{l}(Q_k) \longrightarrow \cdots \longrightarrow \mathsf{l}(Q_1) \longrightarrow \mathsf{l}(Q_0) \longrightarrow \mathsf{Im}\,\mu_P \longrightarrow 0$$

where $\mu\colon \mathsf{le} \longrightarrow \mathrm{Id}_{\mathcal{B}}$ and each Q_j lies in $\mathsf{Proj}\,\mathscr{C}$.

(iii) *The map $\mathsf{e}^n_{\mathsf{Im}\,\mu_P,-}\colon \mathsf{Ext}^n_{\mathscr{B}}(\mathsf{Im}\,\mu_P, -) \longrightarrow \mathsf{Ext}^n_{\mathscr{C}}(\mathsf{e}(\mathsf{Im}\,\mu_P), \mathsf{e}(-))$ is invertible, for every $P \in \mathsf{Proj}\,\mathscr{B}$ and $0 \leq n \leq k-1$.*

(iv) *For all $I \in \mathsf{Inj}\,\mathscr{B}$ the object $\mathsf{Im}\,\nu_I$ has an injective resolution of the form*

$$0 \longrightarrow \mathsf{Im}\,\nu_I \longrightarrow \mathsf{r}(J_0) \longrightarrow \mathsf{r}(J_1) \longrightarrow \cdots \longrightarrow \mathsf{r}(J_k)$$

where $\nu\colon \mathrm{Id}_{\mathscr{B}} \longrightarrow \mathsf{re}$ and each J_i lies in $\mathsf{Inj}\,\mathscr{C}$.

(v) *The map $\mathsf{e}^n_{-,\mathsf{Im}\,\nu_I}\colon \mathsf{Ext}^n_{\mathscr{B}}(-, \mathsf{Im}\,\nu_I) \longrightarrow \mathsf{Ext}^n_{\mathscr{C}}(\mathsf{e}(-), \mathsf{e}(\mathsf{Im}\,\nu_I))$ is invertible, for every $I \in \mathsf{Inj}\,\mathscr{B}$ and $0 \leq n \leq k-1$.*

We now pass to the second part of this subsection. Motivated mainly by the problem of how to compare the finite generation (Fg) condition for Hochschild cohomology for an algebra Λ and $e\Lambda e$, where e is an idempotent of Λ, the following notion was introduced in [**113**]. We will discuss the (Fg) condition as well as more applications of eventually homological isomorphisms in Section 12.

DEFINITION 7.9. Let $F\colon \mathscr{B} \longrightarrow \mathscr{C}$ be a functor between abelian categories and let t be an integer. The functor F is called a t-**homological isomorphism** if there is an isomorphism:

$$\mathsf{Ext}^j_{\mathscr{B}}(B, B') \cong \mathsf{Ext}^j_{\mathscr{C}}(F(B), F(B'))$$

for every pair of objects B and B' in $\mathscr{B}$ and every $j > t$. If F is a t-homological isomorphism for some t, then it is an **eventually homological isomorphism**.

Note that we do not require these isomorphisms to be induced by the functor F, see Remark 7.13. The main problem in this part can be formulated as follows.

PROBLEM 7.10. *Characterize when the quotient functor* $\mathsf{e}\colon \mathscr{B} \longrightarrow \mathscr{C}$ *in a recollement situation* $\mathsf{R_{ab}}(\mathscr{A},\mathscr{B},\mathscr{C})$ *is an eventually homological isomorphism.*

Before we provide a solution, we fix some notation. Let $\mathscr{A}$ be an abelian category with enough projective objects. Let A be an object in $\mathscr{A}$ and $\cdots \longrightarrow P_1 \longrightarrow P_0 \longrightarrow A \longrightarrow 0$ a projective resolution of A in $\mathscr{A}$. The kernel of the morphism $P_{n-1} \longrightarrow P_{n-2}$ is an n-th syzygy of A and is denoted by $\Omega^n(A)$. Following the notation of subsection 2.3, if $\mathcal{X}$ is a class of objects in $\mathscr{B}$, then the right orthogonal subcategory $\{B \in \mathscr{B} \mid \mathsf{Hom}_{\mathscr{B}}(\mathcal{X}, B) = 0\}$ of $\mathcal{X}$ is denoted by $\mathcal{X}^{\circ}$ and the left orthogonal subcategory $\{B \in \mathscr{B} \mid \mathsf{Hom}_{\mathscr{B}}(B,\mathcal{X}) = 0\}$ of $\mathcal{X}$ is denoted by ${}^{\circ}\mathcal{X}$.

Let us consider Theorem 7.7 again. The latter theorem describes when the maps $\mathsf{e}^n_{B,B'}\colon \mathsf{Ext}^n_{\mathscr{B}}(B, B') \longrightarrow \mathsf{Ext}^n_{\mathscr{C}}(\mathsf{e}(B), \mathsf{e}(B'))$, induced by the quotient functor $\mathsf{e}\colon \mathscr{B} \longrightarrow \mathscr{C}$, are isomorphisms in all degrees n. In order to provide an answer to Problem 7.10 we need to give a similar description of when these maps are isomorphisms in almost all degrees. In other words, we need to show a higher degree version of Theorem 7.7. The basic idea is to translate the conditions in Theorem 7.7 to conditions stated for almost all degrees. However, we will see below that these conditions need to be modified.

In the following result we characterize when the functor e in a recollement $\mathsf{R_{ab}}(\mathscr{A},\mathscr{B},\mathscr{C})$ induces isomorphisms of extension groups in almost all degrees.

THEOREM 7.11. ([**113**, Theorem 3.4]) *Let* $\mathsf{R_{ab}}(\mathscr{A},\mathscr{B},\mathscr{C})$ *be a recollement of abelian categories and assume that* $\mathscr{B}$ *and* $\mathscr{C}$ *have enough projective and injective objects. Consider the following statements for an object* B *of* $\mathscr{B}$ *and two integers* n *and* m*:*

(a) *The map*

$$\mathsf{e}^j_{B,B'}\colon \mathsf{Ext}^j_{\mathscr{B}}(B, B') \xrightarrow{\cong} \mathsf{Ext}^j_{\mathscr{C}}(\mathsf{e}(B), \mathsf{e}(B'))$$

is an isomorphism for every object B' *in* $\mathscr{B}$ *and every integer* $j > m+n$.

(b) *The object* B *has a projective resolution of the form*

$$\cdots \longrightarrow \mathsf{l}(Q_1) \longrightarrow \mathsf{l}(Q_0) \longrightarrow P_{n-1} \longrightarrow \cdots \longrightarrow P_0 \longrightarrow B \longrightarrow 0$$

where each Q_j *lies in* $\mathsf{Proj}\,\mathscr{C}$.

(c) $\mathsf{Ext}^j_{\mathscr{B}}(B, \mathsf{i}(A)) = 0$ *for every* A *in* $\mathscr{A}$ *and* $j > n$, *and there exists an* n-*th syzygy of* B *lying in the left orthogonal subcategory* ${}^{\circ}\mathsf{i}(\mathscr{A})$.

(d) $\mathsf{Ext}^j_{\mathscr{B}}(B, \mathsf{i}(I)) = 0$ *for every* I *in* $\mathsf{Inj}\,\mathscr{A}$ *and* $j > n$, *and and there exists an* n-*th syzygy of* B *lying in the left orthogonal subcategory* ${}^{\circ}\mathsf{i}(\mathsf{Inj}\,\mathscr{A})$.

Then the following statements are equivalent: (b) $\Longleftrightarrow$ (c) $\Longleftrightarrow$ (d). *If one of these holds and in addition we have* $\mathsf{pd}_{\mathscr{C}}\,\mathsf{e}(P) \leq m$ *for every projective object* P *in* $\mathscr{B}$, *then statement* (a) *holds.*

Let $\mathsf{R_{ab}}(\mathscr{A},\mathscr{B},\mathscr{C})$ be a recollement of abelian categories. In [**108**] the notion of $\mathscr{A}$-relative global dimension of $\mathscr{B}$ was defined as follows:

$$\mathsf{gl.dim}_{\mathscr{A}}\mathscr{B} := \sup\{\mathsf{pd}_{\mathscr{B}}\,\mathsf{i}(A) \mid A \in \mathscr{A}\}$$

We call this the $\mathscr{A}$-**relative projective global dimension** of $\mathscr{B}$ and we denote it by $\mathsf{pgl.dim}_{\mathscr{A}}\mathscr{B}$. Dually, we define the $\mathscr{A}$-**relative injective global dimension** of $\mathscr{B}$ by

$$\mathsf{igl.dim}_{\mathscr{A}}\mathscr{B} := \sup\{\mathsf{id}_{\mathscr{B}}\mathsf{i}(A) \mid A \in \mathscr{A}\}.$$

We remark that the above relative dimensions are used drastically in [**54**] to investigate the finitistic dimension of finite dimensional algebras, see Section 13.

Recall that an abelian category is a **length category** if it is essentially small and every object has a finite composition series [**42**].

The following result solves Problem 7.10 and generalizes [**113**, Corollary 3.12] from Artin algebras to abelian categories with certain conditions.

THEOREM 7.12. *Let* $\mathsf{R_{ab}}(\mathscr{A},\mathscr{B},\mathscr{C})$ *be a recollement. Assume that* $\mathscr{A}$ *is a length category and that* $\mathscr{B}$ *has projective covers. The following statements are equivalent:*

(i) *There is an integer* t *such that for every pair of objects* X *and* Y *in* $\mathscr{B}$, *and every* $j > t$, *the map*

$$\mathsf{e}^j_{X,Y}\colon \mathsf{Ext}^j_{\mathscr{B}}(X,Y) \xrightarrow{\;\cong\;} \mathsf{Ext}^j_{\mathscr{C}}(\mathsf{e}(X),\mathsf{e}(Y))$$

is an isomorphism.

(ii) *The functor* $\mathsf{e}\colon \mathscr{B} \longrightarrow \mathscr{C}$ *is an eventually homological isomorphism.*

(iii) (α) $\mathsf{igl.dim}_{\mathscr{A}}\mathscr{B} < \infty$ *and* (β) $\sup\{\mathsf{pd}_{\mathscr{C}}\,\mathsf{e}(P) \mid P \in \mathsf{Proj}\,\mathscr{B}\} < \infty$.

(iv) (γ) $\mathsf{pgl.dim}_{\mathscr{A}}\mathscr{B} < \infty$ *and* (δ) $\sup\{\mathsf{id}_{\mathscr{C}}\,\mathsf{e}(I) \mid I \in \mathsf{Inj}\,\mathscr{B}\} < \infty$.

In particular, if the functor e *is a* t*-homological isomorphism, then each of the dimensions in* (iii) *and* (iv) *are at most* t. *The bound* t *in* (i) *is bounded by the sum of the dimensions occurring in* (iii), *and also bounded by the sum of the dimensions occurring in* (iv).

PROOF. Clearly, (i) implies (ii). We first show the implication (ii)$\Longrightarrow$(iii) and similarly it follows that (ii)$\Longrightarrow$(iv). Next, we prove that (iii)$\Longrightarrow$(i) and similarly we get that (iv)$\Longrightarrow$(i). Then we obtain the equivalence of the above four statements.

(ii)$\Longrightarrow$(iii): Let A be an object of $\mathscr{A}$. Then, there is a positive integer t such that for every object B in $\mathscr{B}$ and for all $j > t$, we have the isomorphism

$$\mathsf{Ext}^j_{\mathscr{B}}(B,\mathsf{i}(A)) \cong \mathsf{Ext}^j_{\mathscr{C}}(\mathsf{e}(B),\mathsf{ei}(A))$$

The latter extension group vanishes since $\mathsf{e}\circ\mathsf{i} = 0$ by Remark 2.4 (ii). This implies that $\mathsf{id}_{\mathscr{B}}\,\mathsf{i}(A) \leq t$ and therefore (α) holds. For (β), let P be a projective object of $\mathscr{B}$. For every C in $\mathscr{C}$ and $j > t$, we have the following isomorphism

$$\mathsf{Ext}^j_{\mathscr{C}}(\mathsf{e}(P),C) \cong \mathsf{Ext}^j_{\mathscr{C}}(\mathsf{e}(P),\mathsf{el}(C)) \cong \mathsf{Ext}^j_{\mathscr{B}}(P,\mathsf{l}(C)) = 0$$

since $\mathsf{e}\circ\mathsf{l} \simeq \mathrm{Id}_{\mathscr{C}}$ by Remark 2.4 (iv). We infer that $\mathsf{pd}_{\mathscr{C}}\,\mathsf{e}(P) \leq t$, thus (β) holds.

(iii)$\Longrightarrow$(i): For this implication we need the following isomorphism. Let S be a simple object in $\mathscr{B}$. Then, for any B in $\mathscr{B}$ and $n \geq 1$, there is an isomorphism:

$$\mathsf{Ext}^n_{\mathscr{B}}(B,S) \cong \mathsf{Hom}_{\mathscr{B}}(\Omega^n(B),S) \tag{7.1}$$

Here is where we need to assume that $\mathscr{B}$ has projective covers. For the proof of this isomorphism we refer to [**54**].

Assume that $\mathsf{igl.dim}_{\mathscr{A}}\mathscr{B} = m-1 < \infty$ and let S be a simple object in $\mathscr{A}$. Since by Proposition 2.9 the object $\mathsf{i}(S)$ is simple in $\mathscr{B}$, we have by (7.1) the isomorphism:

$$\mathsf{Ext}^j_{\mathscr{B}}(B,\mathsf{i}(S)) \cong \mathsf{Hom}_{\mathscr{B}}(\Omega^j(B),\mathsf{i}(S))$$

Since $\mathsf{id}_{\mathscr{B}}\,\mathsf{i}(S) \leq m-1$, it follows that $\mathsf{Hom}_{\mathscr{B}}(\Omega^j(B),\mathsf{i}(S)) = 0$ for all $j > m-1$. Note that every object A in $\mathscr{A}$ has a finite composition series. Thus, we infer that $\mathsf{Hom}_{\mathscr{B}}(\Omega^j(B),\mathsf{i}(A)) = 0$ for all $j > m-1$. Hence, so far we have shown that $\mathsf{Ext}^j_{\mathscr{B}}(B,\mathsf{i}(A)) = 0$ for every A in $\mathscr{A}$ and $j > m$, and there exists an m-th syzygy of B lying in ${}^{\circ}\mathsf{i}(\mathscr{A})$. Using now assumption (β), Theorem 7.11 implies (i). □

REMARK 7.13. (i) The implication (ii)$\Longrightarrow$(i) in Theorem 7.12 shows that whenever we know that the quotient functor e is an eventually homological isomorphism, we can obtain the desired isomorphisms from the functor e. This explains why in Definition 7.9 we don't require these isomorphisms to be induced by the involved functor.

(ii) It is natural to consider if any other pair of the four conditions $(\alpha) - (\delta)$ in Theorem 7.12 implies that the functor e is an eventually homological isomorphism. We refer to [**113**, Subsection 8.1] for examples showing that this is not the case.

Finally, we specialize Theorem 7.12 to Artin algebras using the recollement of module categories $(\mathsf{mod}\text{-}\Lambda/\Lambda a\Lambda, \mathsf{mod}\text{-}\Lambda, \mathsf{mod}\text{-}a\Lambda a)$, see Example 3.1. Compare the conditions (α)–(δ) of Theorem 7.12 to the conditions below. We are now ready to characterize when the functor $\mathsf{e} = a(-)\colon \mathsf{mod}\text{-}\Lambda \longrightarrow \mathsf{mod}\text{-}a\Lambda a$ is an eventually homological isomorphism.

COROLLARY 7.14. ([**113**, Corollary 3.12]) *Let Λ be an Artin algebra and let a be an idempotent element of Λ. Consider the following conditions:*

$$(\alpha)\ \mathsf{id}_{\Lambda}\Big(\frac{\Lambda/\langle a\rangle}{\mathsf{rad}\Lambda/\langle a\rangle}\Big) < \infty \qquad\qquad (\beta)\ \mathsf{pd}_{a\Lambda a}\, a\Lambda < \infty$$

$$(\gamma)\ \mathsf{pd}_{\Lambda}\Big(\frac{\Lambda/\langle a\rangle}{\mathsf{rad}\Lambda/\langle a\rangle}\Big) < \infty \qquad\qquad (\delta)\ \mathsf{pd}_{(a\Lambda a)^{\mathrm{op}}}\, \Lambda a < \infty$$

Then the following statements are equivalent:

(i) *(α) and (β) hold.*
(ii) *(γ) and (δ) hold.*
(iii) *The functor e is an eventually homological isomorphism.*

We will further discuss the conditions (α)–(δ) in Section 12.

8. Lifting Recollements of Module Categories to the Derived Level

In this section we investigate when a recollement of module categories lifts to a recollement at the level of (unbounded) derived categories. In general, a useful tool to produce recollements of derived module categories is the concept of **ring epimorphsim**, i.e. an epimorphism in the category of rings. Recall that ring epimorphisms with domain R (up to equivalence) are in bijection with **bireflective** subcategories of $\mathsf{Mod}\text{-}R$, i.e. full subcategories of $\mathsf{Mod}\text{-}R$ whose inclusion functor admits both left and right adjoints. In order for the (exact) restriction of scalars functor to induce a fully faithful functor on the derived level, however, one needs to require more from a ring epimorphism.

THEOREM 8.1. ([**47**, Theorem 4.4]) *Let R be a ring and $f\colon R \longrightarrow S$ a ring homomorphism. The following statements are equivalent.*

(i) *The derived functor $f_*\colon \mathsf{D}(S) \longrightarrow \mathsf{D}(R)$ is fully faithful.*
(ii) *The map f is a ring epimorphism and $\mathsf{Tor}_i^R(S,S) = 0$ for any $i > 0$.*
(iii) *The restriction functor $f_*\colon \mathsf{Mod}\text{-}S \longrightarrow \mathsf{Mod}\text{-}R$ is a homological embedding.*

A ring epimorphism is called **homological** if the above equivalent conditions are satisfied. From Theorem 5.7 it follows that ring epimorphisms do not in general give rise to recollements of module categories. However, a homological ring epimorphism always gives rise to a recollement of triangulated categories, see [**97**].

Let R be a ring, e an idempotent element of R and $f\colon R \longrightarrow R/ReR$ the associated ring epimorphism. By Example 3.1 there is the following recollement induced by the idempotent element e:

$$\xymatrix{\mathsf{Mod}\text{-}R/ReR \ar[r]^{f_*} & \mathsf{Mod}\text{-}R \ar@/_2pc/[l]_{R/ReR\otimes_R -} \ar@/^2pc/[l]^{\mathsf{Hom}_R(R/ReR,-)} \ar[r]^{\mathsf{Hom}_R(Re,-)} & \mathsf{Mod}\text{-}eRe \ar@/_2pc/[l]_{Re\otimes_{eRe}-} \ar@/^2pc/[l]^{\mathsf{Hom}_{eRe}(eR,-)}}$$

Assume that the restriction functor $f_*\colon \mathsf{Mod}\text{-}R/ReR \longrightarrow \mathsf{Mod}\text{-}R$ is a homological embedding (see Definition 7.2). Equivalently, the morphism f is a homological ring epimorphism (i.e. the ideal ReR is stratifying [**27**]). Then it follows from Cline-Parshall-Scott [**27**] that there is a recollement of derived module categories:

$$\xymatrix{\mathsf{D}(R/ReR) \ar[r]^{f_*} & \mathsf{D}(R) \ar@/_2pc/[l]_{R/ReR\otimes^{\mathbb{L}}_R -} \ar@/^2pc/[l]^{\mathbb{R}\,\mathsf{Hom}_R(R/ReR,-)} \ar[r]^{\mathbb{R}\,\mathsf{Hom}_R(Re,-)} & \mathsf{D}(eRe). \ar@/_2pc/[l]_{Re\otimes^{\mathbb{L}}_{eRe}-} \ar@/^2pc/[l]^{\mathbb{R}\,\mathsf{Hom}_{eRe}(eR,-)}} \tag{8.1}$$

In [**111**] we formulated the notion of stratifying recollements.

DEFINITION 8.2. A recollement of a derived module category by derived module categories is called **stratifying** if it is equivalent to one of the form (8.1).

The reason for calling such recollements stratifying is the notion of stratifying ideals [**27**]. Our aim in this section is to reprove the above mentioned result of Cline-Parshall-Scott using the methods of [**111**]. More precisely, we provide necessary and sufficient conditions for a recollement $\mathsf{R}_{\mathsf{ab}}(\mathsf{Mod}\text{-}R/ReR, \mathsf{Mod}\text{-}R, \mathsf{Mod}\text{-}eRe)$ to induce a recollement $\mathsf{R}_{\mathsf{tr}}(\mathsf{D}(R/ReR), \mathsf{D}(R), \mathsf{D}(eRe))$ of unbounded derived categories. The key ingredient of the proof is the equivalence between $\mathsf{D}_{R/ReR}(R)$ and $\mathsf{D}(R)$ established in Theorem 7.6. We remark that the latter result is proved in [**111**] in a more general context but for simplicity we provide a proof below for derived module categories. We encourage the reader to see [**111**, Section 6] for more details. From Theorem 5.7 it suffices to show the next result for a recollement of module categories induced by an idempotent element e.

THEOREM 8.3. ([**27**], [**111**, Theorem 6.10]) *Let* $\mathsf{R}_{\mathsf{ab}}(\mathsf{Mod}\text{-}R/ReR, \mathsf{Mod}\text{-}R, \mathsf{Mod}\text{-}eRe)$ *be a recollement of module categories induced by an idempotent element* e *of* R. *The following statements are equivalent:*

(i) *The functor* $f_*\colon \mathsf{Mod}\text{-}R/ReR \longrightarrow \mathsf{Mod}\text{-}R$ *is a homological embedding.*
(ii) *There is a recollement of derived categories as in* (8.1).

PROOF. (i)$\Longrightarrow$(ii): Consider the recollement $\mathsf{R}_{\mathsf{ab}}(\mathsf{Mod}\text{-}R/ReR, \mathsf{Mod}\text{-}R, \mathsf{Mod}\text{-}eRe)$, see Example 3.1. Then by Remark 2.4 there is an exact sequence of abelian categories:

$$0 \longrightarrow \mathsf{Mod}\text{-}R/ReR \xrightarrow{f_*} \mathsf{Mod}\text{-}R \xrightarrow{e(-)} \mathsf{Mod}\text{-}eRe \longrightarrow 0$$

The fact that $\mathsf{R}_{\mathsf{ab}}(\mathsf{Mod}\text{-}R/ReR, \mathsf{Mod}\text{-}R, \mathsf{Mod}\text{-}eRe)$ is a recollement, means precisely that the restriction and the quotient functor of the above short exact sequence admits left and right adjoint functors. If we derive this picture (the right part) we obtain the following diagram:

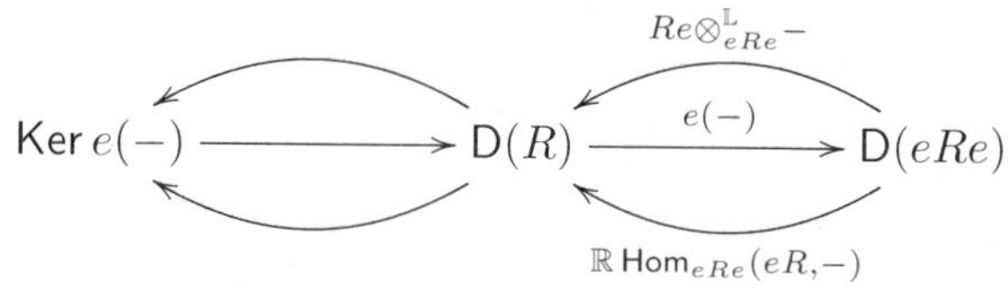

Clearly, this is a recollement of triangulated categories. Then it is easy to compute that the triangulated subcategory $\mathsf{Ker}\, e(-)$ consists of all complexes of R-modules whose cohomology lies in $f_*(\mathsf{Mod}\text{-}R/ReR)$. Following the notation in Theorem 7.6 the latter subcategory is $\mathsf{D}_{R/ReR}(R)$. By assumption, the functor $f_*\colon \mathsf{Mod}\text{-}R/ReR \longrightarrow \mathsf{Mod}\text{-}R$ is a homological embedding, thus from Theorem 7.6 the derived functor $f_*\colon \mathsf{D}(R/ReR) \longrightarrow \mathsf{D}(R)$ is fully faithful and $\mathsf{D}_{R/ReR}(R)$ is equivalent to $\mathsf{D}(R/ReR)$. Putting all the above facts together we obtain the recollement diagram (8.1). Note that the derived functors on the left side of (8.1) are the derived version of the adjoint triple $(R/ReR \otimes_R -, f_*, \mathsf{Hom}_R(R/ReR, -))$ at the level of module categories.

(ii)$\Longrightarrow$(i): Since $\mathsf{R}_{\mathsf{tr}}(\mathsf{D}(R/ReR), \mathsf{D}(R), \mathsf{D}(eRe))$ is a recollement, it follows by definition that the derived functor $f_*\colon \mathsf{D}(R/ReR) \longrightarrow \mathsf{D}(R)$ is fully faithful. Then by Theorem 7.6 we infer that the restriction functor $f_*\colon \mathsf{Mod}\text{-}R/ReR \longrightarrow \mathsf{Mod}\text{-}R$ is a homological embedding. □

For lifting recollement situations $\mathsf{R}_{\mathsf{ab}}(\mathscr{A}, \mathscr{B}, \mathscr{C})$ to the level of bounded derived categories $\mathsf{R}_{\mathsf{tr}}(\mathsf{D}^{\mathsf{b}}(\mathscr{A}), \mathsf{D}^{\mathsf{b}}(\mathscr{B}), \mathsf{D}^{\mathsf{b}}(\mathscr{C}))$ we refer to [**108**, Theorem 8.2].

Closely related to the problem of lifting recollements of module categories to the derived level, is the question of what is happening when the restriction functor $f_*\colon \mathsf{Mod}\text{-}R/ReR \longrightarrow \mathsf{Mod}\text{-}R$ is not a homological embedding. Note that from the proof of Theorem 8.3 we always have the following recollement:

(8.2)

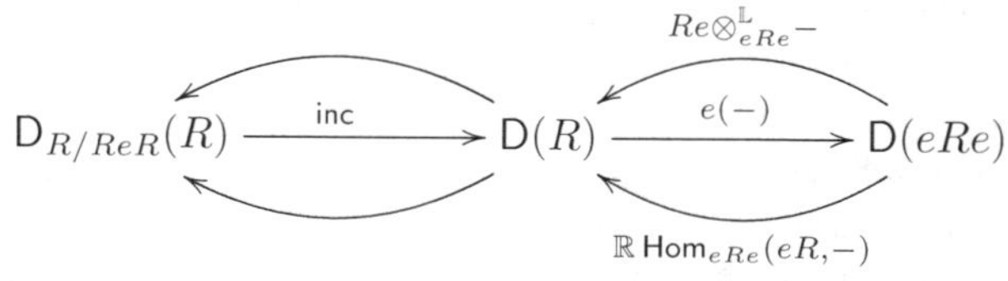

It is natural to consider if the left side of the above recollement can be described as a derived category of a maybe more complicated algebra that keeps track of the underlying module category. For this purpose we need the notion of dg algebras.

Let k be a commutative ring. A differential graded algebra, dg algebra for short, is a $\mathbb{Z}$-graded k-algebra $A = \oplus_{n\in\mathbb{Z}} A^n$ endowed with a differential d_A of degree one such that the graded Leibniz rule: $d(ab) = d(a)b + (-1)^n a d(b)$ for all $a \in A^n$ and $b \in A$, holds. A dg (left) A-module X is a $\mathbb{Z}$-graded (left) A-module $X = \oplus_{n\in\mathbb{Z}} X^n$ endowed with a differential d_X of degree one such that $d_X(ax) = a d_X(x) + (-1)^n d_A(a) x$ for all $x \in X^n$ and $a \in A$. A morphism $f\colon X \longrightarrow Y$ of dg A-modules is a morphism of the underlying graded A-modules which is homogeneous of degree zero and commutes with the differential. Then the category $\mathsf{C}(A)$ of all left dg A-modules is abelian. A morphism f in $\mathsf{C}(A)$ is a quasi-isomorphism if it induces isomorphisms in homology. Moreover, a morphism $f\colon X \longrightarrow Y$ of dg A-modules is null-homotopic if there is a morphism of dg A-modules $h\colon X \longrightarrow Y$ of degree minus one such that $f = hd_X + d_Y h$. Then the homotopy category $\mathsf{H}(A)$ has the same objects as $\mathsf{C}(A)$ and its morphism space consists of the equivalences classes of morphisms in $\mathsf{C}(A)$ modulo the null-homotopic ones. It turns out that the homotopy category $\mathsf{H}(A)$ is triangulated and then the derived category $\mathsf{D}(A)$ is the localization of $\mathsf{H}(A)$ with respect to the quasi-isomorphisms. For more details on dg algebras and their uses we refer to Keller's papers [**69**, **70**].

We are now ready to formulate the last result of this section. It is due to Kalck and Yang [**66**] (see also references in loc. cit.) and shows that we can always lift a recollement of module categories over k-algebras at the derived level but over dg algebras. Note that we view an algebra as a dg algebra concentrated in degree zero.

THEOREM 8.4. ([**66**, Proposition 2.10]) *Let A be a k-algebra which is flat over k and let e be an idempotent element of A. Then there is a dg algebra B with $\mathsf{H}^0(B) \cong A/AeA$ and a recollement of derived categories of dg algebras:*

$$\mathsf{D}(B) \xrightarrow{B\otimes^{\mathbb{L}}_B -} \mathsf{D}(A) \xrightarrow{e(-)} \mathsf{D}(eAe) \qquad (8.3)$$

with left adjoints $B\otimes^{\mathbb{L}}_A - \colon \mathsf{D}(A) \to \mathsf{D}(B)$, $Ae\otimes^{\mathbb{L}}_{eAe} - \colon \mathsf{D}(eAe) \to \mathsf{D}(A)$ and right adjoints $\mathbb{R}\,\mathsf{Hom}_A(B,-)\colon \mathsf{D}(A) \to \mathsf{D}(B)$, $\mathbb{R}\,\mathsf{Hom}_{eAe}(eA,-)\colon \mathsf{D}(eAe) \to \mathsf{D}(A)$.

We finally remark that by the recollement situations (8.2) and (8.3), the triangulated categories $\mathsf{D}(B)$ and $\mathsf{D}_{A/AeA}(A)$ are equivalent.

9. Compatible Torsion Pairs in Recollements

Let $\mathsf{R}_{\mathsf{ab}}(\mathscr{A},\mathscr{B},\mathscr{C})$ be a recollement. In this section we investigate when a torsion pair $(\mathcal{X},\mathcal{Y})$ in $\mathscr{B}$ restricts to $\mathscr{A}$ and $\mathscr{C}$ as well as the converse, i.e. how torsion pairs in $\mathscr{A}$ and $\mathscr{C}$ induce torsion pairs in $\mathscr{B}$. This was also studied in [**83**] for recollements of hearts. For the necessary preliminaries on torsion pairs see subsection 2.3.

Our first result shows that a torsion pair $(\mathcal{X},\mathcal{Y})$ in $\mathscr{B}$ always restricts to $\mathscr{A}$ and under some natural conditions it also restricts to $\mathscr{C}$. We also show how to produce torsion pairs in $\mathscr{B}$ from torsion pairs in $\mathscr{A}$ and $\mathscr{C}$. This should be considered as an abelian analogue of the glueing method of t-structures along a recollement of triangulated categories due to Belinson-Bernstein-Deligne, see Section 4. The proof of the following result follows closely the proofs in [**83**, Lemma 6.2, Proposition 6.6] by Liu-Vitória-Yang, but it is given in detail for clarity.

PROPOSITION 9.1. *Let $\mathsf{R}_{\mathsf{ab}}(\mathscr{A},\mathscr{B},\mathscr{C})$ be a recollement of abelian categories.*

(i) *Let $(\mathcal{X},\mathcal{Y})$ be a torsion pair in $\mathscr{B}$. The following hold.*
   (a) *The pair $(\mathsf{q}(\mathcal{X}),\mathsf{p}(\mathcal{Y}))$ is a torsion pair in $\mathscr{A}$.*
   (b) *The pair $(\mathsf{e}(\mathcal{X}),\mathsf{e}(\mathcal{Y}))$ is a torsion pair in $\mathscr{C}$ if and only if $\mathsf{le}(\mathcal{X}) \subseteq \mathcal{X}$ if and only if $\mathsf{re}(\mathcal{Y}) \subseteq \mathcal{Y}$.*

(ii) *Let $(\mathcal{X}_{\mathscr{A}},\mathcal{Y}_{\mathscr{A}})$ and $(\mathcal{X}_{\mathscr{C}},\mathcal{Y}_{\mathscr{C}})$ be torsion pairs in $\mathscr{A}$ and $\mathscr{C}$ respectively. Then the pair:*

$$\big(\mathcal{X} = \{B \in \mathscr{B} \mid \mathsf{q}(B) \in \mathcal{X}_{\mathscr{A}}, \mathsf{e}(B) \in \mathcal{X}_{\mathscr{C}}\}, \mathcal{Y} = \{B \in \mathscr{B} \mid \mathsf{p}(B) \in \mathcal{Y}_{\mathscr{A}}, \mathsf{e}(B) \in \mathcal{Y}_{\mathscr{C}}\}\big)$$

is a torsion pair in $\mathscr{B}$ compatible with the recollement $\mathsf{R}_{\mathsf{ab}}(\mathscr{A},\mathscr{B},\mathscr{C})$.

PROOF. (i)(a) Let X be an object in $\mathcal{X}$ and Y an object in $\mathcal{Y}$. From Proposition 2.8, see equation (2.4), there is an exact sequence:

$$0 \longrightarrow \mathsf{ip}(Y) \longrightarrow Y \longrightarrow \mathsf{re}(Y) \longrightarrow \mathsf{i}(A) \longrightarrow 0$$

for some A in $\mathscr{A}$. Applying the functor $\mathsf{Hom}_{\mathscr{B}}(X,-)$ we get the exact sequence

$$0 \longrightarrow \mathsf{Hom}_{\mathscr{B}}(X,\mathsf{ip}(Y)) \longrightarrow \mathsf{Hom}_{\mathscr{B}}(X,Y) \longrightarrow \mathsf{Hom}_{\mathscr{B}}(X,\mathsf{re}(Y))$$

and since $\mathsf{Hom}_{\mathscr{B}}(\mathcal{X},\mathcal{Y})=0$ it follows that $\mathsf{Hom}_{\mathscr{B}}(X,\mathsf{ip}(Y))=0$. This implies that $\mathsf{ip}(Y)$ lies in $\mathcal{Y}$ (see subsection 2.3) and similarly we show that the object $\mathsf{iq}(X)$ belongs to $\mathcal{X}$. Since the functor $\mathsf{i}\colon \mathscr{A} \longrightarrow \mathscr{B}$ is fully faithful, we have the isomorphism $\mathsf{Hom}_{\mathscr{A}}(\mathsf{q}(X),\mathsf{p}(Y)) \cong \mathsf{Hom}_{\mathscr{B}}(\mathsf{iq}(X),\mathsf{ip}(Y))$ and therefore $\mathsf{Hom}_{\mathscr{A}}(\mathsf{q}(\mathcal{X}),\mathsf{p}(\mathcal{Y}))=0$. Let A be an object of $\mathscr{A}$. Then, since $(\mathcal{X},\mathcal{Y})$ is a torsion pair in $\mathscr{B}$, for the object $\mathsf{i}(A)$ in $\mathscr{B}$ there exists an exact sequence $0 \longrightarrow X \longrightarrow \mathsf{i}(A) \longrightarrow Y \longrightarrow 0$ with X in $\mathcal{X}$ and Y in $\mathcal{Y}$. Since $\mathscr{A}$ is a Serre subcategory of $\mathscr{B}$ and $\mathsf{i}\colon \mathscr{A} \longrightarrow \mathscr{B}$ is an exact embedding (see Remark 2.4), it follows that there exist objects X' and Y' in $\mathscr{A}$ such that $X \cong \mathsf{i}(X')$, $Y \cong \mathsf{i}(Y')$ and moreover the sequence $0 \longrightarrow X' \longrightarrow A \longrightarrow Y' \longrightarrow 0$ is exact in $\mathscr{A}$ (see also Remark 7.3 (i) for an explanation via Yoneda extensions). We infer that $(\mathsf{q}(\mathcal{X}),\mathsf{p}(\mathcal{Y}))$ is a torsion pair in $\mathscr{A}$.

(i)(b) Let C be an object in $\mathscr{C}$. Since $(\mathcal{X},\mathcal{Y})$ is a torsion pair in $\mathscr{B}$, there exists an exact sequence $0 \longrightarrow X \longrightarrow \mathsf{l}(C) \longrightarrow Y \longrightarrow 0$ with X in $\mathcal{X}$ and Y in $\mathcal{Y}$. Applying the functor $\mathsf{e}\colon \mathscr{B} \longrightarrow \mathscr{C}$ and using Remark 2.4 (iv) we obtain the exact sequence

$$0 \longrightarrow \mathsf{e}(X) \longrightarrow C \longrightarrow \mathsf{e}(Y) \longrightarrow 0$$

with $\mathsf{e}(X)$ in $\mathsf{e}(\mathcal{X})$ and $\mathsf{e}(Y)$ in $\mathsf{e}(\mathcal{Y})$. Furthermore, from the adjoint pairs (l,e) and (e,r) we have the following isomorphisms:

$$\mathsf{Hom}_{\mathscr{C}}(\mathsf{e}(\mathcal{X}),\mathsf{e}(\mathcal{Y})) \cong \mathsf{Hom}_{\mathscr{B}}(\mathsf{le}(\mathcal{X}),\mathcal{Y}) \cong \mathsf{Hom}_{\mathscr{B}}(\mathcal{X},\mathsf{re}(\mathcal{Y}))$$

Then, it follows that $\mathsf{Hom}_{\mathscr{C}}(\mathsf{e}(\mathcal{X}),\mathsf{e}(\mathcal{Y}))=0$ if and only if $\mathsf{le}(\mathcal{X}) \subseteq \mathcal{X}$ if and only if $\mathsf{re}(\mathcal{Y}) \subseteq \mathcal{Y}$, and this completes the proof.

(ii) We show that $\mathcal{X} = {}^{\circ}\mathcal{Y}$. We first claim that $\mathcal{X} \subseteq {}^{\circ}\mathcal{Y}$. Let X be an object in $\mathcal{X}$, i.e. $\mathsf{q}(X)$ lies in $\mathcal{X}_{\mathscr{A}}$ and $\mathsf{e}(X)$ lies in $\mathcal{X}_{\mathscr{C}}$. Then for every object Y in $\mathcal{Y}$, i.e. $\mathsf{p}(Y)$ lies in $\mathcal{Y}_{\mathscr{A}}$ and $\mathsf{e}(Y)$ lies in $\mathcal{Y}_{\mathscr{C}}$, we have as above the exact sequence $0 \longrightarrow \mathsf{ip}(Y) \longrightarrow Y \longrightarrow \mathsf{re}(Y)$. Applying the functor $\mathsf{Hom}_{\mathscr{B}}(X,-)$ we get the following exact sequence:

$$0 \longrightarrow \mathsf{Hom}_{\mathscr{B}}(X,\mathsf{ip}(Y)) \longrightarrow \mathsf{Hom}_{\mathscr{B}}(X,Y) \longrightarrow \mathsf{Hom}_{\mathscr{B}}(X,\mathsf{re}(Y))$$

From the adjunctions (q,i) and (e,r), and since $(\mathcal{X}_{\mathscr{A}},\mathcal{Y}_{\mathscr{A}})$ and $(\mathcal{X}_{\mathscr{C}},\mathcal{Y}_{\mathscr{C}})$ are torsion pairs, we have $\mathsf{Hom}_{\mathscr{B}}(X,\mathsf{ip}(Y)) \cong \mathsf{Hom}_{\mathscr{A}}(\mathsf{q}(X),\mathsf{p}(Y))=0$ and $\mathsf{Hom}_{\mathscr{B}}(X,\mathsf{re}(Y)) \cong \mathsf{Hom}_{\mathscr{C}}(\mathsf{e}(X),\mathsf{e}(Y))=0$. Then we have $\mathsf{Hom}_{\mathscr{B}}(X,Y)=0$ by the above exact sequence. We conclude that $\mathcal{X} \subseteq {}^{\circ}\mathcal{Y}$.

We now show that ${}^{\circ}\mathcal{Y} \subseteq \mathcal{X}$. Let Z be an object in ${}^{\circ}\mathcal{Y}$. It suffices to show that $\mathsf{q}(Z)$ lies in $\mathcal{X}_{\mathscr{A}}$ and $\mathsf{e}(Z)$ lies in $\mathcal{X}_{\mathscr{C}}$. First, using Remark 2.4 observe that $\mathsf{i}(\mathcal{Y}_{\mathscr{A}}) \subseteq \mathcal{Y}$ and $\mathsf{r}(\mathcal{Y}_{\mathscr{C}}) \subseteq \mathcal{Y}$. Let $Y_{\mathscr{A}}$ be an object in $\mathcal{Y}_{\mathscr{A}}$ and $Y_{\mathscr{C}}$ an object in $\mathcal{Y}_{\mathscr{C}}$. From the adjunction (q,i) and since $\mathsf{Hom}_{\mathscr{B}}(Z,\mathcal{Y})=0$, we have $\mathsf{Hom}_{\mathscr{A}}(\mathsf{q}(Z),Y_{\mathscr{A}}) \cong \mathsf{Hom}_{\mathscr{B}}(Z,\mathsf{i}(Y_{\mathscr{A}}))=0$. Moreover, we also have $\mathsf{Hom}_{\mathscr{C}}(\mathsf{e}(Z),Y_{\mathscr{C}}) \cong \mathsf{Hom}_{\mathscr{B}}(Z,\mathsf{r}(Y_{\mathscr{C}}))=0$. Since $(\mathcal{X}_{\mathscr{A}},\mathcal{Y}_{\mathscr{A}})$ and $(\mathcal{X}_{\mathscr{C}},\mathcal{Y}_{\mathscr{C}})$ are torsion pairs, we infer that ${}^{\circ}\mathcal{Y} \subseteq \mathcal{X}$ and thus $\mathcal{X} = {}^{\circ}\mathcal{Y}$. Dually it follows that $\mathcal{Y} = \mathcal{X}^{\circ}$. We conclude that $(\mathcal{X},\mathcal{Y})$ is a torsion pair in $\mathscr{B}$. □

DEFINITION 9.2. Let $\mathsf{R_{ab}}(\mathscr{A},\mathscr{B},\mathscr{C})$ be a recollement of abelian categories. A torsion pair $(\mathcal{X},\mathcal{Y})$ in $\mathscr{B}$ is called **compatible** with respect to $\mathsf{R_{ab}}(\mathscr{A},\mathscr{B},\mathscr{C})$ if the condition of Proposition 9.1 (i) (b) is satisfied.

We continue with the next result which shows that we always have some compatible torsion pairs in a recollement situation.

COROLLARY 9.3. *Let $\mathsf{R_{ab}}(\mathscr{A},\mathscr{B},\mathscr{C})$ be a recollement of abelian categories. Then*

$$(\operatorname{Ker}\mathsf{q},\mathsf{i}(\mathscr{A}),\operatorname{Ker}\mathsf{p})$$

is a TTF-triple in $\mathscr{B}$ and the torsion pairs $(\operatorname{Ker}\mathsf{q},\mathsf{i}(\mathscr{A}))$, $(\mathsf{i}(\mathscr{A}),\operatorname{Ker}\mathsf{p})$ are compatible with respect to $\mathsf{R_{ab}}(\mathscr{A},\mathscr{B},\mathscr{C})$.

PROOF. The statement that $(\operatorname{Ker}\mathsf{q},\mathsf{i}(\mathscr{A}),\operatorname{Ker}\mathsf{p})$ is a TTF-triple in $\mathscr{B}$ was proved in Theorem 6.5. Since the composition of functors $\mathsf{q}\circ\mathsf{l}=0$ and $\mathsf{e}\circ\mathsf{i}=0$, see Remark 2.4 (ii), it follows that $\mathsf{le}(\operatorname{Ker}\mathsf{q})\subseteq\operatorname{Ker}\mathsf{q}$ and $\mathsf{le}(\mathsf{i}(\mathscr{A}))\subseteq\mathsf{i}(\mathscr{A})$. We infer that the torsion pairs $(\operatorname{Ker}\mathsf{q},\mathsf{i}(\mathscr{A}))$ and $(\mathsf{i}(\mathscr{A}),\operatorname{Ker}\mathsf{p})$ are compatible. □

We would like to develop more theory on compatible torsion pairs in recollements of abelian categories. For this purpose, we recall the following important construction of Happel, Reiten and Smalø [**60**] on constructing t-structures in the derived category from torsion pairs of the underlying category.

THEOREM 9.4. (i) ([**60**, Proposition 2.1]) *Let $(\mathcal{X},\mathcal{Y})$ be a torsion pair in an abelian category $\mathscr{A}$. Then the pair:*

$$\begin{aligned}(\mathsf{D}^{\leq 0} &= \{X^\bullet\in\mathsf{D}(\mathscr{A})\mid H^0(X^\bullet)\in\mathcal{X}\ \text{and}\ H^i(X^\bullet)=0,\ \forall i>0\},\\ \mathsf{D}^{\geq 0} &= \{X^\bullet\in\mathsf{D}(\mathscr{A})\mid H^{-1}(X^\bullet)\in\mathcal{Y}\ \text{and}\ H^i(X^\bullet)=0,\ \forall i<-1\})\end{aligned}$$

is a t-structure in $\mathsf{D}(\mathscr{A})$.

(ii) ([**21**, Proposition 2.5]) *Let $\mathcal{T}$ be a triangulated category with a bounded t-structure $(\mathcal{T}^{\leq 0},\mathcal{T}^{\geq 0})$, $\mathscr{A}$ its heart and H^i, $i\in\mathbb{Z}$, the associated cohomology functors. Assume that $(\mathcal{X},\mathcal{Y})$ is a torsion pair in $\mathscr{A}$. Then the pair:*

$$\begin{aligned}(\mathsf{D}^{\leq 0}_{\mathscr{A}} &= \{X\in\mathcal{T}\mid H^0(X)\in\mathcal{X}\ \text{and}\ H^i(X)=0,\ \forall i>0\},\\ \mathsf{D}^{\geq 0}_{\mathscr{A}} &= \{X\in\mathcal{T}\mid H^{-1}(X)\in\mathcal{Y}\ \text{and}\ H^i(X)=0,\ \forall i<-1\})\end{aligned}$$

is a t-structure in $\mathcal{T}$. The t-structure $(\mathsf{D}^{\leq 0},\mathsf{D}^{\geq 0})$ is called **the HRS-tilt** *(or* **the HRS t-structure***) of $(\mathcal{T}^{\leq 0},\mathcal{T}^{\geq 0})$ with respect to $(\mathcal{X},\mathcal{Y})$.*

It would be interesting to know under what conditions a torsion pair is compatible with respect to a recollement of abelian categories. In this direction, we have the following result for recollements of module categories (it can also be formulated for abelian categories using the abelian analogue of Theorem 8.3, i.e. see [**111**, Theorem 6.10]) which is inspired by [**83**, Proposition 6.3]. For the notion of glueing t-structure with respect to a recollement see Section 4.

PROPOSITION 9.5. *Let $\mathsf{R_{ab}}(\mathsf{Mod}\text{-}S,\mathsf{Mod}\text{-}R,\mathsf{Mod}\text{-}T)$ be a recollement of module categories and let $(\mathcal{X},\mathcal{Y})$ be a torsion pair in $\mathsf{Mod}\text{-}R$. The following are equivalent:*

(i) *The torsion pair $(\mathcal{X},\mathcal{Y})$ is compatible with repsect to $(\mathsf{Mod}\text{-}S,\mathsf{Mod}\text{-}R,\mathsf{Mod}\text{-}T)$.*
(ii) *The HRS t-structure $(\mathsf{D}^{\leq 0},\mathsf{D}^{\geq 0})$ is glued with respect to the recollement $\mathsf{R_{tr}}(\mathsf{D}_S(R),\mathsf{D}(R),\mathsf{D}(T))$.*

PROOF. From Theorem 5.7 it suffices to work with a recollement of module categories $\mathsf{R_{ab}}(\mathsf{Mod}\text{-}R/ReR, \mathsf{Mod}\text{-}R, \mathsf{Mod}\text{-}eRe)$ induced by an idempotent element e. Moreover, by the proof of Theorem 8.3 (see also diagram (8.2)) there is a recollement of triangulated categories $\mathsf{R_{tr}}(\mathsf{D}_{R/ReR}(R), \mathsf{D}(R), \mathsf{D}(eRe))$. Let X be an object of $\mathsf{D}(R)$. Then from the standard t-structure we have the next triangle in $\mathsf{D}(R)$:

$$\tau_{\mathsf{st}}^{\leq -1}(X) \longrightarrow X \longrightarrow H^0(X) \longrightarrow \tau_{\mathsf{st}}^{\leq -1}(X)[1]$$

We denote by l the derived functor $Re \otimes^{\mathbb{L}}_{eRe} -\colon \mathsf{D}(eRe) \longrightarrow \mathsf{D}(R)$. Applying the exact functor $\mathsf{l} \circ \mathsf{e}$ we get the following triangle in $\mathsf{D}(R)$:

$$\mathsf{le}(\tau_{\mathsf{st}}^{\leq -1}(X)) \longrightarrow \mathsf{le}(X) \longrightarrow \mathsf{le}(H^0(X)) \longrightarrow \mathsf{le}(\tau_{\mathsf{st}}^{\leq -1}(X))[1]$$

Since $\mathsf{le}(\mathsf{D}_{\mathsf{st}}^{\leq -1}) \subseteq \mathsf{D}_{\mathsf{st}}^{\leq -1}$ and $\mathsf{D}_{\mathsf{st}}^{\leq -1} \subseteq \mathsf{D}^{\leq 0}$, it follows by the above triangle that $\mathsf{le}(X)$ lies in $\mathsf{D}^{\leq 0}$ if and only if the object $\mathsf{le}(H^0(X))$ lies in $\mathsf{D}^{\leq 0}$. This implies that $\mathsf{le}(\mathsf{D}^{\leq 0}) \subseteq \mathsf{D}^{\leq 0}$ if and only if $Re \otimes_{eRe} e(\mathcal{X}) \subseteq \mathcal{X}$, and this completes the proof. □

Important examples of torsion pairs arise in Tilting Theory. We recall the following notions from [**60**].

DEFINITION 9.6. Let $\mathscr{A}$ be an abelian category.

(i) A torsion class $\mathcal{X}$ in $\mathscr{A}$ is called **tilting** if $\mathcal{X}$ is a cogenerator for $\mathscr{A}$, i.e. for all A in $\mathscr{A}$ there is a monomorphism $A \longrightarrow X$ with X in $\mathcal{X}$.
(ii) A torsion-free class $\mathcal{Y}$ in $\mathscr{A}$ is called **cotilting** if $\mathcal{Y}$ is a generator for $\mathscr{A}$, i.e. for all A in $\mathscr{A}$ there is an epimorphism $Y \longrightarrow A$ with Y in $\mathcal{Y}$.

We say that a torsion pair $(\mathcal{X}, \mathcal{Y})$ is **tilting**, resp. **cotilting**, if the torsion class $\mathcal{X}$ is tilting, resp. the torsion-free class $\mathcal{Y}$ is cotilting.

In the next result we examine the above properties for a compatible torsion pair.

LEMMA 9.7. *Let $(\mathcal{X}, \mathcal{Y})$ be a compatible torsion pair in $\mathsf{R_{ab}}(\mathscr{A}, \mathscr{B}, \mathscr{C})$.*

(i) *Assume that $(\mathcal{X}, \mathcal{Y})$ is tilting. Then the torsion pair $(\mathsf{e}(\mathcal{X}), \mathsf{e}(\mathcal{Y}))$ in $\mathscr{C}$ is tilting and $\mathsf{p}(\mathcal{X})$ is a cogenerator for $\mathscr{A}$.*
(ii) *Assume that $(\mathcal{X}, \mathcal{Y})$ is cotilting. Then the torsion pair $(\mathsf{e}(\mathcal{X}), \mathsf{e}(\mathcal{Y}))$ in $\mathscr{C}$ is cotilting and $\mathsf{q}(\mathcal{Y})$ is a generator for $\mathscr{A}$.*

PROOF. We prove only (i), the statement (ii) is dual and is left to the reader. Assume that the torsion pair $(\mathcal{X}, \mathcal{Y})$ is tilting and let C be an object of $\mathscr{C}$. Then for the object $\mathsf{l}(C)$ of $\mathscr{B}$ there exists a monomorphism $\mathsf{l}(C) \longrightarrow X$ with X in $\mathcal{X}$. Then, applying the exact functor $\mathsf{e}\colon \mathscr{B} \longrightarrow \mathscr{C}$ and using Remark 2.4 (iv) we obtain the monomorphism $C \longrightarrow \mathsf{e}(X)$ with $\mathsf{e}(X)$ in $\mathsf{e}(\mathcal{X})$. This implies that $(\mathsf{e}(\mathcal{X}), \mathsf{e}(\mathcal{Y}))$ is a tilting torsion pair in $\mathscr{C}$. If A is an object in $\mathscr{A}$ we have a monomorphism $\mathsf{i}(A) \longrightarrow X$ with X in $\mathcal{X}$. Then applying the left exact functor $\mathsf{p}\colon \mathscr{B} \longrightarrow \mathscr{A}$ and using again Remark 2.4 the result follows. □

Given a recollement $\mathsf{R_{ab}}(\mathscr{A}, \mathscr{B}, \mathscr{C})$ we have from Corollary 9.3 the torsion pairs $(\mathsf{Ker}\,\mathsf{q}, \mathsf{i}(\mathscr{A}))$ and $(\mathsf{i}(\mathscr{A}), \mathsf{Ker}\,\mathsf{p})$ in $\mathscr{B}$. In the next remark, we show that $\mathsf{i}(\mathscr{A})$ is never tilting or cotilting, and with an example we show that $\mathsf{Ker}\,\mathsf{q}$ is in general not tilting and $\mathsf{Ker}\,\mathsf{p}$ is in general not cotilting.

REMARK 9.8. (i) Assume that $\mathsf{i}(\mathscr{A})$ is tilting. Then, for any object B in $\mathscr{B}$ there is a monomorphism $B \longrightarrow \mathsf{i}(A)$, with A in $\mathscr{A}$. Then, if we apply the exact functor $\mathsf{e}\colon \mathscr{B} \longrightarrow \mathscr{C}$ we get by Remark 2.4 (ii) that $\mathsf{e}(B) = 0$ and therefore the object B lies in $\mathsf{i}(\mathscr{A})$. This implies that the functor $\mathsf{i}\colon \mathscr{A} \longrightarrow \mathscr{B}$ is an equivalence of categories and thus from Remark 2.4 (vii) we get that $\mathscr{C} = 0$. Hence, the torsion class $\mathsf{i}(\mathscr{A})$ cannot be tilting. Similar considerations hold if we assume that $\mathsf{i}(\mathscr{A})$ is cotilting.

(ii) Let $\Lambda = \left(\begin{smallmatrix} R & {}_RM_S \\ 0 & S \end{smallmatrix}\right)$ be a triangular matrix ring. Then from Example 3.10, see also diagram (10.4), we have the recollement of module categories $\mathsf{R_{ab}}(\mathsf{Mod}\text{-}R, \mathsf{Mod}\text{-}\Lambda, \mathsf{Mod}\text{-}S)$ and the torsion pair $(\mathsf{Mod}\text{-}R, \mathsf{Mod}\text{-}S)$ in $\mathsf{Mod}\text{-}\Lambda$. Recall that a torsion-free class $\mathcal{Y}$ in an abelian category $\mathscr{A}$ with enough projectives is cotilting if and only if $\mathsf{Proj}\,\mathscr{A} \subseteq \mathcal{Y}$. In our setting, if P is a projective R-module then the triple $\mathsf{Z}_R(P) = (P, 0, 0)$ is projective as a Λ-module and clearly it is not contained in the image of $\mathsf{Mod}\text{-}S$ in $\mathsf{Mod}\text{-}\Lambda$. This shows that the torsion-free class $\mathsf{Mod}\text{-}S$ is not cotilting.

It is natural to consider what the relation between the abelian category $\mathscr{B}$ and the HRS-tilt hearts $\mathcal{H}_{(\mathsf{Ker}\,\mathsf{q}, \mathsf{i}(\mathscr{A}))}$ and $\mathcal{H}_{(\mathsf{i}(\mathscr{A}), \mathsf{Ker}\,\mathsf{p})}$ is. The above remark shows that we cannot apply [**60**, Theorem 3.3] to obtain a triangle equivalence between the bounded derived categories of these abelian categories. However, it is still an interesting (open) problem to try to determine any possible relation.

It is known that every tilting object T in a Grothendieck category $\mathscr{B}$ generates a torsion pair in $\mathscr{B}$, called the **tilting torsion pair**. Our aim in this section is to examine when tilting torsion pairs are compatible with respect to a recollement $\mathsf{R_{ab}}(\mathscr{A}, \mathscr{B}, \mathscr{C})$ of abelian categories, where $\mathscr{B}$ is assumed to be Grothendieck.

Recall from Colpi [**28**] that an object T in a Grothendieck category $\mathscr{B}$ is called **tilting** if it is self-small, i.e. $\mathsf{Hom}_{\mathscr{B}}(T, -)$ commutes with coproducts in $\mathsf{Add}\,T$, and $\mathsf{Gen}\,T = T^{\perp_1}$ where $T^{\perp_1} = \{X \in \mathscr{B} \mid \mathsf{Ext}^1_{\mathscr{B}}(T, X) = 0\}$. Note that if T is a tilting object, then it is finitely presented and therefore the functor $\mathsf{Hom}_{\mathscr{B}}(T, -)$ preserves direct limits, see [**120**, Proposition V.3.4].

In what follows, we characterize when a tilting torsion pair is compatible with respect to a recollement of abelian categories.

PROPOSITION 9.9. *Let $\mathsf{R_{ab}}(\mathscr{A}, \mathscr{B}, \mathscr{C})$ be a recollement of abelian categories. Assume that $\mathscr{B}$ is Grothendieck and let T be a tilting object in $\mathscr{B}$. Then the following statements are equivalent:*

(i) *The torsion pair $(\mathsf{Gen}\,T, T^{\circ})$ is compatible with respect to $\mathsf{R_{ab}}(\mathscr{A}, \mathscr{B}, \mathscr{C})$.*
(ii) $\mathsf{Ext}^1_{\mathscr{B}}(T, \mathsf{le}(T)) = 0$.

PROOF. From Proposition 9.1 the torsion pair $(\mathsf{Gen}\,T, T^{\circ})$ is compatible with respect to $\mathsf{R_{ab}}(\mathscr{A}, \mathscr{B}, \mathscr{C})$ if and only if $\mathsf{le}(\mathsf{Gen}\,T) \subseteq \mathsf{Gen}\,T = T^{\perp_1}$. Thus, if statement (i) holds then we get $\mathsf{Ext}^1_{\mathscr{B}}(T, \mathsf{le}(T)) = 0$.

Conversely, suppose that $\mathsf{Ext}^1_{\mathscr{B}}(T, \mathsf{le}(T)) = 0$ and let M be an object in $\mathsf{Gen}\,T$. Then there is an epimorphism $p\colon T^{(\alpha)} \longrightarrow M$ for some cardinal α. Since left adjoint functors preserve coproducts, by our assumption and since T is finitely presented we have the following relations:

$$\mathsf{Ext}^1_{\mathscr{B}}(T, \mathsf{le}(T^{(\alpha)})) \cong \mathsf{Ext}^1_{\mathscr{B}}(T, (\mathsf{le}(T))^{(\alpha)}) \cong \mathsf{Ext}^1_{\mathscr{B}}(T, \mathsf{le}(T))^{(\alpha)} = 0 \tag{9.1}$$

Since the composition le is right exact, it follows that the map $\mathsf{le}(p)$ is an epimorphism. Since $T^{\perp_1}$ is a torsion class, it follows from Proposition 2.11 that $T^{\perp_1}$ is closed under quotients. Using relation (9.1) we obtain that $\mathsf{Ext}^1_{\mathscr{B}}(T, \mathsf{le}(M)) = 0$. We infer that the object $\mathsf{le}(M)$ belongs to $\mathsf{Gen}\, T$ and this completes the proof. □

Given a recollement of module categories $\mathsf{R_{ab}}(\mathsf{Mod}\text{-}R/ReR, \mathsf{Mod}\text{-}R, \mathsf{Mod}\text{-}eRe)$ and T an R-tilting module, then from Proposition 9.9 we have that $(\mathsf{Gen}\, T, T^{\circ})$ is compatible if and only if $\mathsf{Ext}^1_R(T, Re \otimes_{eRe} eT) = 0$.

We remark that the material of this section was obtained in discussions with Jorge Vitória, and may appear in a forthcoming paper on "*Compatible Torsion Pairs and Tilting Theory*". Note that applications of compatible torsion pairs with respect to recollements of hearts are discussed in Section 10.

10. Recollements of Hearts and HRS-tilt

Let $\mathsf{R}_{\heartsuit}(\mathscr{A}, \mathscr{B}, \mathscr{C})$ be a recollement of hearts. From Section 4, this means that we have a diagram of recollements as follows:

(10.1)

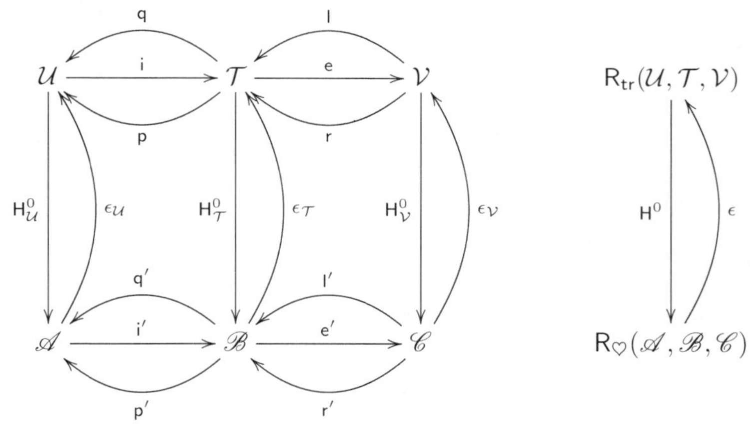

Note that we have slightly changed the notation of a recollement of hearts. From Corollary 9.3 we have the torsion pairs $(\mathsf{Ker}\, \mathsf{q}', \mathsf{i}'(\mathscr{A}))$ and $(\mathsf{i}'(\mathscr{A}), \mathsf{Ker}\, \mathsf{p})$ in $\mathscr{B}$. Then we can apply the HRS-tilt to obtain the hearts $\mathcal{H}_{(\mathsf{Ker}\, \mathsf{q}', \mathsf{i}'(\mathscr{A}))}$ and $\mathcal{H}_{(\mathsf{i}'(\mathscr{A}), \mathsf{Ker}\, \mathsf{p}')}$ inside $\mathsf{D}(\mathscr{B})$. It is, therefore, natural to consider how these abelian categories are related to the abelian categories of the recollement $\mathsf{R}_{\heartsuit}(\mathscr{A}, \mathscr{B}, \mathscr{C})$. In Corollary 10.3 below, we show that there exists a recollement of hearts that connect the abelian categories mentioned above. Motivated by this problem, we show that given a recollement of hearts we can produce infinitely many new recollements of abelian categories.

THEOREM 10.1. *Let $\mathsf{R}_{\heartsuit}(\mathscr{A}, \mathscr{B}, \mathscr{C})$ be a recollement of hearts. Then there exist infinitely many recollements of abelian categories:*

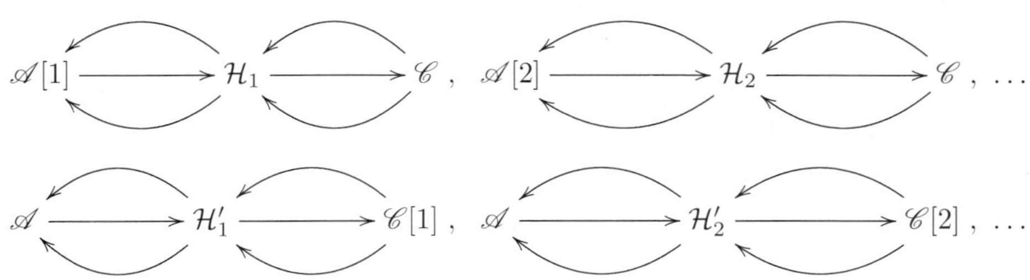

The easy proof of the above result will be given at the end of this section. We start with the next result which shows how from a recollement of hearts with a compatible torsion pair we obtain a new recollement of abelian categories.

PROPOSITION 10.2. *Let* $\mathsf{R}_{\heartsuit}(\mathscr{A},\mathscr{B},\mathscr{C})$ *be a recollement of hearts as in* (10.1) *and let* $(\mathcal{X},\mathcal{Y})$ *be a torsion pair in* $\mathscr{B}$ *which is compatible with respect to* $\mathsf{R}_{\heartsuit}(\mathscr{A},\mathscr{B},\mathscr{C})$. *Then we have the recollement of abelian categories* $\mathsf{R}_{\heartsuit}(\mathcal{H}_{(\mathsf{q}'(\mathcal{X}),\mathsf{p}'(\mathcal{Y}))},\mathcal{H}_{(\mathcal{X},\mathcal{Y})},\mathcal{H}_{(\mathsf{e}'(\mathcal{X}),\mathsf{e}'(\mathcal{Y}))})$:

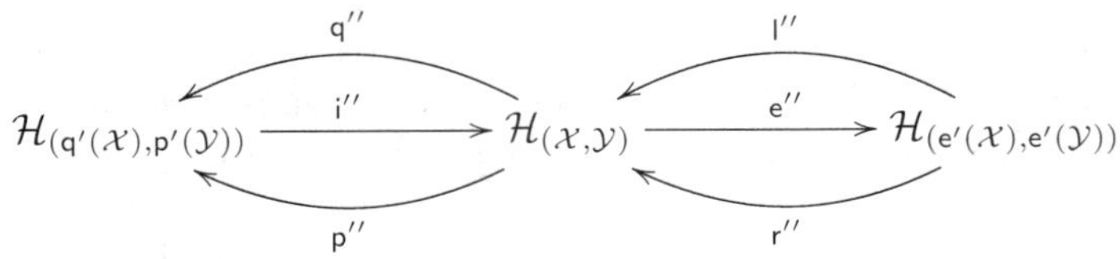

PROOF. Since $(\mathcal{X},\mathcal{Y})$ is a compatible torsion pair in $\mathsf{R}_{\heartsuit}(\mathscr{A},\mathscr{B},\mathscr{C})$, we have by Proposition 9.1 the torsion pairs $(\mathsf{q}'(\mathcal{X}),\mathsf{p}'(\mathcal{Y}))$ and $(\mathsf{e}'(\mathcal{X}),\mathsf{e}'(\mathcal{Y}))$ in $\mathscr{A}$ and $\mathscr{C}$ respectively. From Theorem 9.4 (ii) we have the HRS t-structure $(\mathsf{D}^{\leq 0}_{\mathscr{A}},\mathsf{D}^{\geq 0}_{\mathscr{A}})$ in $\mathcal{U}$, where

$$\begin{aligned}\mathsf{D}^{\leq 0}_{\mathscr{A}} &= \{X\in\mathcal{U} \mid H^0(X)\in\mathsf{q}'(\mathcal{X}) \text{ and } H^i(X)=0,\ \forall i>0\}\\ \mathsf{D}^{\geq 0}_{\mathscr{A}} &= \{X\in\mathcal{U} \mid H^{-1}(X)\in\mathsf{p}'(\mathcal{Y}) \text{ and } H^i(X)=0,\ \forall i<-1\})\end{aligned}$$

and the HRS t-structure $(\mathsf{D}^{\leq 0}_{\mathscr{C}},\mathsf{D}^{\geq 0}_{\mathscr{C}})$ in $\mathcal{V}$, where

$$\begin{aligned}\mathsf{D}^{\leq 0}_{\mathscr{C}} &= \{Y\in\mathcal{V} \mid H^0(Y)\in\mathsf{e}'(\mathcal{X}) \text{ and } H^i(Y)=0,\ \forall i>0\}\\ \mathsf{D}^{\geq 0}_{\mathscr{C}} &= \{Y\in\mathcal{V} \mid H^{-1}(Y)\in\mathsf{e}'(\mathcal{Y}) \text{ and } H^i(Y)=0,\ \forall i<-1\}\end{aligned}$$

We denote by $\mathcal{H}_{(\mathsf{q}'(\mathcal{X}),\mathsf{p}'(\mathcal{Y}))}$ the heart of the t-structure $(\mathsf{D}^{\leq 0}_{\mathscr{A}},\mathsf{D}^{\geq 0}_{\mathscr{A}})$, and by $\mathcal{H}_{(\mathsf{e}'(\mathcal{X}),\mathsf{e}'(\mathcal{Y}))}$ the heart of the t-structure $(\mathsf{D}^{\leq 0}_{\mathscr{C}},\mathsf{D}^{\geq 0}_{\mathscr{C}})$. Then by the glueing method of Beilinson, Bernstein and Deligne, see Section 4, we obtain a t-structure $(\mathcal{T}^{\leq 0},\mathcal{T}^{\geq 0})$ in $\mathcal{T}$ where

$$\begin{aligned}\mathcal{T}^{\leq 0} &= \{X\in\mathcal{T} \mid \mathsf{e}(X)\in\mathsf{D}^{\leq 0}_{\mathscr{C}} \text{ and } \mathsf{q}(X)\in\mathsf{D}^{\leq 0}_{\mathscr{A}}\}\\ \mathcal{T}^{\geq 0} &= \{X\in\mathcal{T} \mid \mathsf{e}(X)\in\mathsf{D}^{\geq 0}_{\mathscr{C}} \text{ and } \mathsf{p}(X)\in\mathsf{D}^{\geq 0}_{\mathscr{A}}\}\end{aligned}$$

Moreover, from [**83**, Theorem 6.4] the glued t-structure $(\mathcal{T}^{\leq 0},\mathcal{T}^{\geq 0})$ in $\mathcal{T}$ is the HRS t-structure $(\mathsf{D}^{\leq 0}_{\mathscr{B}},\mathsf{D}^{\geq 0}_{\mathscr{B}})$ of the torsion pair $(\mathcal{X},\mathcal{Y})$, where

$$\begin{aligned}\mathsf{D}^{\leq 0}_{\mathscr{B}} &= \{X\in\mathcal{T} \mid H^0(X)\in\mathcal{X} \text{ and } H^i(X)=0,\ \forall i>0\}\\ \mathsf{D}^{\geq 0}_{\mathscr{B}} &= \{X\in\mathcal{T} \mid H^{-1}(X)\in\mathcal{Y} \text{ and } H^i(X)=0,\ \forall i<-1\}\end{aligned}$$

Denote by $\mathcal{H}_{(\mathcal{X},\mathcal{Y})}$ the heart of the t-structure $(\mathsf{D}^{\leq 0}_{\mathscr{B}},\mathsf{D}^{\geq 0}_{\mathscr{B}})$. Then Proposition 4.2 implies the recollement of hearts $\mathsf{R}_{\heartsuit}(\mathcal{H}_{(\mathsf{q}'(\mathcal{X}),\mathsf{p}'(\mathcal{Y}))},\mathcal{H}_{(\mathcal{X},\mathcal{Y})},\mathcal{H}_{(\mathsf{e}'(\mathcal{X}),\mathsf{e}'(\mathcal{Y}))})$. □

Recall that a t-structure $(\mathcal{T}^{\leq 0},\mathcal{T}^{\geq 0})$ in a triangulated category $\mathcal{T}$ is called **non-degenerate** if $\cap_{n\in\mathbb{Z}}\mathsf{Ob}\,\mathcal{T}^{\leq n}=0=\cap_{n\in\mathbb{Z}}\mathsf{Ob}\,\mathcal{T}^{\geq n}$. In this case, it is well known ([**15**, Proposition 1.3.7], see also [**49**, Theorem IV.4.11]) that $(\mathcal{T}^{\leq 0},\mathcal{T}^{\geq 0})$ can be cohomologically described as follows:

$$\begin{aligned}\mathcal{T}^{\leq 0} &= \{X\in\mathcal{T} \mid \mathsf{H}^n(X)=0, \text{ for all } n>0\}\\ \mathcal{T}^{\geq 0} &= \{X\in\mathcal{T} \mid \mathsf{H}^n(X)=0, \text{ for all } n<0\}\end{aligned}$$

In the following result we assume that the t-structures $\mathsf{T}_{\mathcal{U}} = (\mathcal{U}^{\leq 0}, \mathcal{U}^{\geq 0})$ in $\mathcal{U}$ and $\mathsf{T}_{\mathcal{V}} = (\mathcal{V}^{\leq 0}, \mathcal{V}^{\geq 0})$ in $\mathcal{V}$, which are used in the construction of diagram 10.1, are nondegenerate.

COROLLARY 10.3. *Let $\mathsf{R}_{\heartsuit}(\mathscr{A}, \mathscr{B}, \mathscr{C})$ be a recollement of hearts. Then we have the following recollements of abelian categories:*

$$\mathscr{A}[1] \longrightarrow \mathcal{H}_{(\mathsf{Ker\,q}', \mathsf{i}'(\mathscr{A}))} \longrightarrow \mathscr{C} \tag{10.2}$$

and

$$\mathscr{A} \longrightarrow \mathcal{H}_{(\mathsf{i}'(\mathscr{A}), \mathsf{Ker\,p}')} \longrightarrow \mathscr{C}[1] \tag{10.3}$$

PROOF. We first show how we obtain from $\mathsf{R}_{\heartsuit}(\mathscr{A}, \mathscr{B}, \mathscr{C})$ the recollement (10.2). From Corollary 9.3 the torsion pairs $(\mathsf{Ker\,q}', \mathsf{i}'(\mathscr{A}))$ and $(\mathsf{i}'(\mathscr{A}), \mathsf{Ker\,p}')$ in $\mathscr{B}$ are compatible with respect to $\mathsf{R}_{\heartsuit}(\mathscr{A}, \mathscr{B}, \mathscr{C})$. Let us first consider the torsion pair $(\mathsf{Ker\,q}', \mathsf{i}'(\mathscr{A}))$. From Remark 2.4 the restriction of $(\mathsf{Ker\,q}', \mathsf{i}'(\mathscr{A}))$ to $\mathscr{A}$ and $\mathscr{C}$ gives the trivial torsion pairs $(0, \mathscr{A})$ and $(\mathscr{C}, 0)$. The HRS t-structure of $(0, \mathscr{A})$ is

$$\begin{aligned} \mathsf{D}^{\leq 0}_{\mathscr{A}} &= \{X \in \mathcal{U} \mid H^i(X) = 0, \forall i > -1\} \\ \mathsf{D}^{\geq 0}_{\mathscr{A}} &= \{X \in \mathcal{U} \mid H^{-1}(X) \in \mathscr{A} \text{ and } H^i(X) = 0, \forall i < -1\} \end{aligned}$$

Since the t-structure $(\mathcal{U}^{\leq 0}, \mathcal{U}^{\geq 0})$ is non-degenerate, from the description above we have $\mathsf{D}^{\leq 0}_{\mathscr{A}} = \mathcal{U}^{\leq 1}$. This implies that the HRS t-structure of $(0, \mathscr{A})$ is $(\mathcal{U}^{\leq 1}, \mathcal{U}^{\geq 1})$ and therefore the heart is the category $\mathscr{A}[1]$. Computing the HRS t-structure of $(\mathscr{C}, 0)$ we get the following:

$$\begin{aligned} \mathsf{D}^{\leq 0}_{\mathscr{C}} &= \{X \in \mathcal{V} \mid H^0(X) \in \mathscr{C} \text{ and } H^i(X) = 0, \forall i > 0\} \\ \mathsf{D}^{\geq 0}_{\mathscr{C}} &= \{X \in \mathcal{V} \mid H^i(X) = 0, \forall i < 0\} \end{aligned}$$

Since $(\mathcal{V}^{\leq 0}, \mathcal{V}^{\geq 0})$ is non-degenerate, it follows that the HRS t-structure of $(\mathscr{C}, 0)$ is precisely $(\mathcal{V}^{\leq 0}, \mathcal{V}^{\geq 0})$. Hence, the heart is the category $\mathscr{C}$. Then from Proposition 10.2 we infer that $\mathsf{R}_{\heartsuit}(\mathscr{A}[1], \mathcal{H}_{(\mathsf{Ker\,q}', \mathsf{i}'(\mathscr{A}))}, \mathscr{C})$ is a recollement of abelian categories. Once more, note that the abelian category $\mathcal{H}_{(\mathsf{Ker\,q}', \mathsf{i}'(\mathscr{A}))}$, which is the heart of the glued t-structure, is the heart of the HRS t-structure of the torsion pair $(\mathsf{Ker\,q}', \mathsf{i}'(\mathscr{A}))$ (see [**83**, Theorem 6.4]). We leave to the reader to describe explicitly the functors in $\mathsf{R}_{\heartsuit}(\mathscr{A}[1], \mathcal{H}_{(\mathsf{Ker\,q}', \mathsf{i}'(\mathscr{A}))}, \mathscr{C})$, see Section 4. Similarly as above, using now the torsion pair $(\mathsf{i}'(\mathscr{A}), \mathsf{Ker\,p}')$ in $\mathscr{B}$, we obtain the recollement (10.3). □

We mention that the recollement (10.2) is obtained by glueing the t-structures $(\mathcal{U}^{\leq 0}[1], \mathcal{U}^{\geq 0}[1])$ and $(\mathcal{V}^{\leq 0}, \mathcal{V}^{\geq 0})$, and the recollement (10.3) is obtained by glueing the t-structures $(\mathcal{U}^{\leq 0}, \mathcal{U}^{\geq 0})$ and $(\mathcal{V}^{\leq 0}[1], \mathcal{V}^{\geq 0}[1])$. Let us go back to diagram (10.1). The starting point of this diagram is the t-structures $\mathsf{T}_{\mathcal{U}} = (\mathcal{U}^{\leq 0}, \mathcal{U}^{\geq 0})$ in $\mathcal{U}$ and $\mathsf{T}_{\mathcal{V}} = (\mathcal{V}^{\leq 0}, \mathcal{V}^{\geq 0})$ in $\mathcal{V}$. From the discussion so far, it is natural to iterate the construction in (10.1) by taking shifts of one of the above t-structures and then repeat the glueing method of Beilinson, Bernstein and Deligne in order to derive more recollements of abelian categories. We are now ready to prove Theorem 10.1.

Proof of Theorem 10.1: Set $\mathcal{H}_1$ for the heart $\mathcal{H}_{(\mathsf{Ker}\,\mathsf{q}',\mathsf{i}'(\mathscr{A}))}$. Then by Corollary 10.3 we have the recollement $\mathsf{R}_{\heartsuit}(\mathscr{A}[1], \mathcal{H}_1, \mathscr{C})$. We fix the t-structure $(\mathcal{V}^{\leq 0}, \mathcal{V}^{\geq 0})$ in $\mathcal{V}$ and we take shifts of the t-structure $(\mathcal{U}^{\leq 0}[n], \mathcal{U}^{\geq 0}[n])$ in $\mathcal{U}$. In particular, if we consider the t-structure $(\mathcal{U}^{\leq 0}[1], \mathcal{U}^{\geq 0}[1])$, as in diagram 10.1 we obtain a recollement of hearts $\mathsf{R}_{\heartsuit}(\mathscr{A}[1], \mathscr{B}_1, \mathscr{C})$. In this recollement of hearts, if we apply Corollary 10.3 we get the recollement $\mathsf{R}_{\heartsuit}(\mathscr{A}[2], \mathcal{H}_2, \mathscr{C})$. We iterate in the same way, and similarly we obtain the remaining recollements by fixing the t-structure $(\mathcal{U}^{\leq 0}, \mathcal{U}^{\geq 0})$ and by taking shifts of the t-structure $(\mathcal{V}^{\leq 0}[n], \mathcal{V}^{\geq 0}[n])$. This completes the proof. $\square$

It is natural to consider how different are the hearts in Theorem 10.1. For more details on this problem we refer to [**112**].

Consider now a triangular matrix ring $\Lambda = \left(\begin{smallmatrix} R & {}_RM_S \\ 0 & S \end{smallmatrix}\right)$. Then from Example 3.10 we have the following two recollements of module categories:

(10.4)

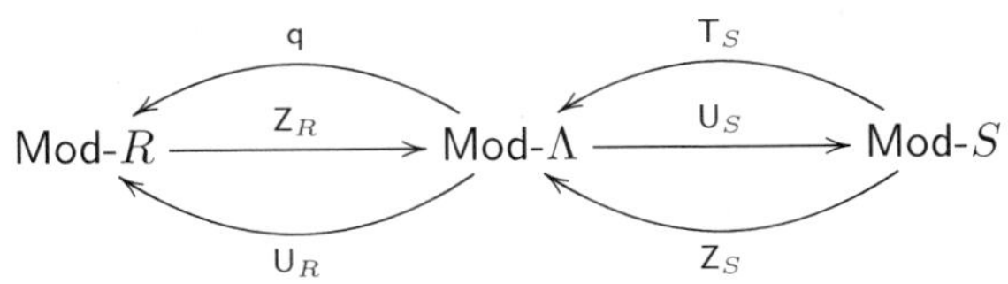

and

(10.5)

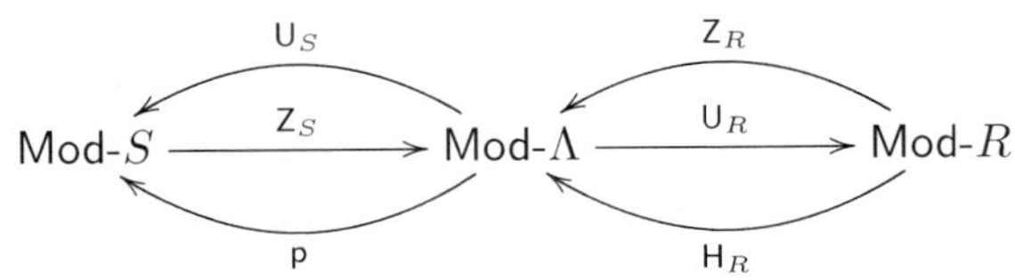

We claim that the above two diagrams are recollements of hearts. Looking at the left side of the recollements, note first that the functors U_R and U_S are exact. Then from Theorem 7.4 (in particular, from statement (viii) and (x) respectively) it follows that the functors Z_R and Z_S are homological embeddings. Hence, by Theorem 8.3 we can lift the recollements (10.4) and (10.5) to derived level. If we consider the standard t-structures and the associated hearts, see Example 4.1, then we obtain the following diagram:

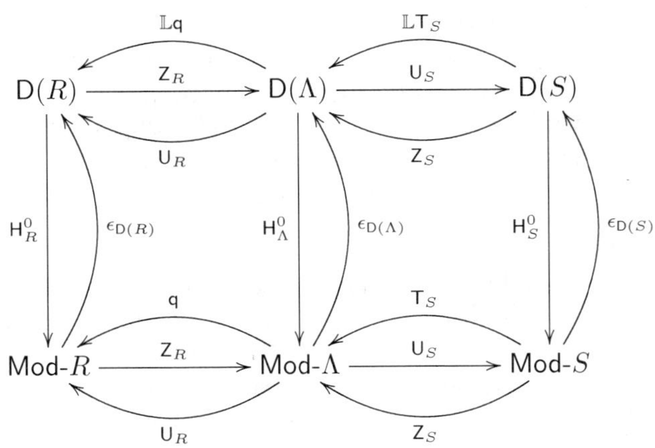

This implies that (10.4) is a recollement of hearts. Similarly we derive that (10.5) is also a recollement of hearts.

Consider now the recollement (10.4) and the torsion pair $(\mathsf{Z}_R(\mathsf{Mod}\text{-}R), \operatorname{Ker}\mathsf{U}_R)$ in $\mathsf{Mod}\text{-}\Lambda$ (see Corollary 9.3). Our aim is to understand in this case the recollement situation (10.3) of Corollary 10.3. In this connection, we obtain in the next result that this recollement splits. We remark that the description of this heart was obtained by Happel, Reiten and Smalø [**60**] using different methods.

COROLLARY 10.4. ([**60**, Proposition 2.3]) *Let* $\Lambda = \left(\begin{smallmatrix} R & {}_RM_S \\ 0 & S \end{smallmatrix}\right)$ *be a triangular matrix ring. Then there is an equivalence of categories:*

$$\mathcal{H}_{(\mathsf{Mod}\text{-}R,\mathsf{Mod}\text{-}S)} \xrightarrow{\simeq} \mathsf{Mod}\text{-}R \times \mathsf{Mod}\text{-}S$$

PROOF. We have shown that $\mathsf{R_{ab}}(\mathsf{Mod}\text{-}R, \mathsf{Mod}\text{-}\Lambda, \mathsf{Mod}\text{-}S)$ is a recollement of hearts. Moreover, the torsion pair $(\mathsf{Z}_R(\mathsf{Mod}\text{-}R), \operatorname{Ker}\mathsf{U}_R)$ in $\mathsf{Mod}\text{-}\Lambda$ is identified with the pair $(\mathsf{Mod}\text{-}R, \mathsf{Mod}\text{-}S)$. Then from Corollary 10.3 we have the recollement of hearts $(\mathsf{Mod}\text{-}R, \mathcal{H}_{(\mathsf{Mod}\text{-}R,\mathsf{Mod}\text{-}S)}, \mathsf{Mod}\text{-}S[1])$. We claim that the latter recollement splits. In particular, we use the characterization of splitting recollements given in Example 3.8 to show that $\mathsf{R_{ab}}(\mathsf{Mod}\text{-}R, \mathcal{H}_{(\mathsf{Mod}\text{-}R,\mathsf{Mod}\text{-}S)}, \mathsf{Mod}\text{-}S[1])$ splits.

Let Y be an S-module. From Proposition 2.8 (iii) we have the exact sequence:

$$0 \longrightarrow \mathcal{H}_{\mathsf{Z}_R}\mathcal{H}_{\mathsf{U}_R}\mathcal{H}_{\mathsf{T}_S}(Y) \longrightarrow \mathcal{H}_{\mathsf{T}_S}(Y) \longrightarrow \mathcal{H}_{\mathsf{Z}_S}(Y) \longrightarrow \mathcal{H}_{\mathsf{Z}_R}\mathcal{H}_q\mathcal{H}_{\mathsf{Z}_S}(Y) \longrightarrow 0$$

We use the above notation for the functors in the induced recollement of hearts (10.3) (note that it is different with the one given in (10.1)). From Example 3.8 it suffices to show that $\mathcal{H}_{\mathsf{Z}_R}\mathcal{H}_{\mathsf{U}_R}\mathcal{H}_{\mathsf{T}_S}(Y) = 0$ and $\mathcal{H}_{\mathsf{Z}_R}\mathcal{H}_q\mathcal{H}_{\mathsf{Z}_S}(Y) = 0$. Note that the module Y is concentrated in degree minus one. We compute:

$$\begin{aligned} \mathcal{H}_{\mathsf{Z}_R}\mathcal{H}_q\mathcal{H}_{\mathsf{Z}_S}(Y) &= (\mathsf{H}^0_\Lambda\mathsf{Z}_R\epsilon_{\mathsf{D}(R)})(\mathsf{H}^0_R\mathbb{L}\mathsf{q}\epsilon_{\mathsf{D}(\Lambda)})(\mathsf{H}^0_\Lambda\mathsf{Z}_S\epsilon_{\mathsf{D}(S)})(Y[1]) \\ &= \mathsf{H}^0_\Lambda\mathsf{Z}_R\epsilon_{\mathsf{D}(R)}\mathsf{H}^0_R(\mathbb{L}\mathsf{q}(0 \longrightarrow (0,Y,0) \longrightarrow 0)) \end{aligned}$$

Since the complex $\mathbb{L}\mathsf{q}(0 \longrightarrow (0,Y,0) \longrightarrow 0)$ lives in degrees minus one and higher and H^0_R gives the zero part, it follows that $\mathcal{H}_{\mathsf{Z}_R}\mathcal{H}_q\mathcal{H}_{\mathsf{Z}_S}(Y) = 0$. Next we compute

$$\mathcal{H}_{\mathsf{Z}_R}\mathcal{H}_{\mathsf{U}_R}\mathcal{H}_{\mathsf{T}_S}(Y) = (\mathsf{H}^0_\Lambda\mathsf{Z}_R\epsilon_{\mathsf{D}(R)})(\mathsf{H}^0_R\mathsf{U}_R\epsilon_{\mathsf{D}(\Lambda)})(\mathsf{H}^0_\Lambda\mathbb{L}\mathsf{T}_S\epsilon_{\mathsf{D}(S)})(Y[1]).$$

Set $X = \mathbb{L}\mathsf{T}_S(Y[1])$. The object X lies in $\mathsf{D}^{\leq -1}_{\mathsf{st}} \subseteq \mathsf{D}^{\leq 0}_{\mathcal{H}}$ and therefore we have $\mathsf{H}^0_\Lambda(X) = \tau^{\geq 0}_{\mathcal{H}}\tau^{\leq 0}_{\mathcal{H}}(X) = \tau^{\geq 0}_{\mathcal{H}}(X)$. Then from the triangle

$$\tau^{\leq -1}_{\mathcal{H}}(X) \longrightarrow X \longrightarrow \tau^{\geq 0}_{\mathcal{H}}(X) \longrightarrow \tau^{\leq -1}_{\mathcal{H}}(X)[1]$$

we get that $\tau^{\geq 0}_{\mathcal{H}}(X)$ lies in $\mathsf{D}^{\leq -1}_{\mathsf{st}}$. Since $\tau^{\geq 0}_{\mathcal{H}}(X)$ lies also in the heart $\mathcal{H}_{(\mathsf{Mod}\text{-}R,\mathsf{Mod}\text{-}S)}$ it follows that $\tau^{\geq 0}_{\mathcal{H}}(X)$ lies in $\mathsf{Mod}\text{-}S$. This implies that $\mathsf{U}_R(\tau^{\geq 0}_{\mathcal{H}}(X)) = 0$ and therefore $\mathcal{H}_{\mathsf{Z}_R}\mathcal{H}_{\mathsf{U}_R}\mathcal{H}_{\mathsf{T}_S}(Y) = 0$. We infer that $\mathsf{R_{ab}}(\mathsf{Mod}\text{-}R, \mathcal{H}_{(\mathsf{Mod}\text{-}R,\mathsf{Mod}\text{-}S)}, \mathsf{Mod}\text{-}S[1])$ is a splitting recollement, i.e. $\mathcal{H}_{(\mathsf{Mod}\text{-}R,\mathsf{Mod}\text{-}S)}$ is equivalent to $\mathsf{Mod}\text{-}R \times \mathsf{Mod}\text{-}S$. □

We remark that there are more hearts in the above triangular matrix case that can be determined using similar methods. In general, it is a natural problem to try to describe the hearts $\mathcal{H}_{(\operatorname{Ker}\mathsf{q}',\mathsf{i}'(\mathscr{A}))}$ and $\mathcal{H}_{(\mathsf{i}'(\mathscr{A}),\operatorname{Ker}\mathsf{p}')}$ considered in Corollary 10.3, but also to investigate what is the relation between $\mathcal{H}_{(\operatorname{Ker}\mathsf{q}',\mathsf{i}'(\mathscr{A}))}$ and $\mathcal{H}_{(\mathsf{i}'(\mathscr{A}),\operatorname{Ker}\mathsf{p}')}$ (for instance, if they are equivalent with the initial category $\mathscr{B}$). Moreover, another observation is that any recollement $\mathsf{R_{ab}}(\mathsf{Mod}\text{-}R/I, \mathsf{Mod}\text{-}R, \mathscr{C})$ such that the functor $\mathsf{i}\colon \mathsf{Mod}\text{-}R/I \longrightarrow \mathsf{Mod}\text{-}R$ is a homological embedding is a recollement of hearts. It would be interesting to know if any recollement of module categories can be viewed as a recollement of hearts. We consider this as a kind of *enhancement* for the underlying recollement of abelian categories that we work with. Finally, we mention that the material of this section is work in progress with Jorge Vitória and it will appear in a forthcoming paper entitled "*Recollements of Hearts*" [**112**].

11. Stratifying Recollements and Hereditary Algebras

The main purpose of this section is to investigate the following question formulated by Xi for derived module categories.

QUESTION 11.1. *When is a recollement of derived categories equivalent to a derived version of a recollement of abelian categories?*

This section consists of three subsections. In the first subsection we characterize when a recollement of derived module categories is stratifying. Using this result we verify in the second subsection Xi's question for hereditary algebras. In the last subsection, we discuss an example by Angeleri Hügel-Koenig-Liu-Yang [**6**] which shows that Question 11.1 is not true in general. The results of the first two subsections are taken from [**111**]. We remark that Theorem 11.2 is proved in [**111**] in the general context of abelian categories using realization functors due to Beilinson-Bernstein-Deligne [**15**]. For simplicity, we outline the proof for the case of derived module categories. We encourage the reader to see [**111**] where we discuss in a great generality recollements and equivalences between them using realisation functors.

11.1. Stratifying Recollements. In subsection 4 we discussed the glueing method of Beilinson, Bernstein and Deligne [**15**] along a recollement of triangulated categories. Trying to characterize a recollement of derived module categories being stratifying, see Definition 8.2, it is natural to consider a statement about glueing derived equivalences and therefore a statement about glueing t-structures which are induced by compact tilting objects (tilting complexes). We now provide an answer to Question 11.1 in terms of the glueing of tilting t-structures.

THEOREM 11.2. ([**111**, Theorem 6.13, Corollary 6.14]) *Let A, B and C be k-algebras over a commutative ring k and assume that A is projective as a k-module. Suppose that there is a recollement of triangulated categories $\mathcal{R}$ of the form*

$$\mathcal{R}:\quad \mathsf{D}(B) \xrightarrow{\ \mathsf{i}\ } \mathsf{D}(A) \xrightarrow{\ \mathsf{e}\ } \mathsf{D}(C), \qquad \mathsf{q},\mathsf{p}\colon \mathsf{D}(A)\to\mathsf{D}(B), \quad \mathsf{l},\mathsf{r}\colon \mathsf{D}(C)\to\mathsf{D}(A).$$

The following statements are equivalent.

(i) *There is a k-algebra S, projective over k, and an idempotent element e of S such that the canonical ring epimorphism $S \longrightarrow S/SeS$ is homological and the associated recollement of $\mathsf{D}(S)$ by derived module categories is equivalent to $\mathcal{R}$.*

(ii) *There are compact tilting objects V in $\mathsf{D}(A)$, U in $\mathsf{D}(B)$ and W in $\mathsf{D}(C)$ such that the associated tilting t-structures in $\mathsf{D}(B)$ and $\mathsf{D}(C)$ glue along $\mathcal{R}$ to the associated tilting t-structure in $\mathsf{D}(A)$ and such that the k-algebra $\mathsf{End}_{\mathsf{D}(A)}(V)$ is projective over k.*

PROOF. (i)$\Longrightarrow$(ii): Since the ring epimorphism $S \longrightarrow S/SeS$ is homological, equivalently the restriction functor $\mathsf{res}\colon \mathsf{Mod}\text{-}S/SeS \longrightarrow \mathsf{Mod}\text{-}S$ is a homological embedding, it follows from Theorem 8.3 that the recollement of module categories $\mathsf{R_{ab}}(\mathsf{Mod}\text{-}S/SeS, \mathsf{Mod}\text{-}S, \mathsf{Mod}\text{-}eSe)$ can be lifted to a recollement of unbounded derived categories. By assumption, there is an equivalence of recollements as follows:

(11.1)

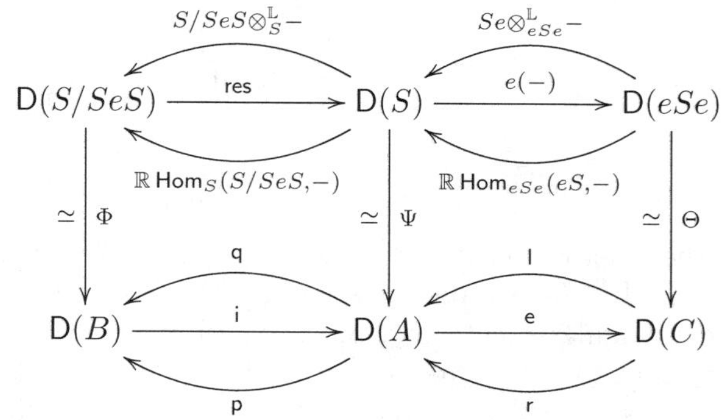

where Φ, Ψ and Θ are restrictable equivalences, i.e. the latter triangle functors induce by restriction equivalences between the bounded derived categories.

We claim that the standard t-structures (Example 4.1) in $\mathsf{D}(S/SeS)$ and $\mathsf{D}(eSe)$ glue to the standard t-structure in $\mathsf{D}(S)$ along the top recollement in diagram (11.1). The functors $\mathsf{res}\colon \mathsf{D}(S/SeS) \longrightarrow \mathsf{D}(S)$ and $e(-)\colon \mathsf{D}(S) \longrightarrow \mathsf{D}(eSe)$ are the derived functors of the underlying exact functors $\mathsf{res}\colon \mathsf{Mod}\text{-}S/SeS \longrightarrow \mathsf{Mod}\text{-}S$ and $e(-)\colon \mathsf{Mod}\text{-}S \longrightarrow \mathsf{Mod}\text{-}eSe$, respectively. This implies that the derived functors res and $e(-)$ are t-exact (see Section 4) with respect to the standard t-structures. Using this, one can show that the standard t-structure $(\mathsf{D}^{\leq 0}_S, \mathsf{D}^{\geq 0}_S)$ in $\mathsf{D}(S)$ coincides with the glued one:

$$\big(\{X \in \mathsf{D}(S) \mid e(X) \in \mathsf{D}^{\leq 0}_{eSe} \text{ and } S/SeS \otimes^{\mathbb{L}}_S (X) \in \mathsf{D}^{\leq 0}_{S/SeS}\},$$

$$\{X \in \mathsf{D}(S) \mid e(X) \in \mathsf{D}^{\geq 0}_{eSe} \text{ and } \mathbb{R}\,\mathsf{Hom}_S(S/SeS, X) \in \mathsf{D}^{\geq 0}_{S/SeS}\}\big).$$

We also have the following description for the standard t-structure $(\mathsf{D}^{\leq 0}_S, \mathsf{D}^{\geq 0}_S)$:

$$\mathsf{D}^{\leq 0}_S = S^{\perp_{>0}} := \big\{X \in \mathsf{D}(S) \mid \mathsf{Hom}_{\mathsf{D}(S)}(S, X[i]) = 0 \ \forall i > 0\big\}$$

$$\mathsf{D}^{\geq 0}_S = S^{\perp_{<0}} := \big\{X \in \mathsf{D}(S) \mid \mathsf{Hom}_{\mathsf{D}(S)}(S, X[i]) = 0 \ \forall i < 0\big\}$$

We leave it to the reader to check the details.

From diagram 11.1 we have the compact tilting objects $U := \Phi(S/SeS)$, $V := \Psi(S)$ and $W := \Theta(eSe)$ in $\mathsf{D}(B)$, $\mathsf{D}(A)$ and $\mathsf{D}(C)$, respectively. Moreover, the standard t-structure $(\mathsf{D}^{\leq 0}_S, \mathsf{D}^{\geq 0}_S)$ is mapped to

$$\big(\mathsf{D}^{\leq 0}_S, \mathsf{D}^{\geq 0}_S\big) \overset{\Psi}{\longmapsto} \big(\Psi(S)^{\perp_{>0}}, \Psi(S)^{\perp_{<0}}\big)$$

and similarly for the other two standard t-structures. These are the associated tilting t-structures mentioned in (ii). We can now finish the proof of (i)$\Longrightarrow$(ii). Since the standard t-structures in the top recollement in (11.1) are sent to the tilting t-structures generated by V, U and W in the bottom recollement, and since the standard t-structures in $\mathsf{D}(S/SeS)$ and $\mathsf{D}(eSe)$ glue to the standard t-structure in $\mathsf{D}(S)$, it follows from the commutativity of the diagram (11.1) that the glueing of the t-structures generated by U and W is the t-structure generated by V.

(ii)$\Longrightarrow$(i): Consider first, as above, the tilting t-structures of the compact tilting objects V in $\mathsf{D}(A)$, U in $\mathsf{D}(B)$ and W in $\mathsf{D}(C)$. Denote by $\mathcal{H}_V$, $\mathcal{H}_U$ and $\mathcal{H}_W$ the associated hearts. By assumption the associated tilting t-structures in $\mathsf{D}(B)$ and $\mathsf{D}(C)$ glue along $\mathcal{R}$ to the associated tilting t-structure in $\mathsf{D}(A)$. Then from Proposition 4.2 there is the following recollement of hearts:

$$\xymatrix{ \mathcal{H}_U \ar[rr]^{\mathsf{H}^0_V\mathsf{i}} && \mathcal{H}_V \ar@/_2pc/[ll]_{\mathsf{H}^0_U\mathsf{q}} \ar@/^2pc/[ll]^{\mathsf{H}^0_U\mathsf{p}} \ar[rr]^{\mathsf{H}^0_W\mathsf{e}} && \mathcal{H}_W \ar@/_2pc/[ll]_{\mathsf{H}^0_V\mathsf{l}} \ar@/^2pc/[ll]^{\mathsf{H}^0_V\mathsf{r}} } \tag{11.2}$$

Note that each of the functors in the above recollement is in fact a triple composition (see Proposition 4.2), but for simplicity we omit the embedding of each heart in the corresponding triangulated category. Since the tilting objects U, V and W are compact, it follows that the above hearts are module categories. In particular, the abelian category $\mathcal{H}_V$ is equivalent to the module category $\mathsf{Mod}\text{-}S$, where S is the endomorphism ring of $\mathsf{H}^0_V(V)$ over $\mathsf{D}(A)$. Note that $\mathsf{H}^0_V(V)$ is a small projective generator of $\mathcal{H}_V$ (see also [**111**, Proposition 4.3]). Similarly descriptions hold for the hearts $\mathcal{H}_U$ and $\mathcal{H}_W$. Then from Theorem 5.7 the recollement of hearts (11.2) is a recollement of module categories induced by an idempotent element.

The next step is to show that the functor $\mathsf{H}^0_V\mathsf{i}$ is a homological embedding. Then we apply Theorem 8.3 to lift the recollement of module categories (11.2) to the derived level. Finally, it remains to show that the latter recollement of derived module categories is equivalent to $\mathcal{R}$. We leave it to the reader to work out the details of this part, in particular, to adapt to our setting the proof of [**111**, Theorem 6.13]. □

We remark that a different criterion for a recollement of derived categories of rings to be stratifying has been independently obtained in [**6**].

11.2. Hereditary Algebras. In this subsection we answer Question 11.1 in the affirmative for hereditary algebras. Before we discuss the proof of this result we recall a series of results that are used in the proof. We start with the following result of Krause-Stovicek.

THEOREM 11.3. ([**77**, Theorem 8.1]) *Let R be a hereditary ring. Then there are bijections between the following sets:*

(i) *Homological ring epimorphisms $R \longrightarrow S$.*
(ii) *Localizing subcategories of $\mathsf{D}(R)$ that are closed under products.*

The above bijection involves universal localizations and Ext-orthogonal pairs. We refer to [**77**] for a thorough discussion.

We continue with the next observation on recollements of triangulated categories $\mathsf{R}_{\mathsf{tr}}(\mathcal{U}, \mathcal{T}, \mathcal{V})$ such that $\mathcal{T}$ admits a Serre functor. Let k be a field. Recall that a k-linear triangulated category $\mathcal{T}$ is Hom-finite if the k-vector space $\mathsf{Hom}_{\mathcal{T}}(X, Y)$ is finite-dimensional for all X and Y in $\mathcal{T}$. Then a **Serre functor** in $\mathcal{T}$ is an autoequivalence $\mathsf{S}_{\mathcal{T}} \colon \mathcal{T} \xrightarrow{\simeq} \mathcal{T}$ with a bifunctorial isomorphism

$$\mathsf{D}\,\mathsf{Hom}_{\mathcal{T}}(X, Y) \cong \mathsf{Hom}_{\mathcal{T}}(Y, \mathsf{S}_{\mathcal{T}}(X))$$

for all X and Y in $\mathcal{T}$. Note that $\mathsf{D} = \mathsf{Hom}_k(-,k)$ is the duality with respect to the base field k. We refer to [**19**, **71**] for more details on Serre functors. In this direction, we have the next result due to Jørgensen.

THEOREM 11.4. ([**64**]) *Let $\mathsf{R}_{\mathsf{tr}}(\mathcal{U},\mathcal{T},\mathcal{V})$ be a recollement of triangulated categories such that $\mathcal{T}$ admits a Serre functor. Then there is an infinite ladder (i.e. any three consecutive rows form a recollement of triangulated categories) as follows:*

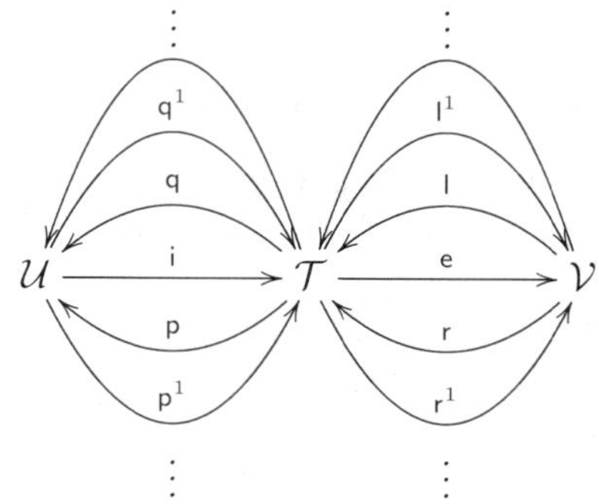

$\mathsf{L}_{\mathsf{tr}}(\mathcal{U},\mathcal{T},\mathcal{V})$

PROOF. Let $\mathsf{S}_{\mathcal{T}}$ be the Serre functor of $\mathcal{T}$ and denote by $\mathsf{S}_{\mathcal{T}}^{-1}$ its quasi-inverse. Then using the recollement situation $\mathsf{R}_{\mathsf{tr}}(\mathcal{U},\mathcal{T},\mathcal{V})$ it follows easily that $\mathsf{S}_{\mathcal{T}}$ induces Serre functors $\mathsf{S}_{\mathcal{U}}$ and $\mathsf{S}_{\mathcal{V}}$ in $\mathcal{U}$ and $\mathcal{V}$, respectively (see [**64**, Proposition 5]). Define the following functors:

$$\mathsf{q}^1 = \mathsf{S}_{\mathcal{T}}^{-1} \circ \mathsf{i} \circ \mathsf{S}_{\mathcal{U}} \quad \text{and} \quad \mathsf{l}^1 = \mathsf{S}_{\mathcal{V}}^{-1} \circ \mathsf{e} \circ \mathsf{S}_{\mathcal{T}}$$

$$\mathsf{p}^1 = \mathsf{S}_{\mathcal{T}} \circ \mathsf{i} \circ \mathsf{S}_{\mathcal{U}}^{-1} \quad \text{and} \quad \mathsf{r}^1 = \mathsf{S}_{\mathcal{V}} \circ \mathsf{e} \circ \mathsf{S}_{\mathcal{T}}^{-1}$$

Then it follows that the following diagrams:

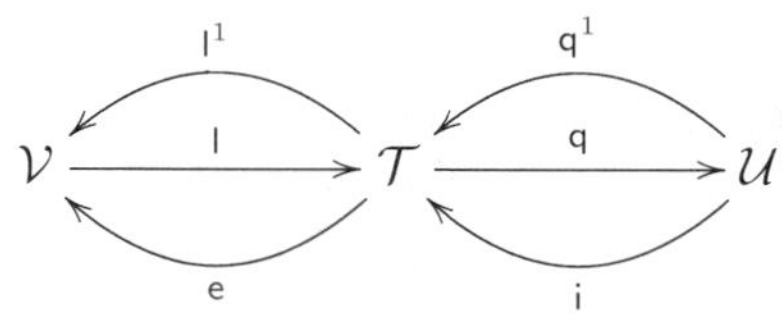

and

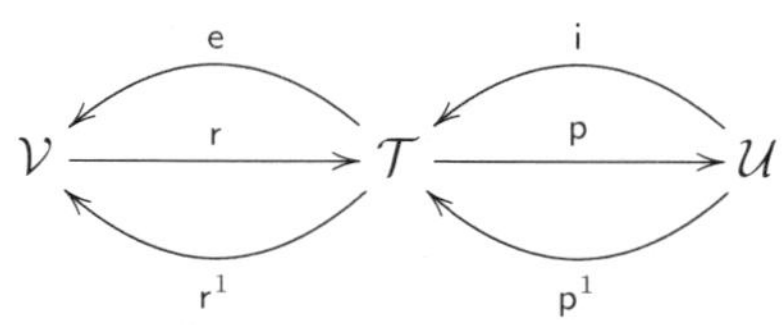

are recollements of triangulated categories. Finally, we iterate the above procedure and in this way we obtain the infinite ladder $\mathsf{L}_{\mathsf{tr}}(\mathcal{U},\mathcal{T},\mathcal{V})$. □

We also need the following results of Angeleri Hügel, Koenig, Liu and Yang.

PROPOSITION 11.5. ([**5**, Lemma 2.10, Proposition 2.14, Corollaries 4.7 and 4.10]) *Consider a recollement of derived categories of rings $\mathsf{R}_{\mathsf{tr}}(\mathsf{D}(B),\mathsf{D}(A),\mathsf{D}(C))$.*

(i) *If A is a finite dimensional k-algebra over a field k, then B and C are also finite dimensional k-algebras.*

(ii) *Assume that A, B and C are finite dimensional k-algebras. Then we have:*

$$\mathsf{gl.dim}\, A < \infty \iff \mathsf{gl.dim}\, B < \infty \ \textit{and}\ \mathsf{gl.dim}\, C < \infty.$$

(iii) *Assume that A is a finite dimensional k-algebra. If the algebra C has finite global dimension, then the recollement $\mathsf{R_{tr}}(\mathsf{D}(B), \mathsf{D}(A), \mathsf{D}(C))$ restricts to $\mathsf{R_{tr}}(\mathsf{D^b}(B), \mathsf{D^b}(A), \mathsf{D^b}(C))$ and to $\mathsf{R_{tr}}(\mathsf{D^b}(\mathsf{mod}\text{-}B), \mathsf{D^b}(\mathsf{mod}\text{-}A), \mathsf{D^b}(\mathsf{mod}\text{-}C))$.*

The following important result is due to Liu, Vitória and Yang. For the explicit description of the object K_X appearing below we refer to [**83**].

THEOREM 11.6. ([**83**, Theorem 4.5]) *Consider a recollement of bounded derived module categories, where A, B and C are finite dimensional k-algebras of finite global dimension:*

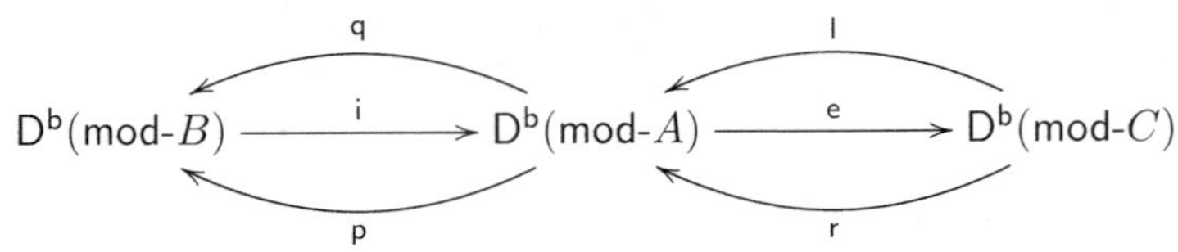

Let X and Y be tilting objects of $\mathsf{D^b}(\mathsf{mod}\text{-}B)$ and $\mathsf{D^b}(\mathsf{mod}\text{-}C)$, respectively. Then there is an object K_X in $\mathsf{D^b}(\mathsf{mod}\text{-}A)$ such that $\mathsf{i}(X) \oplus K_X$ is tilting if and only if the following conditions hold:

(i) $\mathsf{Hom}_{\mathsf{D^b}(\mathsf{mod}\text{-}B)}(X, \mathsf{qr}(Y)[n]) = 0$ *for all* $n < -1$.
(ii) $\mathsf{Hom}_{\mathsf{D^b}(\mathsf{mod}\text{-}B)}(\mathsf{qr}(Y), X[n]) = 0$ *for all* $n < 0$.
(iii) $\mathsf{Hom}_{\mathsf{D^b}(\mathsf{mod}\text{-}B)}(\mathsf{qr}(Y), \mathsf{qr}(Y)[n]) = 0$ *for all* $n < -1$.

We are now ready to show the main result of this subsection. The proof of the following theorem uses all the above mentioned results together with Theorem 11.2. We provide a sketch of the proof.

THEOREM 11.7. ([**111**, Theorem 6.17]) *Let A be a finite dimensional hereditary k-algebra over a field k. Then any recollement of $\mathsf{D}(A)$ (resp. of $\mathsf{D^b}(\mathsf{mod}\text{-}A)$) by (resp. bounded) derived module categories is equivalent to a stratifying one.*

PROOF. Let $\mathcal{R}$ be a recollement of $\mathsf{D}(A)$ of the form

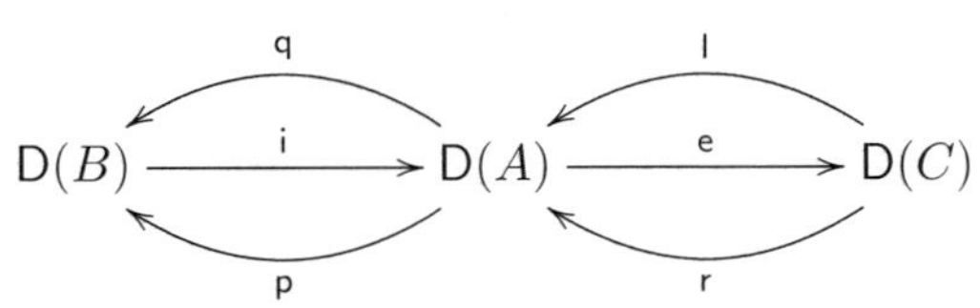

We split the proof into four main steps.
Step 1: From the assignments of Theorem 11.3 for hereditary rings, it follows that $\mathcal{R}$ is equivalent to a recollement induced by a homological ring epimorphism. Note that this equivalence changes only the functors between $\mathsf{D}(B)$ and $\mathsf{D}(A)$, leaving the functors between $\mathsf{D}(A)$ and $\mathsf{D}(C)$ unchanged. Thus, we can assume that $\mathsf{i} = f_*$ for some homological ring epimorphism $f\colon A \longrightarrow B$.
Step 2: Since A is a finite dimensional algebra of finite global dimension, it is well known that the derived Nakayama functor is a Serre functor. Then from Theorem 11.4, see also [**5**, Proposition 3.7], the recollement $\mathcal{R}$ fits in an infinite ladder.

In particular, there is a recollement as follows:

$$\mathcal{R}_u\colon \quad \xymatrix{ \mathsf{D}(C) \ar[r]^{\mathsf{l}} & \mathsf{D}(A) \ar@/_1.5pc/[l]_{\mathsf{l}^1} \ar@/^1.5pc/[l]^{\mathsf{e}} \ar[r]^{\mathsf{q}} & \mathsf{D}(B) \ar@/_1.5pc/[l]_{\mathsf{q}^1} \ar@/^1.5pc/[l]^{f_*} }$$

Applying once again Theorem 11.3, there is a recollement of $\mathsf{D}(A)$ equivalent to $\mathcal{R}_u$, which is induced by a homological ring epimorphism (and the functors between $\mathsf{D}(A)$ and $\mathsf{D}(B)$ remain unchanged). Thus, without loss of generality, once again we assume that $\mathsf{l} = g_*$ for some homological ring epimorphism $g\colon A \longrightarrow C$.
Step 3: Denote by $\mathsf{T}_{\mathsf{D}(A)}$ the glueing of the standard t-structures in $\mathsf{D}(C)$ and $\mathsf{D}(B)$ along $\mathcal{R}$ (see Section 4). Our aim is to check that $\mathsf{T}_{\mathsf{D}(A)}$ is a tilting t-structure in $\mathsf{D}(A)$. The standard t-structures in $\mathsf{D}(C)$ and $\mathsf{D}(B)$ admit left adjacent co-t-structures (see [**20**] for the definition). These co-t-structures, when glued along $\mathcal{R}_u$, yield a left adjacent co-t-structure to $\mathsf{T}_{\mathsf{D}(A)}$ (see [**83**, Remark 2.6]). From Proposition 11.5 the recollement $\mathcal{R}$ restricts to a recollement of $\mathsf{D}^{\mathsf{b}}(A)$. Since the standard t-structures are bounded, then $\mathsf{T}_{\mathsf{D}(A)}$ restricts to a bounded t-structure in $\mathsf{D}^{\mathsf{b}}(A)$ and therefore [**111**, Lemma 4.14 (iii)] implies that $\mathsf{T}_{\mathsf{D}(A)}$ is an intermediate t-structure. From [**7**, Theorem 4.6], we know that any intermediate t-structure that admits a left adjacent co-t-structure is a (bounded) silting t-structure. This implies that there is a bounded silting object T such that $\mathsf{T}_{\mathsf{D}(A)} = \mathsf{T}_T$.
Step 4: From Proposition 11.5 the recollements $\mathcal{R}$ and $\mathcal{R}_u$ restrict to the level of bounded derived categories of finitely generated modules. Then from [**83**] the object T is the glued silting object of B and C along $\mathcal{R}_u$. More precisely, from Theorem 11.6 the object T is tilting if and only if the following conditions hold:

(i) $\mathsf{Hom}_{\mathsf{D}(C)}(C, \mathsf{l}^1 f_* B[n]) = 0$, for all $n < -1$;
(ii) $\mathsf{Hom}_{\mathsf{D}(C)}(\mathsf{l}^1 f_* B, C[n]) = 0$, for all $n < 0$;
(iii) $\mathsf{Hom}_{\mathsf{D}(C)}(\mathsf{l}^1 f_* B, \mathsf{l}^1 f_* B[n]) = 0$, for all $n < -1$.

Using that $\mathsf{i} = f_*$, $\mathsf{l} = g_*$ and that $\mathsf{gl.dim}\, A \leq 1$, it is not so difficult to verify the above conditions (show that $\mathsf{l}^1 f_*(B)$ is a 2-term complex with cohomologies concentrated in degrees 0 and -1). Then the result follows from Theorem 11.2. □

We leave it to the reader to analyse further the proof of Theorem 11.7. For instance, for the unexplained notions used in Step 3 above, we refer to the survey paper on silting theory [**1**] and to the survey paper on co-t-structures [**65**]. We also leave to the reader to check the bounded version of the above result. Clearly, the proof of Theorem 11.7 uses a series of main results from other papers. It would be interesting to find a (shorter) proof avoiding all these theorems. Finally, it is also natural to consider if the class of piecewise hereditary algebras provides another source of stratifying recollements. To our knowledge this problem is open.

11.3. A Counterexample by Angeleri Hügel-Koenig-Liu-Yang. In this subsection we briefly present the example of Angeleri Hügel, Koenig, Liu and Yang [**6**, Example 2.3] which shows that Question 11.1 is not true in general.

Let Q be the quiver with relations ρ given by

Q:

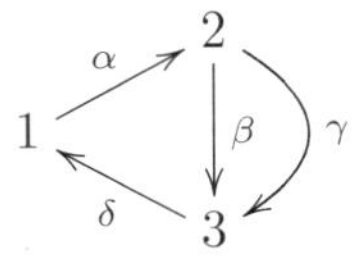

and $\rho = \{\beta\alpha, \alpha\delta, \delta\gamma\}$. Let $\Lambda = kQ/\langle\rho\rangle$ for some field k, and let $a = e_2 + e_3$. Then the recollement of module categories $\mathsf{R_{ab}}(\text{mod-}\Lambda/\Lambda a\Lambda, \text{mod-}\Lambda, \text{mod-}a\Lambda a)$ induces using Theorem 8.3 the recollement of derived categories $\mathsf{R_{tr}}(\mathsf{D}(\Lambda/\Lambda a\Lambda), \mathsf{D}(\Lambda), \mathsf{D}(a\Lambda a))$. Note that $\Lambda/\Lambda a\Lambda \cong k$ and $a\Lambda a$ is isomorphic to the Kronecker algebra. From Proposition 11.5 the algebra Λ has finite global dimension and then by Theorem 11.4 we obtain the recollement $\mathsf{R_{tr}}(\mathsf{D}(a\Lambda a), \mathsf{D}(\Lambda), \mathsf{D}(\Lambda/\Lambda a\Lambda))$ (consider the one going downwards). Using this recollement we have that $\mathsf{l}(k) = S_1$, where l is the left adjoint of the quotient functor. Now from Liu, Vitória and Yang [**83**] the module S_1 is not a direct summand of a tilting complex (S_1 is quasi-isomorphic to the complex $0 \longrightarrow P_2 \longrightarrow P_3 \longrightarrow P_1$). The latter statement contradicts the characterisation given in [**6**, Theorem A]. Hence, the recollement $\mathsf{R_{tr}}(\mathsf{D}(\Lambda/\Lambda a\Lambda), \mathsf{D}(\Lambda), \mathsf{D}(a\Lambda a))$ is not equivalent to a stratifying one. We refer to [**6**, Example 2.3] for more details. It would be interesting to reprove the statement of this example using Theorem 11.2.

12. Applications: Part I

Our aim in this section is to present a common context where we can compare the (Fg) condition for Hochschild cohomology, Gorensteinness and the singularity categories for an algebra Λ and $a\Lambda a$, where a is an idempotent element of Λ. The natural framework to approach the above problem is via recollements of abelian categories. The key idea is to identify assumptions on the idempotent element a of Λ such that we can detect the above properties from the simpler (sub)algebra $a\Lambda a$. The material of this section is taken from [**113**], which is joint work with Øystein Skartsæterhagen and Øyvind Solberg, and consists the first part of our applications.

We remark that this part is mainly motivated from investigations on the finite generation (Fg) condition for Hochschild cohomology. The latter condition is pivotal in the theory of support varieties for finite dimensional algebras as developed by Snashall and Solberg [**117**], see also [**118**]. This condition is necessary in order to have a meaningful theory of support varieties and by this we mean a theory which has similar properties with the theory of support varieties for group algebras [**23**]. In subsection 12.2 we provide the necessary background and references on this topic.

12.1. Gorenstein Algebras.

Let us first recall the notion of Gorenstein abelian categories due to Beligiannis and Reiten [**17**]. Given an abelian category $\mathscr{A}$ with enough projective and injective objects, we consider the following invariants:

$$\mathsf{spli}\,\mathscr{A} = \sup\{\mathsf{pd}_{\mathscr{A}} I \mid I \in \mathsf{Inj}\,\mathscr{A}\} \quad \text{and} \quad \mathsf{silp}\,\mathscr{A} = \sup\{\mathsf{id}_{\mathscr{A}} P \mid P \in \mathsf{Proj}\,\mathscr{A}\}$$

We have the following notion of Gorensteinness for abelian categories.

DEFINITION 12.1. ([**17**]) An abelian category $\mathscr{A}$ with enough projective and injective objects is called **Gorenstein** if $\mathsf{spli}\,\mathscr{A} < \infty$ and $\mathsf{silp}\,\mathscr{A} < \infty$.

Let Λ be an Artin algebra. Recall from [**12, 57**] that Λ is called **Gorenstein** if $\mathsf{id}\,_\Lambda\Lambda < \infty$ and $\mathsf{id}\,\Lambda_\Lambda < \infty$. Equivalently, Λ is Gorenstein if and only if $\mathsf{spli}\,\Lambda = \sup\{\mathsf{pd}\,_\Lambda I \mid I \in \mathsf{inj}\,\Lambda\} < \infty$ and $\mathsf{silp}\,\Lambda = \sup\{\mathsf{id}\,_\Lambda P \mid P \in \mathsf{proj}\,\Lambda\} < \infty$, i.e. $\mathsf{mod}\text{-}\Lambda$ is a Gorenstein abelian category in the sense of Definition 12.1. We refer to [**17**, Chapter VII] for more details on Gorenstein abelian categories.

The main objective of this subsection can be formulated as follows.

PROBLEM 12.2. *When does an additive functor $F\colon \mathscr{A} \longrightarrow \mathscr{B}$ between abelian categories preserve Gorensteinness.*

The key idea here is whether the functor F is an eventually homological isomorphism, see Definition 7.9. In this connection we have the following result.

THEOREM 12.3. ([**113**, Theorem 4.3]) *Let $F\colon \mathscr{A} \longrightarrow \mathscr{B}$ be a functor between abelian categories with enough projectives and injectives. If F is an essentially surjective eventually homological isomorphism, then:*

$$\mathscr{A} \text{ is Gorenstein} \iff \mathscr{B} \text{ is Gorenstein}.$$

PROOF. We provide a very brief sketch of the proof. Assume that F is a t-eventually homological isomorphism. In order to show the desired result, we need to compare the spli and silp dimensions of $\mathscr{A}$ and $\mathscr{B}$. In particular, it can be shown that the following inequalities hold:

$$\mathsf{spli}\,\mathscr{A} \leq \sup\{\mathsf{spli}\,\mathscr{B}, t\}, \quad \mathsf{silp}\,\mathscr{A} \leq \sup\{\mathsf{silp}\,\mathscr{B}, t\}$$

and

$$\mathsf{spli}\,\mathscr{B} \leq \sup\{\mathsf{spli}\,\mathscr{A}, t\}, \quad \mathsf{silp}\,\mathscr{B} \leq \sup\{\mathsf{silp}\,\mathscr{A}, t\}.$$

Then clearly the result follows. To prove the above homological relations, it suffices to show the following:

$$\mathsf{pd}_{\mathscr{A}}\, A \leq \sup\{\mathsf{pd}_{\mathscr{B}}\, F(A), t\}, \quad \mathsf{id}_{\mathscr{A}} A \leq \sup\{\mathsf{id}_{\mathscr{B}} F(A), t\}$$

and

$$\mathsf{pd}_{\mathscr{B}}\, F(A) \leq \sup\{\mathsf{pd}_{\mathscr{A}}\, A, t\}, \quad \mathsf{id}_{\mathscr{B}} F(A) \leq \sup\{\mathsf{id}_{\mathscr{A}} A, t\}.$$

Let us only explain how to show $\mathsf{spli}\,\mathscr{B} \leq \sup\{\mathsf{spli}\,\mathscr{A}, t\}$. Let I be an injective object of $\mathscr{B}$. Since F is essentially surjective, there is an object A in $\mathscr{A}$ such that $F(A) \cong I$. From the relation $\mathsf{id}_{\mathscr{A}} A \leq \sup\{\mathsf{id}_{\mathscr{B}} F(A), t\}$, the object A has finite injective dimension and then it follows that $\mathsf{pd}_{\mathscr{A}}\, A \leq \mathsf{spli}\,\mathscr{A}$. This implies that $\mathsf{pd}_{\mathscr{B}}\, I \leq \sup\{\mathsf{pd}_{\mathscr{A}}\, A, t\} \leq \sup\{\mathsf{spli}\,\mathscr{A}, t\}$ and then the result follows. □

Since the quotient functor $\mathsf{e}\colon \mathscr{B} \longrightarrow \mathscr{C}$ in a recollement situation $\mathsf{R_{ab}}(\mathscr{A}, \mathscr{B}, \mathscr{C})$ is essentially surjective (Remark 2.4 (v)), we can apply the above theorem. We specialize it for the case of Artin algebras, see Example 3.1.

COROLLARY 12.4. *Let Λ be an Artin algebra and e an idempotent element of Λ. Assume that the functor $\mathsf{e} = e(-)\colon \mathsf{mod}\text{-}\Lambda \longrightarrow \mathsf{mod}\text{-}e\Lambda e$ is an eventually homological isomorphism. Then:*

$$\Lambda \text{ is Gorenstein} \iff e\Lambda e \text{ is Gorenstein}.$$

The reader is refereed to [**113**, Section 4] for more details on the proofs of the results of this section and for examples. Moreover, in loc. cit. under certain conditions the authors compare Gorensteinness between $\mathscr{B}$ and $\mathscr{A}$, i.e. the left side of the recollement $\mathsf{R_{ab}}(\mathscr{A}, \mathscr{B}, \mathscr{C})$. Finally, it is natural to consider if the categories

$\mathscr{A}$ and $\mathscr{C}$ being Gorenstein could imply that $\mathscr{B}$ is also Gorenstein. This is not in general true as the example [**113**, Example 4.5] shows.

12.2. The (Fg) Condition. We start this subsection by briefly discussing the terminology we need regarding Hochschild cohomology and the (Fg) condition. For a more detailed exposition of these topics we refer to the survey paper [**118**].

Let Λ be an Artin algebra over a commutative ring k. The **Hochschild cohomology ring** $\mathsf{HH}^*(\Lambda)$ of Λ is defined by

$$\mathsf{HH}^*(\Lambda) = \mathsf{Ext}^*_{\Lambda^e}(\Lambda, \Lambda) = \bigoplus_{i=0}^{\infty} \mathsf{Ext}^i_{\Lambda^e}(\Lambda, \Lambda).$$

The Hochschild cohomology ring $\mathsf{HH}^*(\Lambda)$ is a graded k-algebra with multiplication given by Yoneda product. Note that Hochschild cohomology was originally defined by Hochschild in [**61**] using the bar resolution. If Λ is projective over k, then the above definition coincides with the original one due to Hochschild, see [**24**, IX, §6]. Gerstenhaber showed in [**51**] that the Hochschild cohomology ring as originally defined is graded commutative. This implies that the Hochschild cohomology ring as defined above is graded commutative when Λ is projective over k (for example when Λ is a finite dimensional algebra over a field). It should be noted that Snashall and Solberg [**117**, Theorem 1.1] showed that $\mathsf{HH}^*(\Lambda)$ is graded commutative when Λ is an algebra over a commutative ring k and Λ is flat as a module over k.

Our aim now is to describe the finite generation condition (Fg). For this we first need to define a $\mathsf{HH}^*(\Lambda)$-module structure on the direct sum of all extension groups of a Λ-module with itself. Assume that Λ is flat as k-module, and let M be a Λ-module. The direct sum

$$\mathsf{Ext}^*_{\Lambda}(M, M) = \bigoplus_{i=0}^{\infty} \mathsf{Ext}^i_{\Lambda}(M, M)$$

of all extension groups of M with itself is a graded k-algebra with multiplication given by Yoneda product. Using the graded ring homomorphism

$$\varphi_M \colon \mathsf{HH}^*(\Lambda) \longrightarrow \mathsf{Ext}^*_{\Lambda}(M, M)$$

we give to $\mathsf{Ext}^*_{\Lambda}(M, M)$ a graded $\mathsf{HH}^*(\Lambda)$-module structure. The latter map is defined in the following way. Any homogeneous element of positive degree in $\mathsf{HH}^*(\Lambda)$ can be represented by an exact sequence

$$\eta \colon 0 \longrightarrow \Lambda \longrightarrow X \longrightarrow P_n \longrightarrow \cdots \longrightarrow P_0 \longrightarrow 0$$

of Λ^e-modules, where every P_i is projective. Applying the functor $- \otimes_{\Lambda} M$ gives an exact sequence

$$0 \longrightarrow \Lambda \otimes_{\Lambda} M \longrightarrow X \otimes_{\Lambda} M \longrightarrow P_n \otimes_{\Lambda} M \longrightarrow \cdots \longrightarrow P_0 \otimes_{\Lambda} M \longrightarrow \Lambda \otimes_{\Lambda} M \longrightarrow 0$$

of Λ-modules. Note that the exactness of this sequence follows from the facts that Λ is flat as k-module and that the modules P_i are projective Λ^e-modules. Using the isomorphism $\Lambda \otimes_{\Lambda} M \cong M$, we get an exact sequence of Λ-modules starting and ending in M. We define $\varphi_M([\eta])$ to be the element of $\mathsf{Ext}^*_{\Lambda}(M, M)$ represented by this sequence. For more details on this module structure see [**117**]. We can now formulate the finite generation condition (Fg).

DEFINITION 12.5. Let Λ be an algebra over a commutative ring k such that Λ is flat as a module over k. We say that Λ satisfies the (Fg) **condition** if the following conditions hold:

(i) The ring $\mathsf{HH}^*(\Lambda)$ is noetherian.
(ii) The $\mathsf{HH}^*(\Lambda)$-module $\mathsf{Ext}^*_\Lambda(\Lambda/\mathsf{rad}\Lambda, \Lambda/\mathsf{rad}\Lambda)$ is finitely generated.

As we mentioned already, the above condition is sufficient for the theory of support varieties to have good properties. We remark that the above formulation is given by Solberg who proves in [**118**, Proposition 5.7] that it is equivalent to the initial one given by Erdmann, Holloway, Snashall, Solberg and Taillefer in [**34**].

We mention the following interesting connection between the (Fg) condition and Gorensteinness.

THEOREM 12.6. ([**34**, Theorem 1.5 (a)]) *If an Artin algebra Λ satisfies the* (Fg) *condition, then Λ is Gorenstein.*

Let Λ be an Artin k-algebra and a an idempotent[2] element of Λ. The main issue of this subsection, which was the main problem in [**113**], is as follows.

PROBLEM 12.7. *Identify conditions on a such that we can detect the* (Fg) *condition for Λ from the algebra $a\Lambda a$.*

We start by analysing the main key steps towards the solution of the above problem. First, from Example 3.1 we have the following recollement

$$\xymatrix{ \mathsf{mod}\text{-}\Lambda/\Lambda a\Lambda \ar[rr]^{\mathsf{inc}} && \mathsf{mod}\text{-}\Lambda \ar@/_2pc/[ll]_{\Lambda/\Lambda a\Lambda\otimes_\Lambda -} \ar@/^2pc/[ll]^{\mathsf{Hom}_\Lambda(\Lambda/\Lambda a\Lambda,-)} \ar[rr]^{\mathsf{e}=a(-)} && \mathsf{mod}\text{-}a\Lambda a \ar@/_2pc/[ll]_{\Lambda a\otimes_{a\Lambda a} -} \ar@/^2pc/[ll]^{\mathsf{Hom}_{a\Lambda}(a\Lambda,-)} }$$

Let $\Lambda^{\mathrm{e}} = \Lambda \otimes_k \Lambda^{\mathrm{op}}$ be the enveloping algebra of Λ and consider the idempotent element $\varepsilon = a \otimes a^{\mathrm{op}}$ of Λ^{e}. Then as above we have the following recollement

$$\xymatrix{ \mathsf{mod}\text{-}\Lambda^{\mathrm{e}}/\Lambda^{\mathrm{e}}\varepsilon\Lambda^{\mathrm{e}} \ar[rr]^{\mathsf{inc}} && \mathsf{mod}\text{-}\Lambda^{\mathrm{e}} \ar@/_2pc/[ll]_{\Lambda^{\mathrm{e}}/\Lambda^{\mathrm{e}}\varepsilon\Lambda^{\mathrm{e}}\otimes_{\Lambda^{\mathrm{e}}} -} \ar@/^2pc/[ll]^{\mathsf{Hom}_{\Lambda^{\mathrm{e}}}(\Lambda^{\mathrm{e}}/\Lambda^{\mathrm{e}}\varepsilon\Lambda^{\mathrm{e}},-)} \ar[rr]^{\mathsf{E}=\varepsilon(-)} && \mathsf{mod}\text{-}(a\Lambda a)^{\mathrm{e}} \ar@/_2pc/[ll]_{\Lambda^{\mathrm{e}}\varepsilon\otimes_{(a\Lambda a)^{\mathrm{e}}} -} \ar@/^2pc/[ll]^{\mathsf{Hom}_{(a\Lambda a)^{\mathrm{e}}}(\varepsilon\Lambda^{\mathrm{e}},-)} }$$

Note that $(a\Lambda a)^{\mathrm{e}} \cong \varepsilon\Lambda^{\mathrm{e}}\varepsilon$ as k-algebras and that the functor $\mathsf{E} = (a - a)$, i.e. E is given by multiplication from left and right with a.

Let M be a Λ-module M. Then we have the following diagram

$$\xymatrix{ \mathsf{HH}^*(\Lambda) \ar[rr]^{\varphi_M} \ar[d]_{\mathsf{E}^*_{\Lambda,\Lambda}} && \mathsf{Ext}^*_\Lambda(M,M) \ar[d]^{\mathsf{e}^*_{M,M}} \\ \mathsf{HH}^*(a\Lambda a) \ar[rr]_{\varphi_{\mathsf{e}(M)}} && \mathsf{Ext}^*_{a\Lambda a}(\mathsf{e}(M),\mathsf{e}(M)) }$$

where the horizontal maps φ_M and $\varphi_{\mathsf{e}(M)}$ were defined in the first part of this subsection. The vertical maps $\mathsf{E}^*_{\Lambda,\Lambda}$ and $\mathsf{e}^*_{M,M}$ are graded ring homomorphisms

[2]We denote an idempotent element by a since in this part the letter e is used for the enveloping algebra.

and thinking in terms on Yoneda extensions, it is clear how to define these maps since the functors e and E are exact. In [**113**, Proposition 7.2] it is shown that the above diagram of graded rings is commutative. Our aim is to compare the (Fg) condition for Λ and $a\Lambda a$. Given such a diagram, the key observation for the desired conclusion is that it suffices for the vertical maps to be group isomorphisms for almost all degrees. Here is where the idea of eventually homological isomorphism comes to rescue. The following result provides a solution to Problem 12.7.

THEOREM 12.8. ([**113**, Theorem 7.10]) *Let Λ be a finite dimensional algebra over a field k and let a be an idempotent in Λ. Assume that $(\Lambda/\mathsf{rad}\Lambda) \otimes_k (\Lambda^{\mathrm{op}}/\mathsf{rad}\Lambda^{\mathrm{op}})$ is a semisimple Λ^{e}-module, and that the functor $\mathsf{e}\colon \mathsf{mod}\text{-}\Lambda \longrightarrow \mathsf{mod}\text{-}a\Lambda a$ is an eventually homological isomorphism. Then the algebra Λ satisfies the (Fg) condition if and only if the algebra $a\Lambda a$ satisfies the (Fg) condition.*

Note that the assumption $(\Lambda/\mathsf{rad}\Lambda) \otimes_k (\Lambda^{\mathrm{op}}/\mathsf{rad}\Lambda^{\mathrm{op}})$ being a semisimple Λ^{e}-module is necessary in order to be able to describe the simple Λ^{e}-modules in terms of simple Λ-modules (this is the case when the field k is algebraically closed). For more details on the proof of Theorem 12.8 we refer to [**113**, Sections 6 and 7].

12.3. Cohen-Macaulay Modules and Singularity Categories. Before we start the discussion of the main issues of this subsection, we give some background on singularity categories, and recall some important results.

Singularity categories have been introduced by Buchweitz [**22**] under the name stable derived categories. Later, Orlov introduced the singularity category of an algebraic variety $\mathbb{X}$ as the Verdier quotient

$$\mathsf{D_{sg}}(\mathbb{X}) = \mathsf{D^b}(\mathsf{coh}\mathbb{X})/\mathsf{perf}(\mathbb{X})$$

where $\mathsf{D^b}(\mathsf{coh}\mathbb{X})$ is the bounded derived category of coherent sheaves on $\mathbb{X}$ and $\mathsf{perf}(\mathbb{X})$ is the full subcategory consisting of perfect complexes on $\mathbb{X}$. The singularity category $\mathsf{D_{sg}}(\mathbb{X})$, as one would except, captures many geometric properties of $\mathbb{X}$. If the variety $\mathbb{X}$ is smooth, then the singularity category $\mathsf{D_{sg}}(\mathbb{X})$ is trivial. We remark that $\mathsf{D_{sg}}(\mathbb{X})$ is not only related to the study of the singularities of $\mathbb{X}$ but is also related to the Homological Mirror Symmetry Conjecture due to Kontsevich [**73**]. For more information we refer to [**101**–**103**].

Let R be a Noetherian ring. Then by Buchweitz [**22**] the singularity category of R is defined as the Verdier quotient

$$\mathsf{D_{sg}}(R) = \mathsf{D^b}(\mathsf{mod}\text{-}R)/\mathsf{perf}(R)$$

where $\mathsf{D^b}(\mathsf{mod}\text{-}R)$ is the bounded derived category of finitely generated R-modules and $\mathsf{perf}(R)$ denotes the full subcategory of perfect complexes (i.e. $\mathsf{K^b}(\mathsf{proj}\,R)$). We mention that in this case the singularity category $\mathsf{D_{sg}}(R)$ can be viewed as a categorical measure of the singularities of the spectrum $\mathsf{Spec}(R)$. From a fundamental result of Buchweitz [**22**, Theorem 4.4.1], and independently by Happel [**57**, Theorem 4.6], the singularity category $\mathsf{D_{sg}}(R)$ of a Gorenstein ring R (or a Gorenstein Artin algebra) is equivalent to the stable category of (maximal) Cohen–Macaulay modules $\underline{\mathsf{CM}}(R)$, where the latter is well known to be a triangulated category [**56**]. Recall that the category $\mathsf{CM}(R)$ is defined as follows:

$$\mathsf{CM}(R) = \big\{X \in \mathsf{mod}\text{-}R \mid \mathsf{Ext}^n_R(X,R) = 0 \text{ for all } n \geq 1\big\}$$

Note that the latter equivalence generalizes Rickard's equivalence [**114**] between the singularity category of a selfinjective algebra and the stable module category.

Our aim in this subsection is to study the singularity categories in a recollement $\mathsf{R_{ab}}(\mathscr{A}, \mathscr{B}, \mathscr{C})$. More precisely, we investigate the following problem.

PROBLEM 12.9. *Given a recollement* $\mathsf{R_{ab}}(\mathscr{A}, \mathscr{B}, \mathscr{C})$ *find necessary and sufficient conditions such that the singularity categories of* $\mathscr{B}$ *and* $\mathscr{C}$ *are triangle equivalent.*

First, we can define in the same way as above the singularity category of an abelian category. Let $\mathscr{A}$ be an abelian category with enough projectives. Then the singularity category of $\mathscr{A}$ is the Verdier quotient $\mathsf{D_{sg}}(\mathscr{A}) := \mathsf{D^b}(\mathscr{A})/\mathsf{K^b}(\mathsf{Proj}\,\mathscr{A})$, where $\mathsf{D^b}(\mathscr{A})$ is the bounded derived category of $\mathscr{A}$ and $\mathsf{K^b}(\mathsf{Proj}\,\mathscr{A})$ is the homotopy category of bounded complexes of projective objects of $\mathscr{A}$. We refer to [**113**] and references therein for more details on this quotient. It should be noted that this (big) singularity category was already considered by Beligiannis [**16**] under the notion of the stabilization category. We refer the reader to loc. cit. for more details on this approach.

The following main result of this subsection provides a solution to Problem 12.9.

THEOREM 12.10. ([**113**, Theorem 5.2]) *Let* $\mathsf{R_{ab}}(\mathscr{A}, \mathscr{B}, \mathscr{C})$ *be a recollement of abelian categories. Then the following statements are equivalent:*

(i) *For every* A *in* $\mathscr{A}$ *and* P *in* $\mathsf{Proj}\,\mathscr{B}$ *we have*

$$\mathsf{pd}_{\mathscr{B}}\,\mathsf{i}(A) < \infty \quad \textit{and} \quad \mathsf{pd}_{\mathscr{C}}\,\mathsf{e}(P) < \infty.$$

(ii) *The functor* $\mathsf{e}\colon \mathscr{B} \longrightarrow \mathscr{C}$ *induces a triangle equivalence between the singularity categories of* $\mathscr{B}$ *and* $\mathscr{C}$:

$$\mathsf{D_{sg}}(\mathsf{e})\colon \mathsf{D_{sg}}(\mathscr{B}) \xrightarrow{\simeq} \mathsf{D_{sg}}(\mathscr{C})$$

Let us compare statement (i) with the assumptions in Theorem 7.12. The above statement only states that each object of the form $\mathsf{i}(A)$ and $\mathsf{e}(P)$ has finite projective dimension. However, this doesn't mean that there exists a finite bound for the projective dimensions of all such objects, i.e. that $\mathsf{pgl.dim}_{\mathscr{A}}\mathscr{B} < \infty$ and $\sup\{\mathsf{pd}_{\mathscr{C}}\,e(P) \mid P \in \mathsf{Proj}\,\mathscr{B}\} < \infty$ (conditions (β) and (γ) in Theorem 7.12). Of course the converse implication is true and this is what we use in the last application of this subsection on Cohen-Macaulay modules.

As an immediate consequence of Theorem 12.10 we have the next result. Note that it holds for (left) Noetherian rings, but for completeness we continue to work in the context of Artin algebras.

COROLLARY 12.11. *Let* Λ *be an Artin algebra and* a *an idempotent element of* Λ. *Then the following statements are equivalent:*

(i) *For every* $\Lambda/\Lambda a\Lambda$*-module* X *we have* $\mathsf{pd}_{\Lambda}\,X < \infty$, *and* $\mathsf{pd}_{a\Lambda a}\,a\Lambda < \infty$.

(ii) *The functor* $\mathsf{e} = a(-)\colon \mathsf{mod}\text{-}\Lambda \longrightarrow \mathsf{mod}\text{-}a\Lambda a$ *induces a triangle equivalence between the singularity categories of* Λ *and* $a\Lambda a$:

$$\mathsf{D_{sg}}(\mathsf{e}(-))\colon \mathsf{D_{sg}}(\Lambda) \xrightarrow{\simeq} \mathsf{D_{sg}}(a\Lambda a)$$

For examples illustrating the conditions of the above result we refer to [**113**]. We now discuss quotients of path algebras.

Given a finite dimensional k-algebra $\Lambda = kQ/\langle\rho\rangle$ (i.e. a quotient of a path algebra where k is a field, Q is a quiver and ρ a minimal set of relations in kQ generating an admissible ideal $\langle\rho\rangle$), we present below a way to construct an idempotent element a such that the singularity categories $\mathsf{D_{sg}}(\Lambda)$ and $\mathsf{D_{sg}}(a\Lambda a)$ are triangle equivalent.

COROLLARY 12.12. *Let* $\Lambda = kQ/\langle\rho\rangle$ *be a quotient of a path algebra as above. Choose some vertices in* Q *where no relations start, and let* a *be the sum of all vertices except these. Then the functor* $\mathsf{e} = a(-)\colon \mathsf{mod}\text{-}\Lambda \longrightarrow \mathsf{mod}\text{-}a\Lambda a$ *induces a triangle equivalence between the singularity categories of* Λ *and* $a\Lambda a$:

$$\mathsf{D}_{\mathsf{sg}}(\mathsf{e}(-))\colon \mathsf{D}_{\mathsf{sg}}(\Lambda) \xrightarrow{\simeq} \mathsf{D}_{\mathsf{sg}}(a\Lambda a)$$

PROOF. It is well known that if S is a simple Λ-module corresponding to a vertex v in the quiver Q, then we have $\mathsf{pd}_\Lambda S \leq 1$ if and only if no relation starts in the vertex v. Consider now the recollement of module categories $(\mathsf{mod}\text{-}\Lambda/\Lambda a\Lambda, \mathsf{mod}\text{-}\Lambda, \mathsf{mod}\text{-}a\Lambda a)$. The fact that the idempotent a is the sum of all vertices except the ones where no relations start implies that

$$\mathsf{pd}_\Lambda\Big(\frac{\Lambda/\Lambda a\Lambda}{\mathsf{rad}\,\Lambda/\Lambda a\Lambda}\Big) \leq 1.$$

Then the above condition implies the conditions stated in Corollary 12.11 (i), and therefore we get the desired triangle equivalence between the singularity categories. For more details we refer to [**113**, subsection 8.2]. □

We close this subsection with the next consequence on Cohen-Macaulay modules. The result follows from Corollary 7.14, Corollary 12.4, Corollary 12.11 and the equivalence of Buchweitz [**22**] between the singularity category and the stable category of Cohen-Macaulay modules.

COROLLARY 12.13. ([**113**, Corollary 5.6]) *Let* Λ *be a Gorenstein Artin algebra and* a *an idempotent element of* Λ. *Assume that the functor* $\mathsf{e} = a(-)\colon \mathsf{mod}\text{-}\Lambda \longrightarrow \mathsf{mod}\text{-}a\Lambda a$ *is an eventually homological isomorphism. Then there is a triangle equivalence between the stable categories of Cohen–Macaulay modules of* Λ *and* $a\Lambda a$:

$$\underline{\mathsf{CM}}(\Lambda) \xrightarrow{\simeq} \underline{\mathsf{CM}}(a\Lambda a)$$

13. Applications: Part II

Our purpose in this section is to present three more applications of recollements of abelian categories. In particular, we turn our attention to the behaviour of various homological dimensions of the categories involved in a recollement $\mathsf{R_{ab}}(\mathscr{A},\mathscr{B},\mathscr{C})$. More precisely, we investigate the finitistic dimension, the representation dimension and the dominant dimension. The importance of these homological dimensions is well known in Representation Theory of Finite Dimensional Algebras. The motivation for our investigations in this section is similar to the one given in Section 12. We want to identify conditions in $\mathsf{R_{ab}}(\mathscr{A},\mathscr{B},\mathscr{C})$ such that the finiteness of a homological dimension for $\mathscr{C}$ implies that this homological dimension is also finite for $\mathscr{B}$. In this case the left side of $\mathsf{R_{ab}}(\mathscr{A},\mathscr{B},\mathscr{C})$ is used as well. Moreover, it is always natural to consider cases where the finiteness of a homological dimension for $\mathscr{A}$ and $\mathscr{C}$ implies the same for $\mathscr{B}$. However, this is not a rare phenomenon.

13.1. Finitistic Dimenion. Let Λ be an Artin algebra. The **finitistic dimension** of Λ is defined to be the supremum of the projective dimensions of all finitely generated Λ-modules having finite projective dimension:

$$\mathsf{fin.dim}\,\Lambda = \sup\{\mathsf{pd}_\Lambda X \mid \mathsf{pd}_\Lambda X < \infty \ \text{ and } \ X \in \mathsf{mod}\text{-}\Lambda\}$$

The famous Finitistic Dimension Conjecture due to Bass asserts that the finitistic dimension of an Artin algebra is finite, i.e. $\mathsf{fin.dim}\,\Lambda < \infty$. It is also known that there is an interesting relation between the Finitistic Dimension Conjecture and several other homological conjectures in Represention Theory, like the Gorenstein Symmetry Conjecture, the Wakamatsu-Tilting conjecture and the (Generalized) Nakayama Conjecture. In the next subsection we address an interesting interplay with Auslander's representation dimension. For more details on the finitistic dimension and its relation with several homological conjectures we refer to [**58**, **126**].

We now return to the context of recollemement situations. In this subsection, we assume that the categories $\mathscr{B}$ and $\mathscr{C}$ in a recollement $\mathsf{R_{ab}}(\mathscr{A},\mathscr{B},\mathscr{C})$ have enough projectives (see also Remark 2.7). Recall that the **finitistic projective dimension** $\mathsf{FPD}(\mathscr{B})$ of $\mathscr{B}$ is defined by $\mathsf{FPD}\,(\mathscr{B}) := \sup\{\mathsf{pd}_{\mathscr{B}} B \mid \mathsf{pd}_{\mathscr{B}} B < \infty\}$. In what follows we present a series of results where we relate the finitistic dimension of the categories involved in a recollement $\mathsf{R_{ab}}(\mathscr{A},\mathscr{B},\mathscr{C})$.

For our first result we need the following notion introduced in [**108**]. Consider a right exact functor $\mathsf{F}\colon \mathscr{A} \longrightarrow \mathscr{B}$ between abelian categories, where we assume that $\mathscr{A}$ has enough projectives. We say that F has **locally bounded homological dimension**, if there exists $n \geq 0$ such that whenever the left derived functor $\mathsf{L}_m\mathsf{F}(A) = 0$ for $m \gg 0$ then $\mathsf{L}_m\mathsf{F}(A) = 0$ for every $m \geq n+1$. The minimum such n (if it exists) is called the locally bounded homological dimension of F and is denoted by $\mathsf{l.b.hom.dim}\,\mathsf{F}$. For instance F has locally bounded homological dimension if it has finite homological dimension, i.e. $\mathsf{L}_k\mathsf{F} = 0$ for some $k \geq 1$. In this connection we have the following result, which provides a relation between the finitistic projective dimension of $\mathscr{B}$ and $\mathscr{C}$.

THEOREM 13.1. ([**108**, Theorem 5.1]) *Let* $\mathsf{R_{ab}}(\mathscr{A},\mathscr{B},\mathscr{C})$ *be a recollement. If the functor* $\mathsf{l}\colon \mathscr{C} \longrightarrow \mathscr{B}$ *has locally bounded homological dimension, then:*

$$\mathsf{FPD}\,\mathscr{C} \;\leq\; \mathsf{FPD}\,\mathscr{B} + \mathsf{l.b.hom.dim}\,\mathsf{l}$$

We continue with the next result which provides an upper bound for the finitistic dimension of $\mathscr{B}$ in terms of the finitistic dimension of $\mathscr{C}$ and the $\mathscr{A}$-relative projective global dimension of $\mathscr{B}$ (see Section 7). We suggest to the reader to check again the conditions in Theorem 7.12 in connection with the next result.

THEOREM 13.2. ([**108**, Theorem 5.5]) *Let* $\mathsf{R_{ab}}(\mathscr{A},\mathscr{B},\mathscr{C})$ *be a recollement of abelian categories. If* $\sup\{\mathsf{pd}_{\mathscr{C}}\,\mathsf{e}(P) \mid P \in \mathsf{Proj}\,\mathscr{B}\} < \infty$*, then:*

$$\mathsf{FPD}\,\mathscr{B} \;\leq\; \mathsf{FPD}\,\mathscr{C} + \mathsf{pgl.dim}_{\mathscr{A}}\mathscr{B} + 1$$

When the $\mathscr{A}$-relative projective global dimension of $\mathscr{B}$ is less or equal to one, we can show the following lower and upper bound for the finitistic dimension of $\mathscr{B}$. The next two results are joint work with Edward Green and Øyvind Solberg.

THEOREM 13.3. ([**54**]) *Let* $\mathsf{R_{ab}}(\mathscr{A},\mathscr{B},\mathscr{C})$ *be a recollement of abelian categories. Assume that* $\mathsf{pgl.dim}_{\mathscr{A}}\mathscr{B} \leq 1$*. Then the following hold:*

$$\max\{\mathsf{FPD}\,\mathscr{A}, \mathsf{FPD}\,\mathscr{C}\} \leq \mathsf{FPD}\,\mathscr{B} \leq \mathsf{FPD}\,\mathscr{C} + 2.$$

In particular, we have: $\mathsf{FPD}\,\mathscr{B} < \infty$ *if and only if* $\mathsf{FPD}\,\mathscr{C} < \infty$*.*

Using now the finiteness of the $\mathscr{A}$-relative injective global dimension of $\mathscr{B}$, we have the following upper bound for the finitistic dimension of $\mathscr{B}$.

THEOREM 13.4. ([**54**]) *Let* $\mathsf{R}_{\mathsf{ab}}(\mathscr{A},\mathscr{B},\mathscr{C})$ *be a recollement of abelian categories. Assume that* $\mathscr{B}$ *has projective covers and that* $\operatorname{igl.dim}_{\mathscr{A}}\mathscr{B}<\infty$*. Then we have*

$$\mathsf{FPD}\,\mathscr{B}\leq\mathsf{FPD}\,\mathscr{C}+\operatorname{igl.dim}_{\mathscr{A}}\mathscr{B}$$

In a forthcoming paper with Edward Green and Øyvind Solberg [**54**] the above results are used intensively to provide a reduction technique for the finitistic dimension of Artin algebras. For instance, if Λ is an Artin algebra and e an idempotent element of Λ, then Theorem 13.4 gives the following nice relation:

$$\operatorname{fin.dim}\Lambda\leq\operatorname{fin.dim}e\Lambda e+\sup\{\operatorname{id}_{\Lambda}S\mid S\text{ simple }\Lambda/\Lambda e\Lambda\text{-module}\}$$

This result suggests for quotients of path algebras that we can remove the vertices where the corresponding simples have finite injective dimension and then we hope to iterate this process using one of the above results. The goal is to reduce the problem of computing the finitistic dimension to a possible simpler ("homologically compact") algebra. It is worth to mention that the $\mathscr{A}$-relative injective global dimension plays a key role for the finitistic dimension. We remark that this technique gives a variety of new examples where the finitistic dimension conjecture holds. For more details on these results we refer to the paper [**54**]. We finish this section with an example illustrating Theorem 13.3 and Theorem 13.4. The *slogan* is: remove simples of projective dimension at most one or simples of finite injective dimension.

EXAMPLE 13.5. ([**123**, Example 2]) We consider the algebra Λ defined by the following quiver and relations over a field k.

$\Lambda = k$

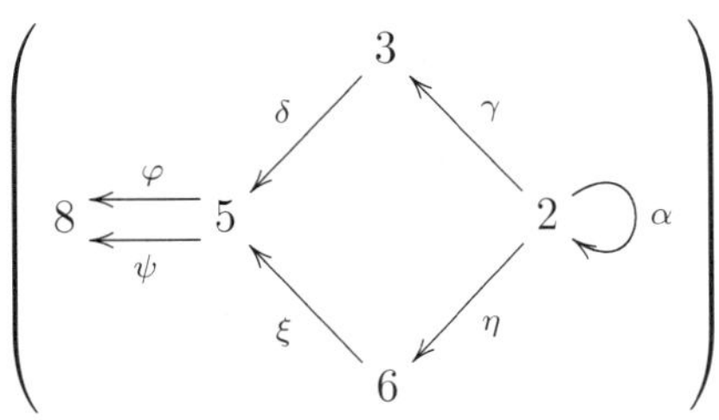

$/\langle\alpha^2,\alpha\eta,\alpha\gamma,\gamma\delta-\eta\xi\rangle$

The injective and the projective dimensions of the simple Λ-modules are given in the following table:

	S_2	S_3	S_6	S_5	S_8
pd	∞	1	1	1	0
id	∞	∞	∞	∞	1

First reduction: Since the projective dimension of the simple modules $\{S_3,S_6,S_5,S_8\}$ are at most 1, we can remove the vertices $\{3,6,5,8\}$ and obtain the local algebra $\Lambda_1=e_2\Lambda e_2$.
Second reduction: Using that the injective dimension of the simple module S_8 is 1, we can remove the vertex 8 and obtain the algebra

$$\Lambda_2=(e_2+e_3+e_6+e_5)\Lambda(e_2+e_3+e_6+e_5),$$

which is given by

$\Lambda_2 = k$

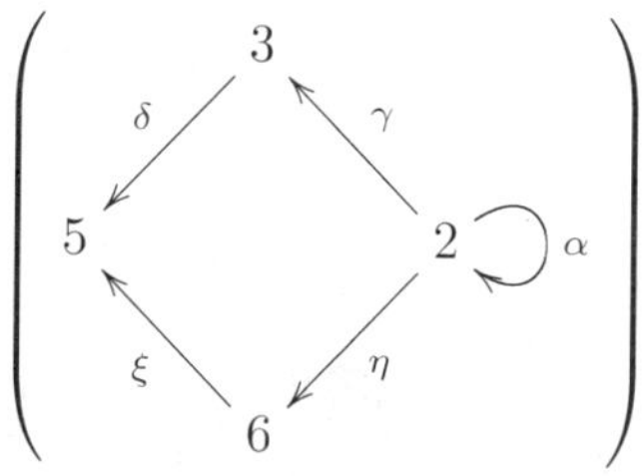

$/\langle \alpha^2, \alpha\eta, \alpha\gamma, \gamma\delta - \eta\xi \rangle$

All simple Λ_2-modules have infinite injective dimension. On the other hand, the simple Λ_2-modules $\{S_3, S_5, S_6\}$ have projective dimension at most 1. Then we can reduce to $e_2\Lambda_2 e_2$, which is a local algebra isomorphic to Λ_1.

Independent of which reduction we carry out, we see that the original algebra has indeed finite finitistic dimension. We remark that either paths of reduction lead to the same algebra up to isomorphism, but this is not a general phenomenon.

In particular, using Theorem 13.3 we obtain that $\mathsf{fin.dim}\,\Lambda \leq 2$. We finally remark that in this example we can also apply triangular reduction (consider the idempotent element $e = e_3 + e_5 + e_6 + e_8$) due to Fossum-Griffith-Reiten [**38**].

13.2. Representation Dimension. In 1971 Auslander introduced in his Queen Mary Notes [**9**] the notion of representation dimension of an Artin algebra. This homological dimension is the minimal possible global dimension of the endomorphism ring of a generator-cogenerator. Auslander's expectation was that this notion should provide a reasonable way of measuring how far an Artin algebra is from being of finite representation type. Let us give Auslander's definition. Let Λ be an Artin algebra. Then the **representation dimension** of Λ is defined as follows:

$$\mathsf{rep.dim}\,\Lambda = \min\{\mathsf{gl.dim}\,\mathsf{End}_\Lambda(\Lambda \oplus \mathsf{D}(\Lambda) \oplus X) \mid X \in \mathsf{mod}\text{-}\Lambda\}$$

Then, Auslander showed that Λ is representation-finite if and only if $\mathsf{rep.dim}\,\Lambda \leq 2$. A Λ-module $Y := \Lambda \oplus \mathsf{D}(\Lambda) \oplus X$ (a module which is a generator and a cogenerator) realizing the minimum above is called **Auslander generator**.

On the other hand, Igusa and Todorov showed the following interesting relation between representation dimension and finitistic dimension. In particular, they proved that the finitistic dimension conjecture holds for Artin algebras of representation dimension at most three.

THEOREM 13.6. ([**62**]) *Let Λ be an Artin algebra with $\mathsf{rep.dim}\,\Lambda \leq 3$. Then:*

$$\mathsf{fin.dim}\,\Lambda < \infty.$$

We now concentrate on the behaviour of representation dimension of the algebras involved in a recollement $\mathsf{R_{ab}}(\mathsf{mod}\text{-}\Lambda/\Lambda e\Lambda, \mathsf{mod}\text{-}\Lambda, \mathsf{mod}\text{-}e\Lambda e)$, see also Example 3.4. We remark that the following result is proved for recollements of abelian categories where in this context the appropriate notion of representation dimension is defined. We refer to [**108**, Section 7] for more details. Part (iii) in the next result follows immediately from statement (i) and Theorem 13.6, see also [**125**, Theorem 2.3].

THEOREM 13.7. ([**108**]) *Let Λ be an Artin algebra. The following hold.*

(i) *Let e be an idempotent element of Λ. If $\mathsf{rep.dim}\,\Lambda \leq 3$, then:*

$$\mathsf{rep.dim}\,e\Lambda e \leq 3$$

(ii) *For any finitely generated projective Λ-module P, we have:*

$$\mathsf{rep.dim}\,\Lambda \leq 3 \iff \mathsf{rep.dim}\,\mathsf{End}_\Lambda(P) \leq 3$$

(iii) *If* $\mathsf{rep.dim}\,\Lambda \leq 3$, *then:*

$$\mathsf{fin.dim}\, e\Lambda e < \infty$$

Among other results, Auslander proved in [**9**] that if Λ is of finite representation type, then so is $\mathsf{End}_\Lambda(P)$. Thus, Theorem 13.7 (ii) generalizes this result.

Let Λ be an Artin algebra. A pair (Γ, e), consisting of an Artin algebra Γ and an idempotent element e in Γ, is called an **Auslander pair** for Λ, if Γ has finite global dimension and $\Lambda \cong e\Gamma e$. By a basic result of Auslander [**9**], any Artin algebra admits an Auslander pair. Moreover, Dlab and Ringel [**32**] showed that Γ can be chosen to be quasi-hereditary. In fact, consider the endomorphism algebra $\Gamma = \mathsf{End}_\Lambda(\oplus_{i=1}^n \Lambda/\mathfrak{r}^i)$, where $\mathfrak{r}$ is the Jacobson radical of Λ and $n = \ell\ell(\Lambda)$ its Loewy length. In this case Auslander proved that $\mathsf{gl.dim}\,\Gamma \leq \ell\ell(\Lambda)$.

We close this section with the following result which provides an interesting interplay between representation dimension and finitistic dimension.

THEOREM 13.8. ([**108**, Theorem 7.13]) *Let Λ be an Artin algebra.*

(i) *Let* $\Gamma = \mathsf{End}_\Lambda(\Lambda \oplus \mathsf{D}\,\Lambda)$ *and assume that the Nakayama functor* $\nu = \mathsf{D}\Lambda \otimes_\Lambda -$ *has locally finite homological dimension. Then:*

$$\mathsf{fin.dim}\,\Lambda \leq \mathsf{fin.dim}\,\Gamma + \mathsf{l.b.hom.dim}\,\nu$$

(ii) *Let* $\Gamma = \mathsf{End}_\Lambda(\oplus_{i=1}^{\ell\ell(\Lambda)} \Lambda/\mathfrak{r}^i)$ *and let e be an idempotent element in Γ such that $\Lambda \cong e\Gamma e$. If the functor* $\Gamma e \otimes_{e\Gamma e} - \colon \mathsf{Mod}\text{-}\Lambda \longrightarrow \mathsf{Mod}\text{-}\Gamma$ *has locally bounded homological dimension, then* $\mathsf{fin.dim}\,\Lambda < \infty$. *More precisely:*

$$\mathsf{fin.dim}\,\Lambda \leq \ell\ell(\Lambda) + \mathsf{l.b.hom.dim}\ \Gamma e \otimes_{e\Gamma e} - \ < \infty$$

(iii) *Let (Γ, e) be an Auslander pair for Λ and assume that the functor* $\Gamma e \otimes_{e\Gamma e} - \colon \mathsf{Mod}\text{-}\Lambda \longrightarrow \mathsf{Mod}\text{-}\Gamma$ *has locally bounded homological dimension. Then* $\mathsf{fin.dim}\,\Lambda < \infty$. *In particular, we have:*

$$\mathsf{fin.dim}\,\Lambda \leq \mathsf{gl.dim}\,\Gamma + \mathsf{l.b.hom.dim}\ \Gamma e \otimes_{e\Gamma e} - \ < \infty$$

(iv) *Let $\Lambda \oplus X$ be an Auslander generator of Λ and assume that the functor* $\mathsf{Hom}_\Lambda(X, \Lambda) \otimes_\Lambda - \colon \mathsf{mod}\text{-}\Lambda \longrightarrow \mathsf{mod}\text{-}\Gamma$ *has locally bounded homological dimension. Then we have:*

$$\mathsf{fin.dim}\,\Lambda \leq \mathsf{rep.dim}\,\Lambda + \mathsf{l.b.hom.dim}\ \mathsf{Hom}_\Lambda(X, \Lambda) \otimes_\Lambda - \ < \infty$$

We remark that the above result presents situations where the finitistic dimension of an Artin algebra Λ is finite. In particular, we have $\mathsf{fin.dim}\,\Lambda < \infty$ provided that one of the following conditions hold:

(i) There exists a summand X of an Auslander generator such that the functor $\mathsf{Hom}_\Lambda(X, \Lambda) \otimes_\Lambda -$ has locally bounded homological dimension.
(ii) The natural quasi-hereditary algebra Γ and the idempotent e in Γ associated to Λ has the property that the functor $\Gamma e \otimes_{e\Gamma e} -$ has locally bounded homological dimension.

13.3. Dominant Dimension. In this subsection we address an interesting homological phenomenon in a recollement of module categories under the appearance of dominant dimension. More precisely, given an algebra Λ of dominant dimension at least two we show that there is a close relationship between certain subcategories of $\mathsf{mod}\text{-}\Lambda$ and $\mathsf{mod}\text{-}e\Lambda e$ and between their extension groups (similarly in some sense to our study in Section 7). The material of this section is taken from the paper of Fang and Koenig [**36**]. Our aim is to present some of their results by giving emphasis on the properties of the recollement situation that we deal with.

Throughout this section we fix a finite dimensional k-algebra Λ over an algebraic closed field k. We start by recalling the notion of dominant dimension.

DEFINITION 13.9. Let M be a Λ-module with a minimal injective resolution:

$$0 \longrightarrow M \longrightarrow I_0 \longrightarrow I_1 \longrightarrow \cdots \longrightarrow I_{t-1} \longrightarrow I_t \longrightarrow \cdots$$

The **dominant dimension** $\mathsf{dom.dim}\, M$ of M is the maximal number t (or ∞) such that I_i is projective for all $i < t$ (or ∞).

The dominant dimension $\mathsf{dom.dim}\,\Lambda$ of Λ is $\mathsf{dom.dim}\,\Lambda = \mathsf{dom.dim}\,{}_{\Lambda}\Lambda$. We set $\mathsf{dom.dim}\, M = 0$ if the injective module I_0 is not projective. In case of ${}_{\Lambda}\Lambda$, this means that $\mathsf{dom.dim}\,\Lambda = 0$ if and only if there is no faithful Λ-module which is both projective and injective.

The following famous conjecture justifies our interest to dominant dimension.
Nakayama's Conjecture: If Λ has infinite dominant dimension, then the algebra Λ is self-injective.

We will see below an interesting class of finite dimensional algebras where the above important conjecture holds. We now continue by recalling a characterization, due to Müller, of dominant dimension over Λ in terms of cohomology over $e\Lambda e$.

THEOREM 13.10. ([**90**]) *Let Λ be a finite dimensional k-algebra over a field k with $\mathsf{dom.dim}\,\Lambda \geq 2$. Assume that there is an idempotent element e of Λ such that the right $e\Lambda e$-module Λe is projective, injective and faithful. For a left Λ-module M the following statements are equivalent:*

(i) $\mathsf{dom.dim}\, M \geq n \geq 2$.
(ii) $\mathsf{Hom}_{e\Lambda e}(e\Lambda, eM) \cong M$ *and* $\mathsf{Ext}^i_{e\Lambda e}(e\Lambda, eM) = 0$ *for all* $1 \leq i \leq n-2$.

We remind the reader that the recollement $\mathsf{R_{ab}}(\mathsf{mod}\text{-}\Lambda/\Lambda e\Lambda, \mathsf{mod}\text{-}\Lambda, \mathsf{mod}\text{-}e\Lambda e)$ was discussed in Example 3.1. Moreover, by Remark 2.4 (iv) the counit of the adjoint pair $(\mathsf{e} = e(-), \mathsf{Hom}_{e\Lambda e}(e\Lambda, -))$ is an isomorphism. However, the condition (ii) above says precisely that the module M admits a special property:the unit map of the adjunction $(\mathsf{e} = e(-), \mathsf{Hom}_{e\Lambda e}(e\Lambda, -))$ in M is an isomorphism. We also remark the interesting interplay between left and right modules, i.e. the conditions for Λe in contrast to the characterization for $\mathsf{dom.dim}\,{}_{\Lambda}M$.

From now on we assume that the finite dimensional k-algebra Λ is quasi-hereditary (over a poset X) with a duality (i.e. an anti-automorphism preserving pointwise a full set of primitive orthogonal idempotents). We refer to [**36**] and the references therein for the precise definitions and more explanations. Assume also that $\mathsf{dom.dim}\,\Lambda \geq 2$ and fix an idempotent element e such that Λe is a basic projective, injective and faithful Λ-module. We mention that the above choice of e together with the rest assumptions on Λ provide us with a special recollement situation. The properties of this recollement will be analysed further in [**109**].

As usual, we denote by $\mathcal{F}(\Delta)$ the full subcategory of $\mathsf{mod}\text{-}\Lambda$ consisting of all modules filtered by the standard Λ-modules. Moreover, $\mathcal{F}(\mathsf{e}\Delta)$ denotes the full subcategory of $\mathsf{mod}\text{-}e\Lambda e$ consisting of all modules filtered by $\{\mathsf{e}\Delta(x) \mid x \in X\}$.

Given a full subcategory $\mathscr{C}$ of $\mathsf{mod}\text{-}\Lambda$, the Hemmer-Nakano dimension of $\mathscr{C}$ is the maximal number $n(\mathscr{C})$ such that the functor $\mathsf{e} = e(-)\colon \mathsf{mod}\text{-}\Lambda \longrightarrow \mathsf{mod}\text{-}e\Lambda e$ induces isomorphisms $\mathsf{Ext}^i_\Lambda(M,N) \cong \mathsf{Ext}^i_{e\Lambda e}(\mathsf{e}(M),\mathsf{e}(N))$ for all objects M,N in $\mathscr{C}$ and $0 \leq i \leq n(\mathscr{C})$.

THEOREM 13.11. ([**36**, Theorem 3.9]) *Let Λ be as above, T the characteristic tilting Λ-module and M a Λ-module. Then for any P in* $\mathsf{proj}\,\Lambda$ *and for any K in $\mathcal{F}(\Delta)$ we have the following isomorphisms:*

$$e^i_{M,P}\colon \mathsf{Ext}^i_\Lambda(M,P) \xrightarrow{\cong} \mathsf{Ext}^i_{e\Lambda e}(eM,eP) \ \ \textit{for all } 0 \leq i \leq \mathsf{dom.dim}\,\Lambda - 2$$

$$e^i_{M,K}\colon \mathsf{Ext}^i_\Lambda(M,K) \xrightarrow{\cong} \mathsf{Ext}^i_{e\Lambda e}(eM,eK) \ \ \textit{for all } 0 \leq i \leq \mathsf{dom.dim}\,T - 2$$

As a consequence of Theorem 13.11 we have the following result.

COROLLARY 13.12. *Let Λ be as above. Assume that* $\mathsf{dom.dim}\,\Lambda \geq 4$ *(or equivalently* $\mathsf{dom.dim}\,T \geq 2$*). Then there is an equivalence of categories:*

$$\mathsf{e}\colon \mathcal{F}(\Delta) \xrightarrow{\simeq} \mathcal{F}(\mathsf{e}\Delta)$$

14. Even More Examples: Recollements of Coherent Functors

In Section 4 we studied recollements of hearts. The latter situations arise from recollements of triangulated categories by considering the hearts of the glueing t-structures. In this section we present another method to obtain recollements of abelian categories from recollements of triangulated categories. This is done via the category of coherent functors, a framework which by the work of Auslander and Reiten is known to be very useful in Representation Theory.

We start by recalling some basics on coherent functors. Let $\mathscr{A}$ be an additive category. An additive functor $F\colon \mathscr{A}^{\mathsf{op}} \longrightarrow \mathfrak{Ab}$ is called **coherent**, if there exists an exact sequence of the form:

$$\mathsf{Hom}_{\mathscr{A}}(-,X) \longrightarrow \mathsf{Hom}_{\mathscr{A}}(-,Y) \longrightarrow F \longrightarrow 0$$

where the objects X and Y lie in $\mathscr{A}$. We denote by $\mathsf{mod}\text{-}\mathscr{A}$ the category of coherent functors over $\mathscr{A}$. Recall that a map $X \longrightarrow Y$ is a weak kernel of $Y \longrightarrow Z$ if the following sequence is exact:

$$\mathsf{Hom}_{\mathscr{A}}(-,X) \longrightarrow \mathsf{Hom}_{\mathscr{A}}(-,Y) \longrightarrow \mathsf{Hom}_{\mathscr{A}}(-,Z)$$

Then the category of coherent functors $\mathsf{mod}\text{-}\mathscr{A}$ is abelian if and only if $\mathscr{A}$ has weak kernels. Moreover, the category $\mathsf{mod}\text{-}\mathscr{A}$ has enough projectives and the Yoneda embedding $\mathsf{Y}_{\mathscr{A}}\colon \mathscr{A} \longrightarrow \mathsf{mod}\text{-}\mathscr{A}$, $A \mapsto \mathsf{Hom}_{\mathscr{A}}(-,A)$, induces an equivalence between $\mathscr{A}$ and $\mathsf{Proj}(\mathsf{mod}\text{-}\mathscr{A})$. Note that in this case $\mathscr{A}$ needs to have split idempotents. For more details on coherent functors we refer to the work of Auslander [**8**, **9**], see also [**39**]. As always we denote by $\mathcal{T}$ a triangulated category. Since $\mathcal{T}$ has weak kernels, the category of coherent functors $\mathsf{mod}\text{-}\mathcal{T}$ is abelian. The latter category is also known as the abelianization of $\mathcal{T}$, see for example [**75**, Appendix A].

Let $\mathsf{e}\colon \mathcal{T} \longrightarrow \mathcal{V}$ be a triangle functor between triangulated categories. Then by the universal property of the Yoneda embedding, there is a unique exact functor $\mathsf{e}_{\mathsf{coh}}\colon \mathsf{mod}\text{-}\mathcal{T} \longrightarrow \mathsf{mod}\text{-}\mathcal{V}$ such that the following diagram is commutative:

$$\begin{array}{ccc} \mathcal{T} & \xrightarrow{\mathsf{Y}_{\mathcal{T}}} & \mathsf{mod}\text{-}\mathcal{T} \\ {\scriptstyle \mathsf{e}}\downarrow & & \downarrow{\scriptstyle \mathsf{e}_{\mathsf{coh}}} \\ \mathcal{V} & \xrightarrow{\mathsf{Y}_{\mathcal{V}}} & \mathsf{mod}\text{-}\mathcal{V} \end{array}$$

Consider now a recollement $\mathsf{R}_{\mathsf{tr}}(\mathcal{U},\mathcal{T},\mathcal{V})$ of triangulated categories:

$$\mathcal{U} \begin{array}{c} \xleftarrow{\ \mathsf{q}\ } \\ \xrightarrow{\ \mathsf{i}\ } \\ \xleftarrow[\ \mathsf{p}\ ]{} \end{array} \mathcal{T} \begin{array}{c} \xleftarrow{\ \mathsf{l}\ } \\ \xrightarrow{\ \mathsf{e}\ } \\ \xleftarrow[\ \mathsf{r}\ ]{} \end{array} \mathcal{V}$$

Then it is not so difficult to show that there is an adjoint triple:

$$\mathsf{mod}\text{-}\mathcal{T} \begin{array}{c} \xleftarrow{\ \mathsf{l}_{\mathsf{coh}}\ } \\ \xrightarrow{\ \mathsf{e}_{\mathsf{coh}}\ } \\ \xleftarrow[\ \mathsf{r}_{\mathsf{coh}}\ ]{} \end{array} \mathsf{mod}\text{-}\mathcal{V}$$

Also, the functors $\mathsf{l}_{\mathsf{coh}}$ and $\mathsf{r}_{\mathsf{coh}}$ are fully faithful since the functors l and r are fully faithful, respectively. From Remark 2.5 it follows that the following diagram, denoted by $\mathsf{R}_{\mathsf{coh}}(\mathsf{Ker}\,\mathsf{e}_{\mathsf{coh}}, \mathsf{mod}\text{-}\mathcal{T}, \mathsf{mod}\text{-}\mathcal{V})$, is a recollement of abelian categories:

(14.1)
$$\mathsf{Ker}\,\mathsf{e}_{\mathsf{coh}} \begin{array}{c} \xleftarrow{\qquad} \\ \xrightarrow{\ \mathsf{inc}\ } \\ \xleftarrow{\qquad} \end{array} \mathsf{mod}\text{-}\mathcal{T} \begin{array}{c} \xleftarrow{\ \mathsf{l}_{\mathsf{coh}}\ } \\ \xrightarrow{\ \mathsf{e}_{\mathsf{coh}}\ } \\ \xleftarrow[\ \mathsf{r}_{\mathsf{coh}}\ ]{} \end{array} \mathsf{mod}\text{-}\mathcal{V}$$

On the other hand, it is natural to ask if the following diagram is a recollement:

(14.2)
$$\mathsf{mod}\text{-}\mathcal{U} \begin{array}{c} \xleftarrow{\ \mathsf{q}_{\mathsf{coh}}\ } \\ \xrightarrow{\ \mathsf{i}_{\mathsf{coh}}\ } \\ \xleftarrow[\ \mathsf{p}_{\mathsf{coh}}\ ]{} \end{array} \mathsf{mod}\text{-}\mathcal{T} \begin{array}{c} \xleftarrow{\ \mathsf{l}_{\mathsf{coh}}\ } \\ \xrightarrow{\ \mathsf{e}_{\mathsf{coh}}\ } \\ \xleftarrow[\ \mathsf{r}_{\mathsf{coh}}\ ]{} \end{array} \mathsf{mod}\text{-}\mathcal{V}$$

It should be noted that the above diagram satisfies all conditions of a recollement of abelian categories (see Definition 2.3) except the "crucial one", i.e. $\mathsf{i}_{\mathsf{coh}} = \mathsf{Ker}\,\mathsf{e}_{\mathsf{coh}}$. In other words, the sequence of abelian categories $0 \longrightarrow \mathsf{mod}\text{-}\mathcal{U} \longrightarrow \mathsf{mod}\text{-}\mathcal{T} \longrightarrow \mathsf{mod}\text{-}\mathcal{V} \longrightarrow 0$ is not, in general, exact — see Remark 2.4 (vii). Let us examine what it really means for the diagram (14.2) to be a recollement.

From Proposition 2.8 and for a functor F in $\mathsf{mod}\text{-}\mathcal{T}$ we have the exact sequence

$$\mathsf{l}_{\mathsf{coh}}\mathsf{e}_{\mathsf{coh}}(F) \longrightarrow F \longrightarrow \mathsf{i}_{\mathsf{coh}}\mathsf{q}_{\mathsf{coh}}(F) \longrightarrow 0\ .$$

For $F = \mathsf{Hom}_{\mathcal{T}}(-,T)$ we have the isomorphisms

$$\mathsf{l}_{\mathsf{coh}}\mathsf{e}_{\mathsf{coh}}(F) \cong \mathsf{Hom}_{\mathcal{T}}(-,\mathsf{le}(T)) \quad \text{and} \quad \mathsf{i}_{\mathsf{coh}}\mathsf{q}_{\mathsf{coh}}(F) \cong \mathsf{Hom}_{\mathcal{T}}(-,\mathsf{iq}(T)).$$

Then we obtain the following exact sequence

$$\text{(14.3)} \qquad \mathsf{Hom}_{\mathcal{T}}(-,\mathsf{le}(T)) \longrightarrow \mathsf{Hom}_{\mathcal{T}}(-,T) \longrightarrow \mathsf{Hom}_{\mathcal{T}}(-,\mathsf{iq}(T)) \longrightarrow 0\,.$$

Since $\mathsf{R}_{\mathsf{tr}}(\mathcal{U},\mathcal{T},\mathcal{V})$ is a recollement of triangulated categories, we have the canonical triangle $\mathsf{le}(T) \longrightarrow T \longrightarrow \mathsf{iq}(T) \longrightarrow \mathsf{le}(T)[1]$. Note that the maps in (14.3) are induced by the adjunction morphisms of the canonical triangle. Since the sequence (14.3) is exact for all X in $\mathcal{T}$, it follows by Yoneda's Lemma that the morphism $\mathsf{iq}(T) \longrightarrow \mathsf{le}(T)[1]$ is zero. This implies that the canonical triangle splits. Similarly, from the exact sequence

$$0 \longrightarrow \mathsf{i}_{\mathsf{coh}}\mathsf{p}_{\mathsf{coh}}(F) \longrightarrow F \longrightarrow \mathsf{r}_{\mathsf{coh}}\mathsf{e}_{\mathsf{coh}}(F)$$

we obtain that the canonical triangle $\mathsf{ip}(T) \longrightarrow T \longrightarrow \mathsf{re}(T) \longrightarrow \mathsf{ip}(T)[1]$ splits. Thus, assuming that the diagram (14.2) is a recollement we showed that the recollement $\mathsf{R}_{\mathsf{tr}}(\mathcal{U},\mathcal{T},\mathcal{V})$ splits. It is not difficult to show that the converse also holds.

To conclude, the abelianization of a recollement $\mathsf{R}_{\mathsf{tr}}(\mathcal{U},\mathcal{T},\mathcal{V})$ produces a recollement situation of abelian categories. In particular, the recollement situation (14.1) gives rise to interesting examples of homological recollements of abelian categories, i.e. recollements where the middle functor on the left side is a homological embedding. Moreover, the latter recollements of coherent functors are studied in connection with ladders of recollements of triangulated categories (see the definition in Theorem 11.4). For more details on the material of this section as well as for further results on recollements of coherent functors we refer to [**46**].

Acknowledgments

I would like to thank Kiyoshi Igusa, Alex Martsinkovsky, and Gordana Todorov, the organisers of the Maurice Auslander International Conference, for the invitation to give a talk and to write this survey paper. I deeply thank Alex Martsinkovsky for encouraging me to write this paper. The visit to Woods Hole, Massachusetts, USA, in 2015 was supported by the Norwegian Research Council (NFR 221893) under the project Triangulated categories in Algebra at the Norwegian University of Science and Technology (NTNU). I would like to thank Petter Andreas Bergh and Steffen Oppermann for this opportunity. Some of the results presented here were obtained in my Ph.D. thesis [**107**] under the supervision of Apostolos Beligiannis. I would like to thank him for having been my teacher and for being a continuous source of inspiration. I would also like to thank Steffen Koenig for numerous discussions on recollements. Furthermore, I am grateful to all my coauthors for the fruitful collaborations. I would also like to thank Julian Külshammer for helpful comments and valuable suggestions on previous versions of the paper. Finally, I would like to thank the referees for the useful comments, remarks and suggestions that improved the exposition of this paper.

References

[1] L. Angeleri Hügel, *On the abundance of silting modules*, in this volume.

[2] Lidia Angeleri Hügel, Steffen Koenig, and Qunhua Liu, *Recollements and tilting objects*, J. Pure Appl. Algebra **215** (2011), no. 4, 420–438, DOI 10.1016/j.jpaa.2010.04.027. MR2738361

[3] Lidia Angeleri Hügel, Steffen Koenig, and Qunhua Liu, *On the uniqueness of stratifications of derived module categories*, J. Algebra **359** (2012), 120–137, DOI 10.1016/j.jalgebra.2012.02.022. MR2914629

[4] Lidia Angeleri Hügel, Steffen Koenig, and Qunhua Liu, *Jordan-Hölder theorems for derived module categories of piecewise hereditary algebras*, J. Algebra **352** (2012), 361–381, DOI 10.1016/j.jalgebra.2011.09.041. MR2862193

[5] Lidia Angeleri Hügel, Steffen Koenig, Qunhua Liu, and Dong Yang, *Ladders and simplicity of derived module categories*, J. Algebra **472** (2017), 15–66, DOI 10.1016/j.jalgebra.2016.10.023. MR3584869

[6] Lidia Angeleri Hügel, Steffen Koenig, Qunhua Liu, and Dong Yang, *Recollements and stratifying ideals*, J. Algebra **484** (2017), 47–65, DOI 10.1016/j.jalgebra.2017.04.003. MR3656712

[7] Lidia Angeleri Hügel, Frederik Marks, and Jorge Vitória, *Silting modules*, Int. Math. Res. Not. IMRN **4** (2016), 1251–1284, DOI 10.1093/imrn/rnv191. MR3493448

[8] Maurice Auslander, *Coherent functors*, Proc. Conf. Categorical Algebra (La Jolla, Calif., 1965), Springer, New York, 1966, pp. 189–231. MR0212070

[9] M. Auslander, *Representation dimension of Artin algebras*, Queen Mary College Mathematics Notes, London (1971).

[10] Maurice Auslander, *Representation theory of Artin algebras. I, II*, Comm. Algebra **1** (1974), 177–268; ibid. 1 (1974), 269–310, DOI 10.1080/00927877408548230. MR0349747

[11] Maurice Auslander, *Representation theory of Artin algebras. I, II*, Comm. Algebra **1** (1974), 177–268; ibid. 1 (1974), 269–310, DOI 10.1080/00927877408548230. MR0349747

[12] Maurice Auslander and Idun Reiten, *Cohen-Macaulay and Gorenstein Artin algebras*, Representation theory of finite groups and finite-dimensional algebras (Bielefeld, 1991), Progr. Math., vol. 95, Birkhäuser, Basel, 1991, pp. 221–245. MR1112162

[13] M. Auslander, M. I. Platzeck, and G. Todorov, *Homological theory of idempotent ideals*, Trans. Amer. Math. Soc. **332** (1992), no. 2, 667–692, DOI 10.2307/2154190. MR1052903

[14] Maurice Auslander, Idun Reiten, and Sverre O. Smalø, *Representation theory of Artin algebras*, Cambridge Studies in Advanced Mathematics, vol. 36, Cambridge University Press, Cambridge, 1995. MR1314422

[15] A. A. Beĭlinson, J. Bernstein, and P. Deligne, *Faisceaux pervers* (French), Analysis and topology on singular spaces, I (Luminy, 1981), Astérisque, vol. 100, Soc. Math. France, Paris, 1982, pp. 5–171. MR751966

[16] Apostolos Beligiannis, *The homological theory of contravariantly finite subcategories: Auslander-Buchweitz contexts, Gorenstein categories and (co-)stabilization*, Comm. Algebra **28** (2000), no. 10, 4547–4596, DOI 10.1080/00927870008827105. MR1780017

[17] Apostolos Beligiannis and Idun Reiten, *Homological and homotopical aspects of torsion theories*, Mem. Amer. Math. Soc. **188** (2007), no. 883, viii+207, DOI 10.1090/memo/0883. MR2327478

[18] A. I. Bondal, *Representations of associative algebras and coherent sheaves* (Russian), Izv. Akad. Nauk SSSR Ser. Mat. **53** (1989), no. 1, 25–44, DOI 10.1070/IM1990v034n01ABEH000583; English transl., Math. USSR-Izv. **34** (1990), no. 1, 23–42. MR992977

[19] A. I. Bondal and M. M. Kapranov, *Representable functors, Serre functors, and reconstructions* (Russian), Izv. Akad. Nauk SSSR Ser. Mat. **53** (1989), no. 6, 1183–1205, 1337, DOI 10.1070/IM1990v035n03ABEH000716; English transl., Math. USSR-Izv. **35** (1990), no. 3, 519–541. MR1039961

[20] M. V. Bondarko, *Weight structures vs. t-structures; weight filtrations, spectral sequences, and complexes (for motives and in general)*, J. K-Theory **6** (2010), no. 3, 387–504, DOI 10.1017/is010012005jkt083. MR2746283

[21] Tom Bridgeland, *t-structures on some local Calabi-Yau varieties*, J. Algebra **289** (2005), no. 2, 453–483, DOI 10.1016/j.jalgebra.2005.03.016. MR2142382

[22] R.-O. Buchweitz, *Maximal Cohen-Macaulay modules and Tate-Cohomology over Gorenstein rings*, unpublished manuscript, (1987), 155 pp.

[23] Jon F. Carlson, *The complexity and varieties of modules*, Integral representations and applications (Oberwolfach, 1980), Lecture Notes in Math., vol. 882, Springer, Berlin-New York, 1981, pp. 415–422. MR646116

[24] Henri Cartan and Samuel Eilenberg, *Homological algebra*, Princeton University Press, Princeton, N. J., 1956. MR0077480

[25] E. Cline, B. Parshall, and L. Scott, *Finite-dimensional algebras and highest weight categories*, J. Reine Angew. Math. **391** (1988), 85–99. MR961165

[26] E. Cline, B. Parshall, and L. Scott, *Algebraic stratification in representation categories*, J. Algebra **117** (1988), no. 2, 504–521, DOI 10.1016/0021-8693(88)90123-8. MR957457

[27] Edward Cline, Brian Parshall, and Leonard Scott, *Stratifying endomorphism algebras*, Mem. Amer. Math. Soc. **124** (1996), no. 591, viii+119, DOI 10.1090/memo/0591. MR1350891

[28] Riccardo Colpi, *Tilting in Grothendieck categories*, Forum Math. **11** (1999), no. 6, 735–759, DOI 10.1515/form.1999.023. MR1725595

[29] William Crawley-Boevey and Julia Sauter, *On quiver Grassmannians and orbit closures for representation-finite algebras*, Math. Z. **285** (2017), no. 1-2, 367–395, DOI 10.1007/s00209-016-1712-z. MR3598816

[30] Spencer E. Dickson, *A torsion theory for Abelian categories*, Trans. Amer. Math. Soc. **121** (1966), 223–235, DOI 10.2307/1994341. MR0191935

[31] Alexandru Dimca, *Sheaves in topology*, Universitext, Springer-Verlag, Berlin, 2004. MR2050072

[32] Vlastimil Dlab and Claus Michael Ringel, *Every semiprimary ring is the endomorphism ring of a projective module over a quasihereditary ring*, Proc. Amer. Math. Soc. **107** (1989), no. 1, 1–5, DOI 10.2307/2048026. MR943793

[33] Vlastimil Dlab and Claus Michael Ringel, *Quasi-hereditary algebras*, Illinois J. Math. **33** (1989), no. 2, 280–291. MR987824

[34] Karin Erdmann, Miles Holloway, Rachel Taillefer, Nicole Snashall, and Øyvind Solberg, *Support varieties for selfinjective algebras*, K-Theory **33** (2004), no. 1, 67–87, DOI 10.1007/s10977-004-0838-7. MR2199789

[35] Carl Faith, *Algebra. I. Rings, modules, and categories*, Grundlehren der Mathematischen Wissenschaften [Fundamental Principles of Mathematical Sciences], vol. 190, Springer-Verlag, Berlin-New York, 1981. Corrected reprint. MR623254

[36] Ming Fang and Steffen Koenig, *Schur functors and dominant dimension*, Trans. Amer. Math. Soc. **363** (2011), no. 3, 1555–1576, DOI 10.1090/S0002-9947-2010-05177-3. MR2737277

[37] Vincent Franjou and Teimuraz Pirashvili, *Comparison of abelian categories recollements*, Doc. Math. **9** (2004), 41–56. MR2054979

[38] Robert M. Fossum, Phillip A. Griffith, and Idun Reiten, *Trivial extensions of abelian categories*, Lecture Notes in Mathematics, Vol. 456, Springer-Verlag, Berlin-New York, 1975. Homological algebra of trivial extensions of abelian categories with applications to ring theory. MR0389981

[39] Peter Freyd, *Stable homotopy*, Proc. Conf. Categorical Algebra (La Jolla, Calif., 1965), Springer, New York, 1966, pp. 121–172. MR0211399

[40] Pierre Gabriel, *Des catégories abéliennes* (French), Bull. Soc. Math. France **90** (1962), 323–448. MR0232821

[41] P. Gabriel and M. Zisman, *Calculus of fractions and homotopy theory*, Ergebnisse der Mathematik und ihrer Grenzgebiete, Band 35, Springer-Verlag New York, Inc., New York, 1967. MR0210125

[42] Peter Gabriel, *Indecomposable representations. II*, Symposia Mathematica, Vol. XI (Convegno di Algebra Commutativa, INDAM, Rome, 1971), Academic Press, London, 1973, pp. 81–104. MR0340377

[43] P. Gabriel and J. A. de la Peña, *Quotients of representation-finite algebras*, Comm. Algebra **15** (1987), no. 1-2, 279–307, DOI 10.1080/00927878708823421. MR876981

[44] Nan Gao and Chrysostomos Psaroudakis, *Gorenstein homological aspects of monomorphism categories via Morita rings*, Algebr. Represent. Theory **20** (2017), no. 2, 487–529, DOI 10.1007/s10468-016-9652-1. MR3638357

[45] N. Gao and C. Psaroudakis, *Ladders of compactly generated triangulated categories and preprojective algebras*, arXiv:1611.09973, to appear in Appl. Categ. Structures, https://doi.org/10.1007/s10485-017-9508-9.

[46] N. Gao, S. Koenig and C. Psaroudakis, *Ladders and Completion of Recollements of Abelian Categories*, preprint in preparation (2015), 45 pages.

[47] Werner Geigle and Helmut Lenzing, *Perpendicular categories with applications to representations and sheaves*, J. Algebra **144** (1991), no. 2, 273–343, DOI 10.1016/0021-8693(91)90107-J. MR1140607

[48] I. M. Gel′fand and V. A. Ponomarev, *Model algebras and representations of graphs* (Russian), Funktsional. Anal. i Prilozhen. **13** (1979), no. 3, 1–12. MR545362

[49] Sergei I. Gelfand and Yuri I. Manin, *Methods of homological algebra*, 2nd ed., Springer Monographs in Mathematics, Springer-Verlag, Berlin, 2003. MR1950475
[50] Ronald Gentle, *T.T.F. theories in abelian categories*, Comm. Algebra **16** (1988), no. 5, 877–908, DOI 10.1080/00927878808823609. MR926328
[51] Murray Gerstenhaber, *The cohomology structure of an associative ring*, Ann. of Math. (2) **78** (1963), 267–288, DOI 10.2307/1970343. MR0161898
[52] Edward L. Green, *On the representation theory of rings in matrix form*, Pacific J. Math. **100** (1982), no. 1, 123–138. MR661444
[53] Edward L. Green and Chrysostomos Psaroudakis, *On Artin algebras arising from Morita contexts*, Algebr. Represent. Theory **17** (2014), no. 5, 1485–1525, DOI 10.1007/s10468-013-9457-4. MR3260907
[54] E.L. Green, C. Psaroudakis and Ø. Solberg, *Reduction techniques for the finitistic dimension*, preprint (2017), 32 pages.
[55] James A. Green, *Polynomial representations of* GL_n, Lecture Notes in Mathematics, vol. 830, Springer-Verlag, Berlin-New York, 1980. MR606556
[56] Dieter Happel, *Triangulated categories in the representation theory of finite-dimensional algebras*, London Mathematical Society Lecture Note Series, vol. 119, Cambridge University Press, Cambridge, 1988. MR935124
[57] Dieter Happel, *On Gorenstein algebras*, Representation theory of finite groups and finite-dimensional algebras (Bielefeld, 1991), Progr. Math., vol. 95, Birkhäuser, Basel, 1991, pp. 389–404, DOI 10.1007/978-3-0348-8658-1_16. MR1112170
[58] D. Happel, *Homological conjectures in representation theory of finite- dimensional algebras*, Sherbrook Lecture Notes Series (1991), 12 pp.
[59] Dieter Happel, *Reduction techniques for homological conjectures*, Tsukuba J. Math. **17** (1993), no. 1, 115–130, DOI 10.21099/tkbjm/1496162134. MR1233117
[60] Dieter Happel, Idun Reiten, and Sverre O. Smalø, *Tilting in abelian categories and quasitilted algebras*, Mem. Amer. Math. Soc. **120** (1996), no. 575, viii+ 88, DOI 10.1090/memo/0575. MR1327209
[61] G. Hochschild, *On the cohomology groups of an associative algebra*, Ann. of Math. (2) **46** (1945), 58–67, DOI 10.2307/1969145. MR0011076
[62] Kiyoshi Igusa and Gordana Todorov, *On the finitistic global dimension conjecture for Artin algebras*, Representations of algebras and related topics, Fields Inst. Commun., vol. 45, Amer. Math. Soc., Providence, RI, 2005, pp. 201–204. MR2146250
[63] J. P. Jans, *Some aspects of torsion*, Pacific J. Math. **15** (1965), 1249–1259. MR0191936
[64] Peter Jørgensen, *Reflecting recollements*, Osaka J. Math. **47** (2010), no. 1, 209–213. MR2666132
[65] P. Jørgensen, *Co-t-structures: the first decade*, in this volume.
[66] Martin Kalck and Dong Yang, *Relative singularity categories I: Auslander resolutions*, Adv. Math. **301** (2016), 973–1021, DOI 10.1016/j.aim.2016.06.011. MR3539395
[67] Masaki Kashiwara and Pierre Schapira, *Categories and sheaves*, Grundlehren der Mathematischen Wissenschaften [Fundamental Principles of Mathematical Sciences], vol. 332, Springer-Verlag, Berlin, 2006. MR2182076
[68] Masaki Kashiwara and Pierre Schapira, *Sheaves on manifolds*, Grundlehren der Mathematischen Wissenschaften [Fundamental Principles of Mathematical Sciences], vol. 292, Springer-Verlag, Berlin, 1994. With a chapter in French by Christian Houzel; Corrected reprint of the 1990 original. MR1299726
[69] Bernhard Keller, *Deriving DG categories*, Ann. Sci. École Norm. Sup. (4) **27** (1994), no. 1, 63–102. MR1258406
[70] Bernhard Keller, *On differential graded categories*, International Congress of Mathematicians. Vol. II, Eur. Math. Soc., Zürich, 2006, pp. 151–190. MR2275593
[71] Bernhard Keller, *Calabi-Yau triangulated categories*, Trends in representation theory of algebras and related topics, EMS Ser. Congr. Rep., Eur. Math. Soc., Zürich, 2008, pp. 467–489, DOI 10.4171/062-1/11. MR2484733
[72] Bernhard Keller and Pedro Nicolás, *Weight structures and simple dg modules for positive dg algebras*, Int. Math. Res. Not. IMRN **5** (2013), 1028–1078, DOI 10.1093/imrn/rns009. MR3031826

[73] Maxim Kontsevich, *Homological algebra of mirror symmetry*, Proceedings of the International Congress of Mathematicians, Vol. 1, 2 (Zürich, 1994), Birkhäuser, Basel, 1995, pp. 120–139. MR1403918

[74] Henning Krause, *Functors on locally finitely presented additive categories*, Colloq. Math. **75** (1998), no. 1, 105–132, DOI 10.4064/cm-75-1-105-132. MR1487973

[75] Henning Krause, *Localization theory for triangulated categories*, Triangulated categories, London Math. Soc. Lecture Note Ser., vol. 375, Cambridge Univ. Press, Cambridge, 2010, pp. 161–235, DOI 10.1017/CBO9781139107075.005. MR2681709

[76] Henning Krause, *Highest weight categories and recollements* (English, with English and French summaries), Ann. Inst. Fourier (Grenoble) **67** (2017), no. 6, 2679–2701. MR3742477

[77] Henning Krause and Jan Šťovíček, *The telescope conjecture for hereditary rings via Ext-orthogonal pairs*, Adv. Math. **225** (2010), no. 5, 2341–2364, DOI 10.1016/j.aim.2010.04.027. MR2680168

[78] Nicholas J. Kuhn, *Generic representations of the finite general linear groups and the Steenrod algebra. II*, K-Theory **8** (1994), no. 4, 395–428, DOI 10.1007/BF00961409. MR1300547

[79] Nicholas J. Kuhn, *The generic representation theory of finite fields: a survey of basic structure*, Infinite length modules (Bielefeld, 1998), Trends Math., Birkhäuser, Basel, 2000, pp. 193–212. MR1789216

[80] Nicholas J. Kuhn, *A stratification of generic representation theory and generalized Schur algebras*, K-Theory **26** (2002), no. 1, 15–49, DOI 10.1023/A:1016357323204. MR1918209

[81] Julian Külshammer, *Pro-species of algebras I: Basic properties*, Algebr. Represent. Theory **20** (2017), no. 5, 1215–1238, DOI 10.1007/s10468-017-9683-2. MR3707912

[82] T. Y. Lam, *Lectures on modules and rings*, Graduate Texts in Mathematics, vol. 189, Springer-Verlag, New York, 1999. MR1653294

[83] Qunhua Liu, Jorge Vitória, and Dong Yang, *Gluing silting objects*, Nagoya Math. J. **216** (2014), 117–151, DOI 10.1215/00277630-2847151. MR3319841

[84] Philippe Loustaunau and Jay Shapiro, *Homological dimensions in a Morita context with applications to subidealizers and fixed rings*, Proc. Amer. Math. Soc. **110** (1990), no. 3, 601–610, DOI 10.2307/2047900. MR1033958

[85] Xiu-Hua Luo and Pu Zhang, *Monic representations and Gorenstein-projective modules*, Pacific J. Math. **264** (2013), no. 1, 163–194, DOI 10.2140/pjm.2013.264.163. MR3079765

[86] Saunders Mac Lane, *Categories for the working mathematician*, 2nd ed., Graduate Texts in Mathematics, vol. 5, Springer-Verlag, New York, 1998. MR1712872

[87] Robert MacPherson and Kari Vilonen, *Elementary construction of perverse sheaves*, Invent. Math. **84** (1986), no. 2, 403–435, DOI 10.1007/BF01388812. MR833195

[88] Barry Mitchell, *Theory of categories*, Pure and Applied Mathematics, Vol. XVII, Academic Press, New York-London, 1965. MR0202787

[89] Barry Mitchell, *Rings with several objects*, Advances in Math. **8** (1972), 1–161, DOI 10.1016/0001-8708(72)90002-3. MR0294454

[90] Bruno J. Müller, *The classification of algebras by dominant dimension*, Canad. J. Math. **20** (1968), 398–409, DOI 10.4153/CJM-1968-037-9. MR0224656

[91] C. Năstăsescu and B. Torrecillas, *Colocalization on Grothendieck categories with applications to coalgebras*, J. Algebra **185** (1996), no. 1, 108–124, DOI 10.1006/jabr.1996.0315. MR1409977

[92] D. Murfet, *Derived categories Part I* (2006), available at therisingsea.org.

[93] Amnon Neeman, *The Grothendieck duality theorem via Bousfield's techniques and Brown representability*, J. Amer. Math. Soc. **9** (1996), no. 1, 205–236, DOI 10.1090/S0894-0347-96-00174-9. MR1308405

[94] Amnon Neeman, *Triangulated categories*, Annals of Mathematics Studies, vol. 148, Princeton University Press, Princeton, NJ, 2001. MR1812507

[95] Amnon Neeman, *Non-left-complete derived categories*, Math. Res. Lett. **18** (2011), no. 5, 827–832, DOI 10.4310/MRL.2011.v18.n5.a2. MR2875857

[96] P. Nicolás, *On torsion torsionfree triples*, Ph.D. thesis, University of Murcia, (2007), 184 pages.

[97] Pedro Nicolás and Manuel Saorín, *Parametrizing recollement data for triangulated categories*, J. Algebra **322** (2009), no. 4, 1220–1250, DOI 10.1016/j.jalgebra.2009.04.035. MR2537682

[98] Koichiro Ohtake, *Colocalization and localization*, J. Pure Appl. Algebra **11** (1977/78), no. 1-3, 217–241, DOI 10.1016/0022-4049(77)90053-6. MR0470027

[99] Hiroyuki Tachikawa and Koichiro Ohtake, *Colocalization and localization in abelian categories*, J. Algebra **56** (1979), no. 1, 1–23, DOI 10.1016/0021-8693(79)90320-X. MR527152

[100] F. Oort, *Yoneda extensions in abelian categories*, Math. Ann. **153** (1964), 227–235, DOI 10.1007/BF01360318. MR0162836

[101] D. O. Orlov, *Triangulated categories of singularities and D-branes in Landau-Ginzburg models* (Russian, with Russian summary), Tr. Mat. Inst. Steklova **246** (2004), no. Algebr. Geom. Metody, Svyazi i Prilozh., 240–262; English transl., Proc. Steklov Inst. Math. **3(246)** (2004), 227–248. MR2101296

[102] D. O. Orlov, *Triangulated categories of singularities, and equivalences between Landau-Ginzburg models* (Russian, with Russian summary), Mat. Sb. **197** (2006), no. 12, 117–132, DOI 10.1070/SM2006v197n12ABEH003824; English transl., Sb. Math. **197** (2006), no. 11-12, 1827–1840. MR2437083

[103] Dmitri Orlov, *Derived categories of coherent sheaves and triangulated categories of singularities*, Algebra, arithmetic, and geometry: in honor of Yu. I. Manin. Vol. II, Progr. Math., vol. 270, Birkhäuser Boston, Inc., Boston, MA, 2009, pp. 503–531, DOI 10.1007/978-0-8176-4747-6_16. MR2641200

[104] C. E. Parra and J. Vitória, *Properties of abelian categories via recollements*, arXiv:1710.04632.

[105] B.J. Parshall and L.L. Scott, *Derived categories, quasi-hereditary algebras, and algebraic groups*, Proceedings of the Ottawa-Moosonee Workshop in Algebra (1987), 105 pp., Carleton Univ., Ottawa, ON, 1988.

[106] T. I. Pirashvili, *Polynomial functors* (Russian, with English summary), Trudy Tbiliss. Mat. Inst. Razmadze Akad. Nauk Gruzin. SSR **91** (1988), 55–66. MR1029007

[107] C. Psaroudakis, *Representation Dimension, Cohen-Macaulay Modules and Triangulated Categories*, Ph.D. thesis, University of Ioannina, Greece, (2013), 201 pages.

[108] Chrysostomos Psaroudakis, *Homological theory of recollements of abelian categories*, J. Algebra **398** (2014), 63–110, DOI 10.1016/j.jalgebra.2013.09.020. MR3123754

[109] C. Psaroudakis, *Recollements of Module Categories and Dominant Dimension*, preprint in preparation (2017).

[110] Chrysostomos Psaroudakis and Jorge Vitória, *Recollements of module categories*, Appl. Categ. Structures **22** (2014), no. 4, 579–593, DOI 10.1007/s10485-013-9323-x. MR3227608

[111] Chrysostomos Psaroudakis and Jorge Vitória, *Realisation functors in tilting theory*, Math. Z. **288** (2018), no. 3-4, 965–1028, DOI 10.1007/s00209-017-1923-y. MR3778987

[112] Chrysostomos Psaroudakis and Jorge Vitória, *Recollements of module categories*, Appl. Categ. Structures **22** (2014), no. 4, 579–593, DOI 10.1007/s10485-013-9323-x. MR3227608

[113] Chrysostomos Psaroudakis, Øystein Skartsæterhagen, and Øyvind Solberg, *Gorenstein categories, singular equivalences and finite generation of cohomology rings in recollements*, Trans. Amer. Math. Soc. Ser. B **1** (2014), 45–95, DOI 10.1090/S2330-0000-2014-00004-6. MR3274657

[114] Jeremy Rickard, *Derived categories and stable equivalence*, J. Pure Appl. Algebra **61** (1989), no. 3, 303–317, DOI 10.1016/0022-4049(89)90081-9. MR1027750

[115] Manuel Saorin, *The structure of commutative rings with monomorphic flat envelopes*, Comm. Algebra **23** (1995), no. 14, 5383–5394, DOI 10.1080/00927879508825539. MR1363610

[116] A. H. Schofield, *Representation of rings over skew fields*, London Mathematical Society Lecture Note Series, vol. 92, Cambridge University Press, Cambridge, 1985. MR800853

[117] Nicole Snashall and Øyvind Solberg, *Support varieties and Hochschild cohomology rings*, Proc. London Math. Soc. (3) **88** (2004), no. 3, 705–732, DOI 10.1112/S002461150301459X. MR2044054

[118] Øyvind Solberg, *Support varieties for modules and complexes*, Trends in representation theory of algebras and related topics, Contemp. Math., vol. 406, Amer. Math. Soc., Providence, RI, 2006, pp. 239–270, DOI 10.1090/conm/406/07659. MR2258047

[119] *The Stacks Project*, http://stacks.math.columbia.edu/browse.

[120] Bo Stenström, *Rings of quotients*, Springer-Verlag, New York-Heidelberg, 1975. Die Grundlehren der Mathematischen Wissenschaften, Band 217; An introduction to methods of ring theory. MR0389953

[121] Roger Ware, *Endomorphism rings of projective modules*, Trans. Amer. Math. Soc. **155** (1971), 233–256, DOI 10.2307/1995475. MR0274511

[122] Alfred Wiedemann, *On stratifications of derived module categories*, Canad. Math. Bull. **34** (1991), no. 2, 275–280, DOI 10.4153/CMB-1991-044-0. MR1113309

[123] Changchang Xi, *On the finitistic dimension conjecture. III. Related to the pair $eAe \subseteq A$*, J. Algebra **319** (2008), no. 9, 3666–3688, DOI 10.1016/j.jalgebra.2008.01.021. MR2407846

[124] Dongyuan Yao, *On equivalence of derived categories*, K-Theory **10** (1996), no. 3, 307–322, DOI 10.1007/BF00538187. MR1394382

[125] Aiping Zhang and Shunhua Zhang, *On the finitistic dimension conjecture of Artin algebras*, J. Algebra **320** (2008), no. 1, 253–258, DOI 10.1016/j.jalgebra.2007.12.011. MR2417987

[126] Birge Zimmermann Huisgen, *The finitistic dimension conjectures—a tale of* 3.5 *decades*, Abelian groups and modules (Padova, 1994), Math. Appl., vol. 343, Kluwer Acad. Publ., Dordrecht, 1995, pp. 501–517. MR1378224

Institute of Algebra and Number Theory, University of Stuttgart, Pfaffenwaldring 57, 70569 Stuttgart, Germany

Email address: chrysostomos.psaroudakis@mathematik.uni-stuttgart.de

Contemporary Mathematics
Volume **716**, 2018
https://doi.org/10.1090/conm/716/14428

Going relative with Maurice—A survey

Øyvind Solberg

To the memory of Maurice Auslander.

Contents

1. Introduction

Homological algebra is extensively used in the representation theory of artin algebras and elsewhere. Relative homological algebra has played a much less prominent role. The aim of the series of the three papers [**6**–**9**] was to encourage a more systematic use of this theory. Well, one might discuss how successful we were in doing so, but I was very happy to get the chance to speak at the Maurice Auslander distinguished lectures and international conference and to present a survey on these papers. Of course, it was only possible to include a small part of what we did in those papers, and the presentation here follows to a large degree the survey lecture given there. However I include some of the things I did not get time to present in the lecture and some new results to illustrate further how one can use relative theory to prove statements about objects that are not in the outset defined by relative homological algebra.

The more mathematical structure an object of study has, the more we can say about that mathematical object. Relative homological algebra is sometimes an extra ingredient that brings more structure to a maybe well-known construction or object. For instance, take the endomorphism ring of a cogenerator T for an artin algebra Λ. You might say boring, there is not much structure there. But using

2010 *Mathematics Subject Classification.* Primary 16S37, 16E05, 16W50.

relative homological algebra $\operatorname{End}_\Lambda(T)$ and $\operatorname{End}_\Lambda(\Lambda \oplus D\operatorname{Tr}(T))$ are derived equivalent and even more is true. So this paper is a continuation of the motivation/the attempt of the papers [**6**–**9**] to promote the use of relative homological algebra in the representation theory of artin algebras and more general. Here we concentrate on the former as there we have even more structure.

Let us review the paper section by section. In Section 2 we explain the starting point of relative homological algebra and give the basic results for artin algebras from [**6**] with additional clarifying results from [**10**] and [**14**]. Section 3 is devoted to constructing the relative derived functors of Hom and the tensor product. Homologically finite subcategories is an important topic and received a lot of attention when the papers [**6**–**9**] were written. This topic is discussed in Section 4 by reviewing/generalizing the following three results: Wakamatsu's Lemma, functorially finiteness of the collection of all n-th syzygy modules and the Auslander-Reiten Lemma. In Section 5 we recall the definition of relative cotilting modules and review some of their most important properties. Section 6 is devoted to reviewing the connection between relative cotilting modules and the Wedderburn correspondence. In Section 7 we show that when absolute cotilting modules have so-called dualizing direct summands they are constructed from relative cotilting modules, and we recall how this lead to the notion of n-precluster tilting modules introduced first for $n = 1$ in [**9**] and in general in [**21**]. In the final section we point out two new applications of relative homological algebra. One which one could have seen long ago, and another which is naturally inspired by cluster and precluster tilting modules [**18**–**21**]. In particular, we show that the endomorphism rings of all relative cotilting modules are derived equivalent and that the subcategory $M^{\perp_{n-1}}$ occurring in the definition of a cluster or a precluster tilting module is always constructed from a relative cotilting module.

Our setting throughout this paper would be an artin algebra Λ and the category of finitely generated left Λ-modules denoted by $\operatorname{mod}\Lambda$. We denote by D the usual duality $D\colon \operatorname{mod}\Lambda \to \operatorname{mod}\Lambda^{\mathrm{op}}$ and by Tr the transpose $\operatorname{Tr}\colon \operatorname{mod}\Lambda \to \operatorname{mod}\Lambda^{\mathrm{op}}$.

2. Relative homological algebra

The basic ideas of relative homological algebra for the category of modules over a ring were given by Hochschild [**17**]. Heller [**16**] and Buchsbaum [**11**] generalized and developed these to more abstract categories. These were further discussed and developed by Butler and Horrocks in [**12**]. Common for them all is that relative homological algebra starts with choosing the extensions $\mathcal{E}$ you want to consider in your category. A very natural condition for $\mathcal{E}$ to satisfy, is to be closed under pullbacks and pushouts. It is easy to see that this correspond to choosing a sub-bifunctor

$$F = F_{\mathcal{E}} \subseteq \operatorname{Ext}^1_\Lambda(-,-)\colon (\operatorname{mod}\Lambda)^{\mathrm{op}} \times \operatorname{mod}\Lambda \to \mathrm{Sets}.$$

This section is devoted to discussing the elementary properties of such sub-bifunctors $F \subseteq \operatorname{Ext}^1_\Lambda(-,-)$ and to develop the theory of relative injective and projective objects for these in order to be able to define relative derived functors in the next section.

Let F be a sub-bifunctor of $\operatorname{Ext}^1_\Lambda(-,-)\colon (\operatorname{mod}\Lambda)^{\mathrm{op}} \times \operatorname{mod}\Lambda \to \mathrm{Sets}$. An exact sequence

$$\eta\colon 0 \to A \to B \to C \to 0$$

in $F(C, A)$ is said to be *F-exact*. So having a sub-bifunctor

$$F \subseteq \operatorname{Ext}^1_\Lambda(-,-)\colon (\operatorname{mod}\Lambda)^{\mathrm{op}} \times \operatorname{mod}\Lambda \to \mathrm{Sets},$$

is the same as saying that the collection $\mathcal{E}_F$ of F-exact sequences are closed under pullbacks and pushouts.

A sub-bifunctor $F \subseteq \operatorname{Ext}^1_\Lambda(-,-)\colon (\operatorname{mod}\Lambda)^{\mathrm{op}} \times \operatorname{mod}\Lambda \to \mathrm{Sets}$ need not to have its image in Ab. Reviewing how Baer sum can be constructed and that the collection of F-exact sequences is closed under pullbacks and pushouts, the following is not too hard to prove.

LEMMA 2.1 ([**6**, Lemma 1.1]). *Let F be a sub-bifunctor of $\operatorname{Ext}^1_\Lambda(-,-)\colon (\operatorname{mod}\Lambda)^{\mathrm{op}} \times \operatorname{mod}\Lambda \to \mathrm{Sets}$. Then F is an additive sub-bifunctor of $\operatorname{Ext}^1_\Lambda(-,-)\colon (\operatorname{mod}\Lambda)^{\mathrm{op}} \times \operatorname{mod}\Lambda \to \mathrm{Ab}$ if and only if F is closed under direct sums of F-exact sequences.*

For the rest of the section let $F \subseteq \operatorname{Ext}^1_\Lambda(-,-)\colon (\operatorname{mod}\Lambda)^{\mathrm{op}} \times \operatorname{mod}\Lambda \to \mathrm{Ab}$ be an additive sub-bifunctor. To construct the derived functor $\operatorname{Ext}^1_\Lambda(-,-)$ the injective and projective modules are crucial. Next we define the corresponding objects for F.

To F we associate two subcategories of F-injective and F-projective modules, $\mathcal{I}(F)$ and $\mathcal{P}(F)$, respectively. A module I is said to be *F-injective* if for all F-exact sequences

$$0 \to A \to B \to C \to 0,$$

the sequence

$$0 \to \operatorname{Hom}_\Lambda(C, I) \to \operatorname{Hom}_\Lambda(B, I) \to \operatorname{Hom}_\Lambda(A, I) \to 0$$

is exact. Dually, we define an *F-projective* module. Clearly, we have that the standard injective modules $\mathcal{I}(\Lambda)$ and the standard projective modules $\mathcal{P}(\Lambda)$ are contained in $\mathcal{I}(F)$ and $\mathcal{P}(F)$, respectively.

Given an F-exact sequence $\eta\colon 0 \to A \to B \to C \to 0$ it is easy to see that

$$0 \to \operatorname{Hom}_\Lambda(-, A) \to \operatorname{Hom}_\Lambda(-, B) \to \operatorname{Hom}_\Lambda(-, C) \to F(-, A)$$

and

$$0 \to \operatorname{Hom}_\Lambda(C, -) \to \operatorname{Hom}_\Lambda(B, -) \to \operatorname{Hom}_\Lambda(A, -) \to F(C, -)$$

are exact. It follows from this that a module I is F-injective if and only if $F(-, I) = (0)$, and that a module P is F-projective if and only if $F(P, -) = (0)$ (see [**6**, Corollary 1.4]).

The definition of relative injective and relative projective modules motivates us to look at the following constructions for a subcategory $\mathcal{X}$ of $\operatorname{mod}\Lambda$:

$$F_{\mathcal{X}}(C, A) = \{0 \to A \to B \to C \to 0 \text{ exact} \mid \operatorname{Hom}_\Lambda(-, B)|_{\mathcal{X}} \to \operatorname{Hom}_\Lambda(-, C)|_{\mathcal{X}} \to 0 \text{ is exact}\}$$

and

$$F^{\mathcal{X}}(C, A) = \{0 \to A \to B \to C \to 0 \text{ exact} \mid \operatorname{Hom}_\Lambda(B, -)|_{\mathcal{X}} \to \operatorname{Hom}_\Lambda(A, -)|_{\mathcal{X}} \to 0 \text{ is exact}\}$$

A natural question is then: Are $F_{\mathcal{X}}$ and $F^{\mathcal{X}}$ additive sub-bifunctors of $\operatorname{Ext}^1_\Lambda(-,-)$? This is answered in the affirmative in [**6**, Proposition 1.7].

PROPOSITION 2.2. *Let $\mathcal{X}$ be any subcategory of* $\operatorname{mod}\Lambda$. *Then* $F_{\mathcal{X}}$ *and* $F^{\mathcal{X}}$ *are additive sub-bifunctors of* $\operatorname{Ext}^1_\Lambda(-,-)\colon (\operatorname{mod}\Lambda)^{\mathrm{op}} \times \operatorname{mod}\Lambda \to \mathrm{Ab}$.

Using [**2**, III, Corollary 4.2] we immediately obtain the following.

PROPOSITION 2.3 ([**6**, Corollary 1.6 and Proposition 1.8]).

(a) (i) $\mathcal{P}(F) = \mathcal{P}(\Lambda) \cup \operatorname{Tr} D\mathcal{I}(F)$.
(i) $\mathcal{I}(F) = \mathcal{I}(\Lambda) \cup D\operatorname{Tr}\mathcal{P}(F)$.

(b) *For a subcategory $\mathcal{X}$ of* $\operatorname{mod}\Lambda$ *we have* $F_{\mathcal{X}} = F^{D\operatorname{Tr}\mathcal{X}}$.

The sub-bifunctor F is said to have *enough injectives* if for A in $\operatorname{mod}\Lambda$ there exists an F-exact sequence

$$0 \to A \to I \to A' \to 0$$

with I in $\mathcal{I}(F)$. To have *enough projectives* is defined dually. By definition we have that $F \subseteq F_{\mathcal{P}(F)}$ and $F \subseteq F^{\mathcal{I}(F)}$. If F has enough injectives or enough projectives, then any F-exact sequence

$$0 \to A \to B \to C \to 0$$

is a pullback of an F-exact sequence $0 \to A \to I \to C' \to 0$ with I in $\mathcal{I}(F)$ or a pushout of an F-exact sequence $0 \to A' \to P \to C \to 0$ with P in $\mathcal{P}(F)$, respectively. In either case, it follows that $F^{\mathcal{I}(F)} \subseteq F$ and $F_{\mathcal{P}(F)} \subseteq F$, respectively. To characterize when a sub-bifunctor F has enough injectives or projectives we need to recall the notion of homologically finite subcategories.

Let $\mathcal{X}$ be a subcategory of $\operatorname{mod}\Lambda$. For a module C in $\operatorname{mod}\Lambda$ a homomorphism $f\colon X \to C$ with X in $\mathcal{X}$ is a *right $\mathcal{X}$-approximation* if for all homomorphisms $g\colon X' \to C$ with X' in $\mathcal{X}$, there exists a homomorphism $h\colon X' \to X$ such that $fh = g$. If all modules in $\operatorname{mod}\Lambda$ have a right $\mathcal{X}$-approximation, then $\mathcal{X}$ is a *contravariantly finite* in $\operatorname{mod}\Lambda$. A left $\mathcal{X}$-approximation of a module and a covariantly finite subcategory of $\operatorname{mod}\Lambda$ are defined dually. A covariantly and contravariantly finite subcategory $\mathcal{X}$ of $\operatorname{mod}\Lambda$ is called a *functorially finite* subcategory of $\operatorname{mod}\Lambda$.

Using these notions and the above we infer the following.

THEOREM 2.4 ([**6**, Theorem 1.12]). *Let F be an additive sub-bifunctor of* $\operatorname{Ext}^1_\Lambda(-,-)$.

(a) *The sub-bifunctor F has enough injectives if and only if $\mathcal{I}(F)$ is covariantly finite in* $\operatorname{mod}\Lambda$ *and* $F = F^{\mathcal{I}(F)}$.

(b) *The sub-bifunctor F has enough projectives if and only if $\mathcal{P}(F)$ is contravariantly finite in* $\operatorname{mod}\Lambda$ *and* $F = F_{\mathcal{P}(F)}$.

A standard fact from absolute homological algebra, is, given any short exact sequence $0 \to A \to B \to C \to 0$, it induces long exact sequences

$$0 \to \operatorname{Hom}_\Lambda(-,A) \to \operatorname{Hom}_\Lambda(-,B) \to \operatorname{Hom}_\Lambda(-,C) \to \operatorname{Ext}^1_\Lambda(-,A) \to \operatorname{Ext}^1_\Lambda(-,B) \to \operatorname{Ext}^1_\Lambda(-,C)$$

and

$$0 \to \operatorname{Hom}_\Lambda(C,-) \to \operatorname{Hom}_\Lambda(B,-) \to \operatorname{Hom}_\Lambda(A,-) \to \operatorname{Ext}^1_\Lambda(C,-) \to \operatorname{Ext}^1_\Lambda(B,-) \to \operatorname{Ext}^1_\Lambda(A,-).$$

Does F satisfy a similar property? If $0 \to A \to B \to C \to 0$ is an F-exact sequence, then it is straightforward to see that

$$(*)\quad 0 \to \mathrm{Hom}_\Lambda(-,A) \to \mathrm{Hom}_\Lambda(-,B) \to \mathrm{Hom}_\Lambda(-,C) \to F(-,A) \to F(-,B) \to F(-,C)$$

and

$$(**)\quad 0 \to \mathrm{Hom}_\Lambda(C,-) \to \mathrm{Hom}_\Lambda(B,-) \to \mathrm{Hom}_\Lambda(A,-) \to F(C,-) \to F(B,-) \to F(A,-)$$

are complexes. However, in general they are not exact. If F has the property that $(*)$ or $(**)$ is exact for all F-exact sequences $0 \to A \to B \to C \to 0$, then F is said to be *closed on the left* or *closed on the right*, respectively. A sub-bifunctor F that is closed on the left and the right, is called *closed.*

Next we give an example of an additive sub-bifunctor of $\mathrm{Ext}^1_\Lambda(-,-)$ which is not closed.

EXAMPLE 2.5. Let Λ be an artin algebra. Let $\mathcal{E}$ be the collection of extensions given by the split short exact sequences and the almost split sequences taking the closure under direct sum. This collection is closed under pullbacks, pushouts and finite direct sums, so it correspond to an additive sub-bifunctor $F = F_\mathcal{E} \subseteq \mathrm{Ext}^1_\Lambda(-,-)\colon (\mathrm{mod}\,\Lambda)^{\mathrm{op}} \times \mathrm{mod}\,\Lambda \to \mathrm{Ab}$. The relative projective modules $\mathcal{P}(F)$ and the relative injective modules are only the absolute projective modules $\mathcal{P}(\Lambda)$ and the absolute injective modules $\mathcal{I}(\Lambda)$, respectively. The sub-bifunctor F is not closed in general. For instance, let $0 \to A \to B \to C \to 0$ be an almost split sequence in $\mathrm{mod}\,\Lambda$. To prove that the above complexes are not exact, consider the complex

$$F(-,A) \to F(-,B) \to F(-,C).$$

For an indecomposable module X, the functor $F(-,X)$ is only supported in $\mathrm{Tr}\,D(X)$. Using this and the fact that the Auslander-Reiten quiver of Λ does not contain any loops, the above complex evaluated at $\mathrm{Tr}\,D(B)$ gives the following complex

$$0 \to F(\mathrm{Tr}\,D(B),B) \to 0,$$

where $F(\mathrm{Tr}\,D(B),B) \neq (0)$.

In [**12**, Theorem 1.1] it is pointed out that any exact structure on $\mathrm{mod}\,\Lambda$ is of the form $\mathcal{E}_F$ for some closed additive sub-bifunctor F of $\mathrm{Ext}^1_\Lambda(-,-)$, recalling that $\mathcal{E}_F$ is the collection of all F-exact sequences in $\mathrm{mod}\,\Lambda$. We recall that all the different types of being closed are equivalent (in fact for all abelian categories).

PROPOSITION 2.6 ([**14**, Proposition 1.4]). *Let F be an additive sub-bifunctor of* $\mathrm{Ext}^1_\Lambda(-,-)$.

(a) *$\mathcal{E}_F$ defines an exact structure on* $\mathrm{mod}\,\Lambda$.
(b) *F is closed on the left.*
(c) *F is closed.*
(d) *F is closed on the right.*

Closed sub-bifunctors are maybe hard to understand just from the definition, but the following characterization by Buan ([**10**, Lemma 2.2.1]) is useful. Recall that a sub-bifunctor F is said to have the 3×3-*lemma property* if for every exact commutative 3×3-diagram short exact sequences with 3 rows and 3 columns where the first and third row and all the columns are F-exact, also the second row is F-exact.

LEMMA 2.7 ([**10**, Lemma 2.2.1]). *A sub-bifunctor has the* 3×3*-lemma property if and only if it is closed.*

Using this, it is easy to see that the additive sub-bifunctors $F_{\mathcal{X}}$ and $F^{\mathcal{X}}$ for any subcategory $\mathcal{X}$ in $\operatorname{mod}\Lambda$ are closed.

Given a closed additive sub-bifunctor F of $\operatorname{Ext}^1_\Lambda(-,-)$, then $\operatorname{mod}\Lambda$ can be regarded as an exact category with the exact structure given by $\mathcal{E}_F$. Then we can define the relative derived category of $\operatorname{mod}\Lambda$ as the derived category of the exact category $(\operatorname{mod}\Lambda, \mathcal{E}_F)$.

3. Relative homology

As above, let Λ be an artin algebra and F an additive sub-bifunctor of $\operatorname{Ext}^1_\Lambda(-,-)$ with enough projectives and injectives. In this section we explain how to construct the relative derived functors of $\operatorname{Hom}_\Lambda(-,-)$ and the tensor product $-\otimes_\Lambda -$. This section can be summarized by saying, to construct the relative derived functors we use relative injective and/or relative projective resolutions instead of standard injective and/or standard projective resolutions and make exactly the same constructions as we are used to.

For ease of notation we call a long exact sequence

$$\cdots \to M_{n+1} \xrightarrow{d_{n+1}} M_n \xrightarrow{d_n} M_{n-1} \to \cdots$$

F*-exact* if the sequence

$$0 \to \operatorname{Ker} d_n \to M_n \to \operatorname{Im} d_n \to 0$$

is F-exact for each n in $\mathbb{Z}$. Since F has enough projectives, there exists for each C in $\operatorname{mod}\Lambda$ an F-projective resolution, that is, an F-exact sequence of the form

$$\cdots \to P_2 \to P_1 \to P_0 \to C \to 0$$

such that P_i is in $\mathcal{P}(F)$ for each i in $\mathbb{N}$. Similarly, since F has enough injectives, there exists for each A in $\operatorname{mod}\Lambda$ an F-injective resolution, that is, an F-exact sequence of the form

$$0 \to A \to I^0 \to I^1 \to I^2 \to \cdots$$

such that I^i is in $\mathcal{I}(F)$ for each non-negative integer i. If i is a non-negative integer, A and C are in $\operatorname{mod}\Lambda$, and

$$\cdots \to P_2 \to P_1 \to P_0 \to C \to 0$$

is an F-projective resolution of C, then we denote by $\operatorname{Ext}^i_F(C,A)$ the i-th homology of the complex

$$0 \to \operatorname{Hom}_\Lambda(P_0,A) \to \operatorname{Hom}_\Lambda(P_1,A) \to \operatorname{Hom}_\Lambda(P_2,A) \to \cdots.$$

As in the standard case, this definition does not depend on the choice of a projective resolution of C and $\operatorname{Ext}^i_F(C,A)$ is isomorphic to the i-th homology of the complex

$$0 \to \operatorname{Hom}_\Lambda(C,I^0) \to \operatorname{Hom}_\Lambda(C,I^1) \to \operatorname{Hom}_\Lambda(C,I^2) \to \cdots,$$

where

$$0 \to A \to I^0 \to I^1 \to I^2 \to \cdots$$

is an F-injective resolution of A. One easily checks that

$$\mathrm{Ext}_F^0(C,A)=\mathrm{Hom}_\Lambda(C,A) \qquad \text{and} \qquad \mathrm{Ext}_F^1(C,A)=F(C,A).$$

Given an F-projective resolution of C as above and B in $\operatorname{mod}\Lambda^{\mathrm{op}}$, we define $\mathrm{Tor}_i^F(B,C)$ as the i-th homology of the complex

$$\cdots\to M\otimes_\Lambda P_2\to M\otimes_\Lambda P_1\to M\otimes_\Lambda P_0\to 0.$$

If

$$0\to A\to B\to C\to 0$$

is an F-exact sequence, then for each X in $\operatorname{mod}\Lambda$ we have long exact sequences

$$\begin{aligned}0\to \mathrm{Hom}_\Lambda(X,A)\to \mathrm{Hom}_\Lambda(X,B)\to \mathrm{Hom}_\Lambda(X,C)\to&\\ F(X,A)\to F(X,B)\to F(X,C)\to&\\ \mathrm{Ext}_F^2(X,A)\to \mathrm{Ext}_F^2(X,B)\to \mathrm{Ext}_F^2(X,C)\to\cdots&\end{aligned}$$

and

$$\begin{aligned}0\to \mathrm{Hom}_\Lambda(C,X)\to \mathrm{Hom}_\Lambda(B,X)\to \mathrm{Hom}_\Lambda(A,X)\to&\\ F(C,X)\to F(B,X)\to F(A,X)\to&\\ \mathrm{Ext}_F^2(C,X)\to \mathrm{Ext}_F^2(B,X)\to \mathrm{Ext}_F^2(A,X)\to\cdots.&\end{aligned}$$

By the *F-projective dimension* $\mathrm{pd}_F\, C$ of C in $\operatorname{mod}\Lambda$ we mean the minimal non-negative integer n, such that there exists a finite F-projective resolution of C of the form

$$0\to P_n\to P_{n-1}\to\cdots\to P_1\to P_0\to C\to 0,$$

and it is infinite otherwise. As in the standard case, for a non-negative integer n we have $\mathrm{pd}_F\, C\le n$ if and only if $\mathrm{Ext}_F^{n+1}(C,-)=0$. Dually, by the *$F$-injective dimension* $\mathrm{id}_F\, A$ of A in $\operatorname{mod}\Lambda$ we mean the minimal non-negative integer n, such that there exists an F-injective resolution of A of the form

$$0\to A\to I^0\to I^1\to\cdots\to I^{n-1}\to I^n\to 0,$$

and otherwise infinite. Again, if n is a positive integer, then $\mathrm{id}_F\, A\le n$ if and only if $\mathrm{Ext}_F^{n+1}(-,A)=0$. Finally, we define

$$\mathrm{gldim}_F\,\Lambda=\sup\{\mathrm{pd}_F\, C\mid C\in\operatorname{mod}\Lambda\}.$$

Obviously,

$$\mathrm{gldim}_F\,\Lambda=\sup\{\mathrm{id}_F\, A\mid A\in\operatorname{mod}\Lambda\}.$$

For A and C in $\operatorname{mod}\Lambda$, let $\mathcal{P}(F)(C,A)$ be the subspace of $\mathrm{Hom}_\Lambda(C,A)$ consisting of the homomorphisms which factor through a module from $\mathcal{P}(F)$, and

$$\underline{\mathrm{Hom}}_{\mathcal{P}(F)}(C,A)=\mathrm{Hom}_\Lambda(C,A)/\mathcal{P}(F)(C,A).$$

We end with a summary of the basic properties of the relative derived functors and the relative derived category associated to an additive sub-bifunctor F of $\mathrm{Ext}_\Lambda^1(-,-)$ with enough projectives and injectives.

THEOREM 3.1 ([**6**, Lemma 2.1 and Proposition 2.3], [**10**, Proposition 2.2.3], [**24**, Proposition 1.3]). *Let A and C be in $\operatorname{mod}\Lambda$, and let B be in $\operatorname{mod}\Lambda^{\mathrm{op}}$. Then we have the following.*

(a) $F(C,A)=\mathrm{Ext}_F^1(C,A)$.

(b) $\mathrm{Ext}_F^i(A,D(B))\simeq D\,\mathrm{Tor}_i^F(B,A)$ *for all* $i>0$.

(c) $\operatorname{Ext}^1_F(C, D\operatorname{Tr}(A)) \simeq D(\underline{\operatorname{Hom}}_{\mathcal{P}(F)}(A, C))$.

(d) $D_F(\operatorname{mod}\Lambda) = \mathcal{K}(\operatorname{mod}\Lambda)/K^F_{\mathrm{ac}}(\operatorname{mod}\Lambda)$, *where the construction is a Verdier quotient of the homotopy category* $\mathcal{K}(\operatorname{mod}\Lambda)$ *of* $\operatorname{mod}\Lambda$ *and the homotopy category* $K^F_{\mathrm{ac}}(\operatorname{mod}\Lambda)$ *of* F*-exact complexes in* $\operatorname{mod}\Lambda$.

(e) *For* X *and* Y *in* $\operatorname{mod}\Lambda$ *we have*

$$\operatorname{Hom}_{D_F(\operatorname{mod}\Lambda)}(X, Y[1]) \simeq F(X, Y),$$

when X *and* Y *are viewed as stalk complexes concentrated in the same degree.*

(f) *Let* $\mathcal{X}$ *be a subcategory of* $\operatorname{mod}\Lambda$ *with* $\mathcal{P}(\Lambda) \subseteq \mathcal{X}$. *For a* Λ*-module* A *with* $\operatorname{Ext}^i_\Lambda(\mathcal{X}, A) = (0)$ *for* $i = 1, 2, \ldots, n$, *then*

$$\operatorname{Ext}^i_{F\mathcal{X}}(C, A) \simeq \operatorname{Ext}^i_\Lambda(C, A)$$

for all C *in* $\operatorname{mod}\Lambda$ *and all* $i = 1, 2, \ldots, n$.

4. Homologically finite subcategories

When the papers [**6**–**9**] were written, there was a great interest in homologically finite subcategories. So an important part of them was to discuss this topic. The general approach was to take an absolute/standard (that is, $F = \operatorname{Ext}^1_\Lambda(-,-)$) result on homologically finite subcategories and go relative. There is a fair number of results in the above papers on this, but we only concentrate on three of them: Wakamatsu's Lemma, $\Omega^n_\Lambda(\operatorname{mod}\Lambda) \cup \mathcal{P}(\Lambda)$ is functorially finite in $\operatorname{mod}\Lambda$ and the Auslander-Reiten Lemma. They are all important, but in addition the first and the second we use in the last section to prove new results using relative homological algebra.

Throughout this section let F be an additive sub-bifunctor of $\operatorname{Ext}^1_\Lambda(-,-)$ with enough injectives and projectives.

We first discuss Wakamatsu's Lemma. It was first discoved by Takayoshi Wakamatsu in a special case [**25**], and proved in general by Auslander and Reiten observing that the same proof worked (see [**4**, Lemma 1.3]). According to Maurice Auslander I got the impression he thought that Wakamatsu did not think much of the result, but it has later proven its importance.

Wakamatsu's Lemma states the following: Given an extension closed subcategory $\mathcal{X}$ of $\operatorname{mod}\Lambda$ and

$$0 \to Y_C \xrightarrow{f} X_C \xrightarrow{g} C \to 0$$

is exact with g a minimal right $\mathcal{X}$-approximation of C, then $\operatorname{Ext}^1_\Lambda(\mathcal{X}, Y_C) = (0)$. Another way of saying this is that all minimal right $\mathcal{X}$-approximations for an extension closed subcategory $\mathcal{X}$ are *special*, namely, by definition that $\operatorname{Ext}^1_\Lambda(\mathcal{X}, Y_C) = (0)$.

One essential assumption in Wakamatsu's Lemma is that $\mathcal{X}$ is extension closed. However, not all subcategories are extension closed, for example, $\operatorname{Fac}M$ for some module M in $\operatorname{mod}\Lambda$. Recall that $\operatorname{Fac}M$ is the full subcategory of $\operatorname{mod}\Lambda$ consisting of all quotients of a finite direct sum of copies of M. Can we define a relative homological theory in which $\operatorname{Fac}M$ always is extension closed? The answer here is yes. Let $F = F_{\operatorname{add}M}$ be the additive sub-bifunctor of $\operatorname{Ext}^1_\Lambda(-,-)$ associated to M. If the sequene

$$\eta\colon 0 \to A \xrightarrow{f} B \xrightarrow{g} C \to 0$$

is F-exact with A and C in $\mathrm{Fac}M$ and $M_A \xrightarrow{p_A} A$ and $M_C \xrightarrow{p_C} C$ are right $\mathrm{add}\,M$-approximations of A and C, respectively, then we can construct the following commutative diagram

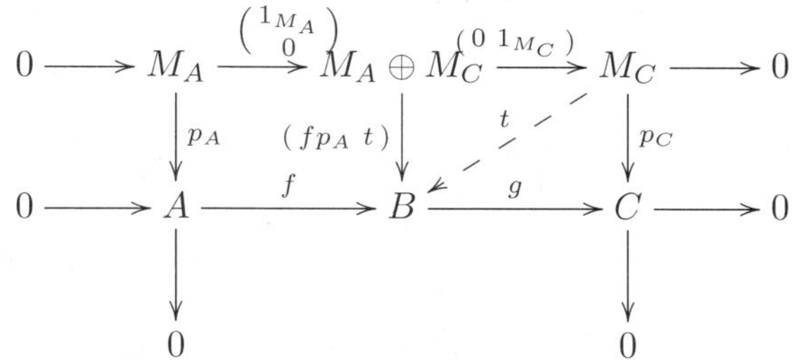

Since η is F-exact, there exists $t\colon M_C \to B$ such that $gt = p_C$. By the Five Lemma the module B is in $\mathrm{Fac}M$. Hence $\mathrm{Fac}M$ is always F-extension closed. If $\mathcal{X} = \mathrm{Fac}M$ and $F = F_{\mathrm{add}\,M}$ above, then $\mathrm{Ext}^1_F(\mathrm{Fac}M, Y_C) = (0)$ assuming that the relative version of Wakamatsu's Lemma would be true. Indeed, it is true and it is formulated as follows.

LEMMA 4.1 ([**6**, Lemma 4.3]). *Supoose $\mathcal{X}$ is an F-extension closed subcategory of* $\mathrm{mod}\,\Lambda$. *If C is a finitely generated module and*

$$0 \to Y_C \to X_C \to C \to 0$$

is a minimal right $\mathcal{X}$-approximation of C which is F-exact, then $\mathrm{Ext}^1_F(\mathcal{X}, Y_C) = (0)$.

Next we discuss the relative version of the fact that $\Omega^n_\Lambda(\mathrm{mod}\,\Lambda) \cup \mathcal{P}(\Lambda)$ is a functorially finite subcategory of $\mathrm{mod}\,\Lambda$ for all $n \geq 1$ (see [**3**, Corollary 1.8] and [**4**]). The relative version of this result reads as follows.

THEOREM 4.2 ([**6**, Proposition 3.1]). *Let F be additive sub-bifunctor of* $\mathrm{Ext}^1_\Lambda(-,-)$ *as above, and let $\mathcal{X}$ be a functorially finite subcategory in* $\mathrm{mod}\,\Lambda$ *with* $\mathcal{P}(\Lambda) \subseteq \mathcal{X}$. *Then* $\mathcal{X} \cup \Omega^n_F(\mathrm{mod}\,\Lambda)$ *is functorially finite in* $\mathrm{mod}\,\Lambda$ *for all* $n \geq 1$.

For an n-cluster tilting (see [**18**, Section 2]) or an n-precluster tilting (see [**21**, Section 3]) module M the subcategory

$$M^{\perp_{n-1}} = \{X \in \mathrm{mod}\,\Lambda \mid \mathrm{Ext}^i_\Lambda(M, X) = (0) \text{ for } i = 1, 2, \ldots, n-1\}$$

is part of the definition. For an n-cluster tilting module M it is equal to $\mathrm{add}\,M$, while for an n-precluster tilting module M it is equal to ${}^{\perp_{n-1}}M$. We use the above result to show that $M^{\perp_{n-1}}$ is functorially finite when M is a generator-cogenerator in $\mathrm{mod}\,\Lambda$ in Section 8.

The last result we relativize is the Auslander-Reiten Lemma [**4**, Proposition 1.8 and the remark after]. It says that if $\mathcal{Z}$ is an extension closed covariantly finite subcategory of $\mathrm{mod}\,\Lambda$, then

$$\mathcal{X}(\mathcal{Z}) = \{X \in \mathrm{mod}\,\Lambda \mid \mathrm{Ext}^1_\Lambda(X, \mathcal{Z}) = (0)\}$$

is an extension closed contravariantly finite generator in $\mathrm{mod}\,\Lambda$ and all kernels of minimal right $\mathcal{X}(\mathcal{Z})$-approximations are in $\mathcal{Z}$.

THEOREM 4.3 ([**6**, Lemma 4.8]). *Let $\mathcal{Z}$ be an F-extension closed subcategory of* $\mathrm{mod}\,\Lambda$. *If $\mathcal{Z}$ is a covariantly finite subcategory of* $\mathrm{mod}\,\Lambda$, *then*

$$\mathcal{X}_F(\mathcal{Z}) = \{X \in \mathrm{mod}\,\Lambda \mid \mathrm{Ext}^1_F(X, \mathcal{Z}) = (0)\}$$

is an F-extension closed contravariantly finite subcategory of $\operatorname{mod}\Lambda$ *with* $\mathcal{P}(F) \subseteq \mathcal{X}_F(\mathcal{Z})$ *and all kernels of minimal right* $\mathcal{X}_F(\mathcal{Z})$*-approximations are in* $\mathcal{Z}$.

5. Relative cotilting modules

The theory of tilting and cotilting has been one of the success stories of representation theory of artin algebras. So it furnish one natural motivation for generalizing the notion to the relative setting. In standard homological algebra, selfinjective algebras have no non-trivial tilting or cotilting modules. This was yet another motivation for studying relative tilting and cotilting modules, as there exist interesting such modules over selfinjective artin algebras. One final inspiration was to use the theory of relative tilting and cotilting modules to further explore the Wedderburn correspondence. This section is devoted to reviewing some of the most important results on relative cotilting theory. The Wedderburn correspondence is discussed in the next section.

We start by defining a relative cotilting module. To that end, assume throughout that F is an additive sub-bifunctor of $\operatorname{Ext}^1_\Lambda(-,-)$ with enough injectives and enough projectives.

DEFINITION 5.1 ([**7**, Section 3]). A Λ-module T is an *F-cotilting module* if

(i) $\operatorname{Ext}^i_F(T,T) = (0)$ for $i > 0$,
(ii) $\operatorname{id}_F T < \infty$,
(iii) for all modules I in $\mathcal{I}(F)$ there exists an F-exact sequence

$$0 \to T_n \to T_{n-1} \to \cdots \to T_1 \to T_0 \to I \to 0$$

for some non-negative integer n with T_i in $\operatorname{add} T$ for all i.

We see that this is literary a direct transfer of the standard definition to the relative setting. So it is clear that any standard cotilting module is an F-cotilting module for $F = \operatorname{Ext}^1_\Lambda(-,-)$.

We mentioned above that selfinjective artin algebras have interesting relative cotilting modules. The definition, when Λ is a selfinjective artin algebra, is forcing all the projective modules to be a direct summand of any relative cotilting module. On the other hand, any module $T = \Lambda \oplus X$ for some Λ-module X is a relative cotilting module, as we explain next.

EXAMPLE 5.2. Let Λ be a selfinjectve artin algebra, and let X be finitely generated Λ-module. Denote by F the additive sub-bifunctor given by $F = F^{\operatorname{add}(\Lambda\oplus X)}$. Let $T = \Lambda\oplus X$. Then T is an F-cotilting module: As we have $\mathcal{I}(F) = \operatorname{add}(\Lambda\oplus X) = \operatorname{add} T$, it is clear that

(i) $\operatorname{Ext}^i_F(T,T) = (0)$ for $i > 0$,
(ii) $\operatorname{id}_F T = 0 < \infty$,
(iii) the exact sequence $0 \to T \to \Lambda \oplus X \to 0$ is F-exact,

and T is an F-cotilting module.

In fact, for the same reasons, any cogenerator T for $\operatorname{mod}\Lambda$ is an F-cotilting module when $F = F^{\operatorname{add} T}$. So it seems very "easy" to be an relative cotilting module. However, they satisfy the following nice properties. This is a summary of the results [**7**, Corollary 3.4, Theorem 3.2, Proposition 3.7, Theorem 3.13 (d), Proposition 3.19, Corollary 3.6 (a)].

THEOREM 5.3. *Let Λ be an artin algebra with an F-cotilting module T for some additive sub-bifunctor F of $\operatorname{Ext}^1_\Lambda(-,-)$. Let $\Gamma = \operatorname{End}_\Lambda(T)$.*

(a) $\Lambda \simeq \operatorname{End}_\Gamma(T)$.

(b) *For any C in ${}^\perp T = \{X \in \operatorname{mod}\Lambda \mid \operatorname{Ext}^i_F(X,T) = (0)$ for all $i > 0\}$ there is an F-exact sequence*

$$0 \to C \to T^0 \xrightarrow{f^0} T^1 \xrightarrow{f^1} T^2 \xrightarrow{f^2} \cdots$$

with T^i in $\operatorname{add} T$ and $\operatorname{Im} f^i$ in ${}^\perp T$ for all $i \geq 0$.

(c)

$$\operatorname{Ext}^i_F(C,A) \simeq \operatorname{Ext}^i_\Gamma(\operatorname{Hom}_\Lambda(A,T), \operatorname{Hom}_\Lambda(C,T))$$

for all $i \geq 0$ and for all A and C in ${}^\perp T$.

(d) *There is a Γ-module U such that $\operatorname{add} U = \operatorname{Hom}_\Lambda(\mathcal{P}(F),T)$ and U is a cotilting Γ-module with*

$$\operatorname{id}_F T \leq \operatorname{id}_\Gamma U \leq \operatorname{id}_F T + 2.$$

In particular, the number of indecomposable non-isomorphic direct summands of U, in $\mathcal{P}(F)$ and in ${}_\Lambda T$ all coincide.

(e) *The functors*

$$\operatorname{Hom}_\Lambda(-,T)\colon \operatorname{mod}\Lambda \to \operatorname{mod}\Gamma$$

and

$$\operatorname{Hom}_\Gamma(-,T)\colon \operatorname{mod}\Gamma \to \operatorname{mod}\Lambda$$

induce quasi-inverse dualities between

$${}^\perp_\Lambda T \quad \text{and} \quad {}^\perp_\Gamma U = \{Y \in \operatorname{mod}\Gamma \mid \operatorname{Ext}^i_\Gamma(Y,U) = (0) \textit{ for all } i > 0\}.$$

Note that Theorem 5.3 (d) is a correction of the orginal statement in [**7**, Theorem 3.13 (d)] given in [**21**, Theorem 2.9 (e)].

6. Wedderburn correspondence

Here we recall the notion of the Wedderburn correspondence, and we show that it is intimately related to relative cotilting modules.

First we recall the definition of a Wedderburn pair from [**1**, Section 8]. Let Σ be an artin algebra and M a finitely generated Σ-module. We can construct another pair of an artin algebra and a module in the following may:

$$(\Sigma, M) \overset{\operatorname{End}}{\longmapsto} (\operatorname{End}_\Sigma(M)^{\mathrm{op}}, \operatorname{Hom}_\Sigma(M,\Sigma)).$$

Then we can apply the same construction again to obtain $\operatorname{End}^2(\Sigma, M)$. But even though there are natural maps from the pair (Σ, M) to $\operatorname{End}^2(\Sigma, M)$, they are not isomorphic in general. A natural question is, when do they coincide? If M is a generator in $\operatorname{mod}\Sigma$, then it is easy to see that (Σ, M) is isomorphic to $\operatorname{End}^2(\Sigma, M)$. When M is a generator, then in the corresponding pair $(\operatorname{End}_\Sigma(M)^{\mathrm{op}}, \operatorname{Hom}_\Sigma(M,\Sigma))$, the module $\operatorname{Hom}_\Sigma(M,\Sigma)$ is a projective $\operatorname{End}_\Sigma(M)^{\mathrm{op}}$-module and it has the property that

$$\operatorname{End}(\operatorname{End}_\Sigma(M)^{\mathrm{op}}, \operatorname{Hom}_\Sigma(M,\Sigma)) \simeq (\Sigma, M).$$

A projective Σ-module P in $\operatorname{mod}\Sigma$ such that $\operatorname{End}^2(\Sigma, P) \simeq (\Sigma, P)$ is called a *Wedderburn projective module.* A pair (Σ, M) where M is a generator in $\operatorname{mod}\Sigma$ or a Wedderburn projective module, is called a *Wedderburn pair* and M is called a *Wedderburn module.* Hence the map End is an involution on the Wedderburn pairs, and it sets up a correspondence between the pairs (algebras, generators) and (algebras, Wedderburn projectives). This correspondence it is called the *Wedderburn correspondence.* The classical example is the following.

EXAMPLE 6.1. Let $\Sigma = M_n(D)$ for a division ring D and $P = \Sigma e_{11}$, where e_{11} is the natural idempotent matrix with zero entries except for the $(1,1)$-position where it is 1. Then

$$\mathrm{End}(\Sigma, P) = (\mathrm{End}_\Sigma(P) \simeq D, \mathrm{Hom}_\Sigma(P, \Sigma) \simeq D^n)$$

and

$$\mathrm{End}^2(\Sigma, P) \simeq (\Sigma, P).$$

This is the classical Wedderburn correspondence between simple artin rings and divisions rings, hence this is why the above correspondence got it's name.

The notion of Wedderburn correspondence might seem unrelated to relative cotilting modules, but they are intimately connected as the Wedderburn correspondence can be viewed as the composition of a relative cotilting functor and an ordinary cotilting functor as given in the next result.

PROPOSITION 6.2 ([**8**, Proposition 3.5]). *Let Λ be an artin algebra with a generator G for* $\mathrm{mod}\,\Lambda$. *Denote by $F = F_{\mathrm{add}\,G}$ the induced sub-bifunctor of* $\mathrm{Ext}^1_\Lambda(-,-)$, *and let T be an F-cotilting module. Furthermore, let $\Gamma = \mathrm{End}_\Lambda(T)$ and $\Sigma = \mathrm{End}_\Lambda(G)^{\mathrm{op}}$. Then the Wedderburn correspondence functor*

$$\mathrm{Hom}_\Lambda(G, -)\colon \mathrm{mod}\,\Lambda \to \mathrm{mod}\,\Sigma$$

is isomorphic to the composition of the relative cotilting functor

$$\mathrm{Hom}_\Lambda(-, T)\colon \mathrm{mod}\,\Lambda \to \mathrm{mod}\,\Gamma$$

and the standard cotilting functor

$$\mathrm{Hom}_\Gamma(-, \mathrm{Hom}_\Lambda(G, T))\colon \mathrm{mod}\,\Gamma \to \mathrm{mod}\,\Sigma.$$

We apply this result in the last section to show that all endomorphism rings of all F-cotilting modules are derived equivalent (F fixed).

7. Dualizing direct summands of cotilting modules

The results we reviewed about relative cotilting modules in Section 5 show that any relative cotilting module T induces an absolute cotilting module over the endomorphism ring of T. However it is not clear if and when we can make a construction in the other direction. This section is devoted to characterizing when a converse construction is possible.

First we take a closer look at the absolute cotilting module U a relative cotilting module T induces over it's endomorphism ring. In particular, we show that whenever an absolute cotilting module has a special direct summand called dualizing, it is constructed from a relative cotilting module.

Let Λ be an artin algebra and $F = F_{\mathrm{add}\,G}$ for a generator G for $\mathrm{mod}\,\Lambda$. In addition, let T be an F-cotilting module and $\Gamma = \mathrm{End}_\Lambda(T)$. Then $U = \mathrm{Hom}_\Lambda(G, T)$ is a Γ-cotilting module and ${}_\Gamma T = \mathrm{Hom}_\Lambda(\Lambda, T)$ is in $\mathrm{add}\,U$ since is $\mathcal{P}(\Lambda)$ is in $\mathrm{add}\,G$. Furthermore, if

$$P_1 \xrightarrow{d_1} P_0 \xrightarrow{d_0} G \to 0$$

is a projective presentation of the generator G, the sequence

$$0 \to \operatorname{Hom}_\Lambda(G,T) \xrightarrow{d_0^*} \operatorname{Hom}_\Lambda(P_0,T) \xrightarrow{d_1^*} \operatorname{Hom}_\Lambda(P_1,T)$$

is exact with $\operatorname{Hom}_\Lambda(P_0,T)$ and $\operatorname{Hom}_\Lambda(P_1,T)$ in $\operatorname{add}_\Gamma T$ and d_0^* is a left $\operatorname{add}_\Gamma T$-approximation. This captures the essense of the definition of a dualizing direct summand, which is explained by the following result.

PROPOSITION 7.1 ([**8**, Proposition 2.1]). *Let M be an arbitrary module in* $\operatorname{mod}\Lambda$, *and let* $\Gamma = \operatorname{End}_\Lambda(M)$. *For a module X in* $\operatorname{mod}\Lambda$ *the natural homomorphism*

$$\alpha_X : X \to \operatorname{Hom}_\Gamma(\operatorname{Hom}_\Lambda(X,M),M) \tag{7.1}$$

is an isomorphism if and only if there exists an exact sequence

$$0 \to X \xrightarrow{d_0} M_0 \xrightarrow{d_1} M_1$$

with M_0 and M_1 in $\operatorname{add}M$ *and d_0 a left* $\operatorname{add}M$*-approximation.*

DEFINITION 7.2. We keep the notation from Proposition 7.1.

(1) A module X is said be *dualized* by M if it satisfies (7.1).
(2) If $X \simeq M \oplus X'$, then M is called a *dualizing direct summand of X*.

Going back to the setup before Proposition 7.1 we can for short say that ${}_\Gamma T$ is a dualizing direct summand of U if $G = \Lambda \oplus G'$. Maybe the surprizing thing is that we also have the converse.

THEOREM 7.3 ([**8**, Theorem 2.8]). *Let Γ be an artin algebra with a cotilting module $U = T \oplus U'$, and let $\Lambda = \operatorname{End}_\Gamma(T)$ and $F = F_{\operatorname{add}\operatorname{Hom}_\Gamma(U,T)} \subseteq \operatorname{Ext}^1_\Lambda(-,-)$. Then T is a dualizing direct summand of U if and only if T is an F-cotilting module in* $\operatorname{mod}\Lambda$.

This notion of being a dualizing direct summand might seem special, but it occurs frequently as the following example explains.

EXAMPLE 7.4. Let Γ be a Gorenstein artin algebra with $\operatorname{domdim}\Gamma \geq 2$. This means that in the minimal injective resolution

$$0 \to \Gamma \to I^0 \to I^1 \to \cdots$$

of Γ, at least the two first terms I^0 and I^1 are also projective. Let T be a basic Γ-module such that $\operatorname{add}T = \operatorname{add}(I^0 \oplus I^1)$. Then $\operatorname{add}T$ is the additive closure of the maximal injective direct summand of Γ. Since Γ is Gorenstein, Γ is both a tilting and a cotilting module. The above exact sequence then immediately tells us that T is a dualizing direct summand of Γ. Hence by Theorem 7.3, if $\Lambda = \operatorname{End}_\Gamma(T)$, the sub-bifunctor $F = F_{\operatorname{add}\operatorname{Hom}_\Gamma(\Gamma,T)}$ and ${}_\Lambda T = \operatorname{Hom}_\Gamma(\Gamma,T)$, then ${}_\Lambda T$ is an F-cotilting module with $\operatorname{End}_\Lambda(T) \simeq \Gamma$.

Given this example, then it is natural to ask how different properties of Γ are reflected in Λ and the F-cotilting module T. One class of algebras considered in [**9**] was the artin Gorenstein algebras Γ satisfying

$$\operatorname{domdim}\Gamma \geq 2 \geq \operatorname{id}_\Gamma\Gamma.$$

The more general class of artin Gorenstein algebras Γ satisfying the property

$$\operatorname{domdim}\Gamma \geq n+1 \geq \operatorname{id}_\Gamma\Gamma.$$

with $n \geq 1$ was studied in [**21**]. Among other things the following was proved about these algebras in [**9**, **21**].

THEOREM 7.5. *Let* Γ *be an artin Gorenstein algebra with* $\operatorname{domdim}\Gamma \geq 2$*, and let* T *be the maximal injective direct summand of* Γ*. Let* $\Lambda = \operatorname{End}_\Gamma(T)$*, the* Λ*-module* $M = \operatorname{Hom}_\Gamma(\Gamma, T)$*, and the sub-bifunctor* $F = F_{\operatorname{add} G}$ *of* $\operatorname{Ext}^1_\Lambda(-,-)$.

(a) Γ *satifies* $\operatorname{domdim}\Gamma = 2 = \operatorname{id}_\Gamma\Gamma$ *if and only if* $\mathcal{P}(F) = \mathcal{I}(F)$ *with* $\mathcal{P}(\Lambda)$ *a proper subcategory, that is,* $\operatorname{add} M = \mathcal{I}(\Lambda) \cup \operatorname{add} D\operatorname{Tr}(M)$ *with* $\mathcal{P}(\Lambda) \subsetneq \operatorname{add} M$.

(b) *Let* $n \geq 1$*. Then* Γ *satifies* $\operatorname{domdim}\Gamma \geq n+1 \geq \operatorname{id}_\Gamma\Gamma$ *if and only if*
   (i) $D\operatorname{Tr}\Omega^{n-1}_\Lambda(M)$ *and* $\operatorname{Tr} D\Omega^{-(n-1)}_\Lambda(M)$ *are in* $\operatorname{add} M$.
   (ii) $\operatorname{Ext}^i_\Lambda(M, M) = (0)$ *for* $0 < i < n$.
   (iii) M *is a generator-cogenerator for* $\operatorname{mod}\Lambda$ *with* $\Gamma = \operatorname{End}_\Lambda(M)$.

Theorem 7.5 (a) was proved in [**9**, Proposition 5.3 and Proposition 5.5], and it was generalized in [**21**], where a module M satisfying (i) – (iii) is called a *precluster tilting module*. Which in turn is a generalization of a cluster tilting module from [**18**]. For related results see [**13**]. Theorem 7.5 (b) has been independently observed in [**23**]. Furthermore, the algebras in Theorem 7.5 (a) are characterized as the artin algebras where the subcategory of finitely generated Gorenstein projective modules is an abelian category [**22**].

8. Applications

This final section is devoted to giving two applications of relative homological algebra for artin algebras. The applications are quite different as the first one we could have observed already when the papers [**6**–**9**] were written, while the other one only became natural to think about after the introduction of n-cluster and n-precluster tilting modules in the papers [**18**, **21**]. In particular, we show that the endomorphism rings of all relative cotilting modules (for a fixed relative theory) are derived equivalent. In addition, we prove that the subcategory $M^{\perp_{n-1}}$ for a generator-cogenerator M such that M is in $M^{\perp_{n-1}}$ is

$$^{\perp}T = \{X \in \operatorname{mod}\Lambda \mid \operatorname{Ext}^i_F(X, T) = (0) \text{ for all } i > 0\}$$

for an F_M-cotilting module T.

Our first application can be seen as a generalization of the following result about standard (co)tilting modules: Using the fundamental result of Happel about (co)tilting modules over artin algebras, that the endomorphism ring $\operatorname{End}_\Lambda(T)$ of a (co)tilting module T and Λ are derived equivalent [**15**]. Consequently, the endomorphism rings of all (co)tilting modules over Λ are all derived equivalent. This generalizes to relative cotilting modules applying Proposition 6.2.

COROLLARY 8.1. *Let* Λ *be an artin algebra. Let* G *be a generator for* $\operatorname{mod}\Lambda$*, and let* $F = F_G$ *be the induced sub-bifunctor of* $\operatorname{Ext}^1_\Lambda(-,-)$.

(a) *If* T *and* T' *are two* F*-cotilting modules, then* $\operatorname{End}_\Lambda(T)$ *and* $\operatorname{End}_\Lambda(T')$ *are derived equivalent.*

(b) *Let* T *be such that* $\operatorname{add} T = \mathcal{I}(\Lambda) \cup \operatorname{add} D\operatorname{Tr}(G)$*. Then* $\operatorname{End}_\Lambda(G)$ *and* $\operatorname{End}_\Lambda(T)$ *are derived equivalent.*

PROOF. (a) The claim follows directly from Proposition 6.2, as $\operatorname{End}_\Lambda(T)$ and $\operatorname{End}_\Lambda(T')$ are derived dual to $\operatorname{End}_\Lambda(G)^{\mathrm{op}}$.

(b) Let T be such that $\operatorname{add} T = \mathcal{I}(\Lambda) \cup \operatorname{add} D\operatorname{Tr}(G)$. Then T is an injective F-cotilting module. By Proposition 6.2 the algebras $\operatorname{End}_\Lambda(G)$ and $\operatorname{End}_\Lambda(T)$ are derived equivalent. □

Here we can make an example in the spirit of Maurice Auslander.

EXAMPLE 8.2. Let Λ be a selfinjective algebra, and let M be an indecomposable non-projective module. Then $\operatorname{End}_\Lambda(\Lambda \oplus M)$ and $\operatorname{End}_\Lambda(\Lambda \oplus D\operatorname{Tr}(M))$ are derived equivalent.

We end this section with the application to the subcategory $M^{\perp_{n-1}}$ for a generator-cogenerator of $\operatorname{mod}\Lambda$ announced in Section 4.

To this end, we recall the notion of resolution dimension. Let $\mathcal{X}$ be a contravariantly finite subcategory of $\operatorname{mod}\Lambda$ with $\mathcal{P}(F) \subseteq \mathcal{X}$ for some additive sub-bifunctor F of $\operatorname{Ext}^1_\Lambda(-,-)$. Then the resolution dimension of a Λ-module M with respect to $\mathcal{X}$ is the minimum of n including infinity such that there exists an F-exact sequence

$$0 \to X_n \to X_{n-1} \to \cdots \to X_1 \to X_0 \to M \to 0,$$

where X_i is in $\mathcal{X}$. We denote this dimension by $\mathcal{X}\text{-resdim}_F M$. If $\mathcal{Y}$ is a subcategory of $\operatorname{mod}\Lambda$, then $\mathcal{X}\text{-resdim}_F \mathcal{Y} = \sup\{\mathcal{X}\text{-resdim}_F Y \mid Y \in \mathcal{Y}\}$.

In [**21**] we introduced the notion of an n-precluster tilting module M. Among other properties, M is a generator-coenerator satisfying $\operatorname{Ext}^i_\Lambda(M,M) = (0)$ for $i = 1,2,\ldots,n-1$. One property of an n-precluster tilting module is that the subcategories

$$M^{\perp_{n-1}} = \{X \in \operatorname{mod}\Lambda \mid \operatorname{Ext}^i_\Lambda(M,X) = (0) \text{ for } i = 1,2,\ldots,n-1\}$$

and

$$^{\perp_{n-1}}M = \{X \in \operatorname{mod}\Lambda \mid \operatorname{Ext}^i_\Lambda(X,M) = (0) \text{ for } i = 1,2,\ldots,n-1\}$$

are equal, functorially finite and has n-fold almost split extensions (see [**21**] for a definition). A natural question is then, what properties does the subcategory $M^{\perp_{n-1}}$ have for an arbitrary generator-cogenerator M? We show how we can use relative homological algebra to analyze this subcategory given in terms of standard homological algebra.

PROPOSITION 8.3. *Let Λ be a artin algebra, and let M be a generator for* $\operatorname{mod}\Lambda$ *with* $\operatorname{Ext}^i_\Lambda(M,M) = (0)$ *for* $i = 1,2,\ldots,n-1$. *Let* $F = F_M$ *be the induced sub-bifunctor of* $\operatorname{Ext}^1_\Lambda(-,-)$. *Then the following assertions hold.*

(a) *Let* $\mathcal{X} = M^{\perp_{n-1}} = \{X \in \operatorname{mod}\Lambda \mid \operatorname{Ext}^i_\Lambda(M,X) = (0)$ *for* $i = 1,2,\ldots,n-1\}$. *Then* $\mathcal{X}\text{-resdim}_F(\operatorname{mod}\Lambda) \leq n-1$.
(b) $\mathcal{X}$ *is an F-resolving subcategory of* $\operatorname{mod}\Lambda$.

In the following, assume in addition that M is a generator-cogenerator for $\operatorname{mod}\Lambda$.

(c) *Then*

$$\mathcal{X} = \Omega^{n-1}_F(\operatorname{mod}\Lambda) \cup \mathcal{P}(F)$$

and $\mathcal{X}$ *is functorially finite in* $\operatorname{mod}\Lambda$.
(d) *The subcategory* $\mathcal{X} \cap \mathcal{X}^\perp$ *of* $\operatorname{mod}\Lambda$ *is the additive closure of an F-cotilting module T such that* $^\perp T = \mathcal{X}$.

PROOF. (a) Suppose that C satisfies $\operatorname{Ext}^i_\Lambda(M,C) = (0)$ for $i = 1,2,\dots,s \leq n-1$. Let $M_C \xrightarrow{p} C$ be a minimal right $\operatorname{add} M$-approximation of C. Wakamatsu's Lemma implies that $\operatorname{Ext}^1_\Lambda(M, \operatorname{Ker} p) = (0)$, and the exact sequences

$$\operatorname{Ext}^i_\Lambda(M, M_C) \to \operatorname{Ext}^i_\Lambda(M,C) \to \operatorname{Ext}^{i+1}_\Lambda(M, \operatorname{Ker} p) \to \operatorname{Ext}^{i+1}_\Lambda(M, M_C)$$

implies that $\operatorname{Ext}^i_\Lambda(M, \operatorname{Ker} p) = (0)$ for $i = 1,2,\dots,s+1 \leq n-1$. Using this we infer that in a sequence of $n-1$ minimal right $\operatorname{add} M$-approximations of a module X, or equivalently a minimal F-projective resolution of X,

$$0 \to \Omega^{n-1}_F(X) \to M_{n-1} \to \cdots \to M_1 \to M_0 \to X \to 0,$$

the module $\Omega^{n-1}_F(X)$ is in $\mathcal{X}$. Hence $\mathcal{X}\text{-resdim}_F(\operatorname{mod}\Lambda) \leq n-1$.

(b) It is clear by the assumptions that $\mathcal{P}(F) = \operatorname{add} M \subseteq \mathcal{X}$. Suppose that $\eta\colon 0 \to A \to B \to C \to 0$ is F-exact. Then the long exact sequence in absolute homology induced by η shows that $\mathcal{X}$ is closed under F-extensions (even all extensions). Suppose that B and C are in $\mathcal{X}$. Using the same long exact sequence and that η is F-exact, we obtain that $\mathcal{X}$ is closed under kernels of epimorphisms occurring in F-exact sequences. This proves that $\mathcal{X}$ is F-resolving.

(c) In general it follows from (a) that $\Omega^{n-1}_F(\operatorname{mod}\Lambda)$ is contained in $\mathcal{X}$. Clearly, $\mathcal{P}(F) = \operatorname{add} M$ is also contained in $\mathcal{X}$, and we infer that $\Omega^{n-1}_F(\operatorname{mod}\Lambda) \cup \mathcal{P}(F)$ is contained in $\mathcal{X}$.

Now assume in addition that M is a generator-cogenerator for $\operatorname{mod}\Lambda$. Let X be in $\mathcal{X}$, and let $f\colon U \to X$ be any Λ-homomorphism with U in $\mathcal{X}$. Construct the following diagram:

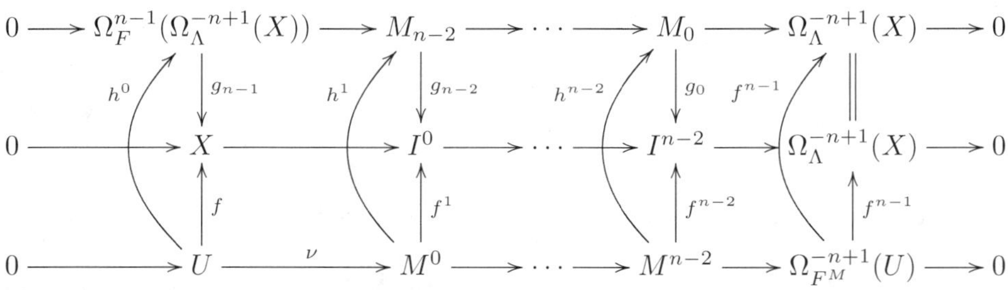

where

(1) the middle row is an injective resolution of X, which is F-exact, since X is in $\mathcal{X}$,

(2) the bottom row is an F^M-injective resolution of U, which is F-exact, since U and M are in $\mathcal{X}$,

(3) the top row is an F-projective resolution of $\Omega^{-n+1}_\Lambda(X)$, by definition F-exact.

The homomorphisms $\{g_i\}_{i=0}^{n-1}$ exist by applying the relative version of the Comparision Theorem, and they make the diagram consisting of the two upper rows commutative. The homomorphisms $\{f^i\}_{i=1}^{n-1}$ exist due to the (relative) Comparision Theorem, and they make the diagram consisting of the two lower rows commutative. The homomorphisms $\{h^i\}_{i=0}^{n-2}$ exist, since the bottom row is an F-projective resolution of $\Omega^{-n+1}_{F^M}(U)$ and the Comparision Theorem, and they make the diagram consisting of the upper and the bottom rows commutative. Since the bottom row is an F-projetive resolution of $\Omega^{-n+1}_{F^M}(U)$ and we have two liftings $\{g_{n-i-1}h^i\}_{i=0}^{n-2}$ and $\{f^i\}_{i=0}^{n-2}$ with $f^0 = f$ of f^{n-1}. It follows that these are homotopic. In particular, we obtain that $f - g_{n-1}h^0 = t\nu$ for some homomorphism $t\colon M^0 \to X$. The homomorphism t factors through the minimal right add M-approximation $g\colon M_X \to X$ of X, so that the homomorphism $(g_{n-1}, g)\colon \Omega_F^{n-1}(\Omega_\Lambda^{n-1}(X)) \oplus M_X \to X$ is a right $\mathcal{X}$-approximation of X. Since we can choose $U = X$ is in $\mathcal{X}$ and $f = 1_X$, we infer that X is a direct summand of $\Omega_F^{n-1}(\Omega_\Lambda^{n-1}(X)) \oplus M_X$. This shows that $\mathcal{X}$ is contained in $\Omega_F^{n-1}(\operatorname{mod}\Lambda) \cup \mathcal{P}(F)$, and we therefore have equality.

By [**6**, Proposition 3.2] or Theorem 4.2 the subcategory $\mathcal{X}$ of $\operatorname{mod}\Lambda$ is functorially finite.

(d) Since $\mathcal{X}$ is an F-resolving contravariantly finite subcategory of $\operatorname{mod}\Lambda$ with $\mathcal{X}\text{-resdim}_F(\operatorname{mod}\Lambda) \leq n-1$, we have by [**7**, Theorem 3.24] that $\mathcal{X} \cap \mathcal{X}^\perp$ is the additive closure of an F-cotilting module. □

REMARK 8.4. Keep the setting and the assumptions from Proposition 8.3 (c) and (d).

(1) The F-cotilting module shown to exist in Proposition 8.3 (d) can be constructed as follows: Let I be such that add $I = \mathcal{I}(F)$. Then consider the sequence of minimal right $\mathcal{X}$-approximations

$$0 \to X_{n-1} \xrightarrow{f_{n-1}} X_{n-2} \xrightarrow{f_{n-2}} \cdots \to X_1 \xrightarrow{f_1} X_0 \xrightarrow{f_0} I \to 0$$

with X_i in $\mathcal{X}$ for all i. Since $\mathcal{X}$ is F-resolving and by the Wakamatsu's Lemma, $\operatorname{Ker} f_i$ is in $\mathcal{X}^\perp$ for $i = 0, 1, \ldots, n-1$. Since I is in $\mathcal{X}^\perp$ and $\mathcal{X}\text{-resdim}_F(\operatorname{mod}\Lambda) \leq n-1$, the last kernel X_{n-1} is also in $\mathcal{X}$. Hence, $T = \oplus_{i=0}^{n-1} X_i$ is in $\mathcal{X} \cap \mathcal{X}^\perp$ and T is an F-cotilting module with add $T = \mathcal{X} \cap \mathcal{X}^\perp$.

(2) Let $\Gamma = \operatorname{End}_\Lambda(T)$, and let $\Sigma = \operatorname{End}_\Lambda(M)^{\mathrm{op}}$. We have that $\mathcal{P}(F) = \operatorname{add} M$. By Proposition 6.2 the functor $\operatorname{Hom}_\Lambda(M, -)\colon \operatorname{mod}\Lambda \to \operatorname{mod}\Sigma$ is isomorphic to the composition of the two functors $\operatorname{Hom}_\Lambda(-, T)\colon \operatorname{mod}\Lambda \to \operatorname{mod}\Gamma$ and $\operatorname{Hom}_\Sigma(-, \operatorname{Hom}_\Lambda(M, T))\colon \operatorname{mod}\Gamma \to \operatorname{mod}\Sigma$. Recall that $\operatorname{Hom}_\Lambda(M, T)$ is a standard cotilting module over Γ. Hence Γ and Σ are derived dual of each other, or equivalently, $\operatorname{End}_\Lambda(T)$ and $\operatorname{End}_\Lambda(M)$ are derived equivalent.

References

[1] Maurice Auslander, *Representation theory of Artin algebras. I, II*, Comm. Algebra **1** (1974), 177–268; ibid. 1 (1974), 269–310, DOI 10.1080/00927877408548230. MR0349747 Math., vol. 37, Dekker, New York, 1978, 1–244.

[2] Maurice Auslander, *Functors and morphisms determined by objects*, Representation theory of algebras (Proc. Conf., Temple Univ., Philadelphia, Pa., 1976), Dekker, New York, 1978, pp. 1–244. Lecture Notes in Pure Appl. Math., Vol. 37. MR0480688

[3] Maurice Auslander and Idun Reiten, *Homologically finite subcategories*, Representations of algebras and related topics (Kyoto, 1990), London Math. Soc. Lecture Note Ser., vol. 168, Cambridge Univ. Press, Cambridge, 1992, pp. 1–42. MR1211476

[4] Maurice Auslander and Idun Reiten, *Applications of contravariantly finite subcategories*, Adv. Math. **86** (1991), no. 1, 111–152, DOI 10.1016/0001-8708(91)90037-8. MR1097029

[5] Maurice Auslander and Idun Reiten, *Syzygy modules for Noetherian rings*, J. Algebra **183** (1996), no. 1, 167–185, DOI 10.1006/jabr.1996.0212. MR1397392

[6] M. Auslander and Ø. Solberg, *Relative homology and representation theory. I. Relative homology and homologically finite subcategories*, Comm. Algebra **21** (1993), no. 9, 2995–3031, DOI 10.1080/00927879308824717. MR1228751

[7] ———, *Relative homology and representation theory* II, *Relative cotilting theory*, Comm. in Alg., 21 (9), 3033–3079.

[8] ———, *Relative homology and representation theory* III, *Cotilting modules and Wedderburn correspondence*, Comm. in Alg., 21 (9), 3081–3097.

[9] ______, *Gorenstein algebras and algebras with dominant at least* 2, Comm. in Alg., 21 (11), 3897–3934.

[10] Aslak Bakke Buan, *Closed subbifunctors of the extension functor*, J. Algebra **244** (2001), no. 2, 407–428, DOI 10.1006/jabr.2001.8858. MR1857752

[11] David A. Buchsbaum, *A note on homology in categories*, Ann. of Math. (2) **69** (1959), 66–74, DOI 10.2307/1970093. MR0140556

[12] M. C. R. Butler and G. Horrocks, *Classes of extensions and resolutions*, Philos. Trans. Roy. Soc. London Ser. A **254** (1961/1962), 155–222, DOI 10.1098/rsta.1961.0014. MR0188267 Math. Res. Not. IMRN 2016, no. 22, 6979–7037.

[13] Hongxing Chen and Steffen Koenig, *Ortho-symmetric modules, Gorenstein algebras, and derived equivalences*, Int. Math. Res. Not. IMRN **22** (2016), 6979–7037. MR3632073

[14] Peter Dräxler, Idun Reiten, Sverre O. Smalø, and Øyvind Solberg, *Exact categories and vector space categories*, Trans. Amer. Math. Soc. **351** (1999), no. 2, 647–682, DOI 10.1090/S0002-9947-99-02322-3. With an appendix by B. Keller. MR1608305

[15] Dieter Happel, *Triangulated categories in the representation theory of finite-dimensional algebras*, London Mathematical Society Lecture Note Series, vol. 119, Cambridge University Press, Cambridge, 1988. MR935124

[16] Alex Heller, *Homological algebra in abelian categories*, Ann. of Math. (2) **68** (1958), 484–525, DOI 10.2307/1970153. MR0100622

[17] G. Hochschild, *Relative homological algebra*, Trans. Amer. Math. Soc. **82** (1956), 246–269, DOI 10.2307/1992988. MR0080654

[18] Osamu Iyama, *Higher-dimensional Auslander-Reiten theory on maximal orthogonal subcategories*, Adv. Math. **210** (2007), no. 1, 22–50, DOI 10.1016/j.aim.2006.06.002. MR2298819 210 (2007), no. 1, 51–82.

[19] Osamu Iyama, *Auslander correspondence*, Adv. Math. **210** (2007), no. 1, 51–82, DOI 10.1016/j.aim.2006.06.003. MR2298820

[20] Osamu Iyama, *Cluster tilting for higher Auslander algebras*, Adv. Math. **226** (2011), no. 1, 1–61, DOI 10.1016/j.aim.2010.03.004. MR2735750

[21] Osamu Iyama and Øyvind Solberg, *Auslander-Gorenstein algebras and precluster tilting*, Adv. Math. **326** (2018), 200–240, DOI 10.1016/j.aim.2017.11.025. MR3758429 Represent. Theory 17 (2014), no. 4, 1289–1301.

[22] Fan Kong, *Characterizing when the category of Gorenstein projective modules is an abelian category*, Algebr. Represent. Theory **17** (2014), no. 4, 1289–1301, DOI 10.1007/s10468-013-9448-5. MR3228489

[23] Kong, F., *Generalization of the Correspondence about* $D\,\mathrm{Tr}$*-selfinjective algebras*, arXiv:1204.0967.

[24] Magdalini Lada, *Relative homology and maximal l-orthogonal modules*, J. Algebra **321** (2009), no. 10, 2798–2811, DOI 10.1016/j.jalgebra.2009.02.015. MR2512627

[25] Takayoshi Wakamatsu, *Stable equivalence for self-injective algebras and a generalization of tilting modules*, J. Algebra **134** (1990), no. 2, 298–325, DOI 10.1016/0021-8693(90)90055-S. MR1074331

Institutt for matematiske fag, NTNU, N–7491 Trondheim, Norway
Email address: `oyvind.solberg@ntnu.no`

Contemporary Mathematics
Volume **716**, 2018
https://doi.org/10.1090/conm/716/14429

Coxeter groups and quiver representations

Hugh Thomas

In this expository note, I will attempt to showcase the relevance of Coxeter groups to quiver representations. The subjects which I will discuss are:

(1) real and imaginary roots,
(2) reflection functors,
(3) torsion free classes and c-sortable elements.

The first two of these topics are classical, going back respectively to Kac [**Kac**] and Bernstein-Gelfand-Ponomarev [**BGP**]. Let Q be a quiver with n vertices having no oriented cycles. Associated to Q is a Coxeter group W which acts on $\mathbb{Z}^n$. Real and imaginary roots in $\mathbb{Z}^n_{\geq 0} \subseteq \mathbb{Z}^n$ are defined using W, and characterize the dimension vectors of the indecomposable representations of Q. This W-action on a set containing the dimension vectors of representations of Q suggests the idea that W might act directly on representations. Such an action does not exist. Reflection functors are, in a sense, the best we can do towards defining such an action. For this classical material I will not give proofs, but I will provide pointers to the textbooks [**DDPW**] and [**Kir**]. These are particularly convenient for us since they take quiver representations as their starting point.

The third topic which I will cover is a more recent development. Torsion free classes of $\operatorname{rep} Q$ are full subcategories closed under extensions and subrepresentations. It is natural to want to classify them. It turns out that the torsion free classes that contain only finitely many isomorphism classes of indecomposables correspond to c-sortable elements of W, a notion which was introduced by Reading [**Re**]. This correspondence was first established in Dynkin type in [**IT**]. The general result was shown by [**AIRT**] as a consequence of their study of preprojective algebras. In this note, we will give a more elementary proof, essentially following the argument of [**IT**], but without restricting to the Dynkin case.

This note is based on an expository talk given at the Maurice Auslander international conference in 2013. I am very grateful to the organizers for having been invited to speak there, and for their patience in waiting for me to write up the talk. I am also happy to acknowledge financial support from NSERC and the Canada Research Chairs program. I thank the referees for their helpful comments which improved the paper.

1. Getting started

Let Q be a quiver without oriented cycles. (It would be possible to work over a general hereditary Artin algebra, but for concreteness, I prefer not to. For parts of this note, e.g. in Section 2, one could also relax the prohibition on oriented cycles in Q.) Suppose that Q has n vertices, numbered 1 to n.

We write $\operatorname{rep} Q$ for the representations of Q over a fixed ground field k. Of our main references, note that [**DDPW**] works over an arbitrary ground field, while [**Kir**] works only over an algebraically closed ground field. Since for the topics we present, the choice of a general ground field does not introduce any additional complications, we do not assume that k is algebraically closed.

Write S_i for the simple representation which has a copy of k at vertex i and the 0 vector space elsewhere.

For $V \in \operatorname{rep} Q$, write V_i for the vector space over vertex i. Write $\underline{\dim} V$ for the n-tuple $(\dim V_1, \ldots, \dim V_n) \in \mathbb{Z}^n$.

1.1. The Euler-Ringel form. There is an extremely useful bilinear form on $\mathbb{Z}^n$.

THEOREM 1.1 ([**Kir**, Theorem 1.25], [**DDPW**, Proposition 1.9(2)]). *The following equation defines a bilinear form on $\mathbb{Z}^n$, known as the Euler-Ringel form:*

$$\langle \underline{\dim} V, \underline{\dim} W \rangle = \dim \operatorname{Hom}(V, W) - \dim \operatorname{Ext}^1(V, W)$$

It is not obvious that this defines a bilinear form! In order for it to do so, the righthand side must depend only on $\underline{\dim} V$ and $\underline{\dim} W$, and not on the specific representations V and W. On the face of it, one would not expect this to be true. Indeed, the dimensions of $\operatorname{Hom}(V, W)$ and of $\operatorname{Ext}^1(V, W)$ *do* depend on the specific representations, rather than on their dimension vectors. However, taking the difference, we get a quantity which only depends on $\underline{\dim} V$ and $\underline{\dim} W$.

Once we know that this the righthand side only depends on $\underline{\dim} V$, $\underline{\dim} W$, it is easy to see that it is linear, and to write down an explicit formula for the pairing, because we can choose the most convenient representations of the given dimension vectors. If we want to evaluate $\langle \beta, \gamma \rangle$, the simplest choice is to take semisimple representations, i.e., representations which are direct sums of simple representations. Since $\operatorname{Hom}(S_i, S_j)$ is one-dimensional if $i = j$ and zero-dimensional otherwise, and the dimension of $\operatorname{Ext}^1(S_i, S_j)$ is the number of arrows from i to j, we find that

$$\langle \beta, \gamma \rangle = \sum_{i=1}^{n} \beta_i \gamma_i - \sum_{(a:i \to j) \in Q} \beta_i \gamma_j.$$

Note that this bilinear form is not symmetric. We should view the fact that we are naturally given a non-symmetric bilinear form as a piece of very good luck. Sometimes, though, it is useful to have a symmetric bilinear form. When we want one, it is easy to get one:

$$(\beta, \gamma) = \langle \beta, \gamma \rangle + \langle \gamma, \beta \rangle$$

If, on the contrary, we only had the symmetric form, we would not know how to unsymmetrize it in a meaningful way. (This is not a hypothetical situation! I got into representation theory from Coxeter groups, and, working on Coxeter groups, I became convinced that I needed to break the symmetry of the symmetric bilinear form, but, not knowing anything about quiver representations, I had no idea how to do it.)

1.2. The Weyl group associated to Q. For $\beta \in \mathbb{Z}^n$ with $(\beta, \beta) \neq 0$, we can define a reflection on $\mathbb{Q}^n$, which we denote t_β, by:

$$t_\beta(\gamma) = \gamma - 2\frac{(\beta,\gamma)}{(\beta,\beta)}\beta$$

This preserves the symmetric form, fixes all vectors orthogonal to β, and sends β to $-\beta$. If $|(\beta,\beta)|$ divides 2, then t_β restricts to a map from $\mathbb{Z}^n$ to $\mathbb{Z}^n$.

It is convenient to let e_i denote the standard basis vector consisting of all zeros, with a 1 in the i-th position. Equivalently, we can say $e_i = \underline{\dim}\, S_i$.

Let us define $s_i = t_{e_i}$. Note that, by our definition of the symmetric form, $(e_i, e_i) = 2$, so $s_i = t_{e_i}$ is well-defined as a map from $\mathbb{Z}^n$ to $\mathbb{Z}^n$. We refer to the elements s_i as *simple reflections.*

Define W to be the group generated by the elements s_i. Since this is a group generated by reflections, it is what is called a *reflection group.* The generators satisfy some relations which it is easy to verify:

- $s_i^2 = e$,
- if i and j are not adjacent, then $s_i s_j = s_j s_i$,
- if there is exactly one arrow between i and j, then $s_i s_j s_i = s_j s_i s_j$,
- for clarity, let me add that if there is more than one arrow between i and j, then there is no corresponding relation.

It is natural to wonder if there are any other relations. It turns out that there are not.

THEOREM 1.2 ([**DDPW**, Theorem 4.2(1)], [**Kir**, Theorem A.11]). *The group W can be defined abstractly as the group generated by n elements $s_1, \dots, s_n$ subject to the above relations.*

The group W is therefore a Coxeter group. For background on Coxeter groups, see [**BB**, **Hu**].

Let us remark at this point that the group W does not depend on the orientation of the arrows of the quiver; that is to say, it depends only on the underlying graph structure. This is in marked contrast to $\operatorname{rep} Q$, which depends highly on the orientation. Nonetheless, as we shall see, there are connections between the representations of different orientations of the same quiver.

EXAMPLE 1.1. An example to bear in mind always is the case that the quiver is a path with the vertices numbered 1 to n in order along it, and the arrows oriented arbitrarily. We refer to such a quiver as being of type A_n. The corresponding Weyl group is isomorphic to the symmetric group on $n+1$ letters; we can take the isomorphism to send s_i to the adjacent transposition $(i\ i+1)$. In examples, we represent an element w of the symmetric group in *one-line notation*, by writing down the sequence $w(1), \dots, w(n+1)$.

By definition, each element of W can be expressed as a product of the simple reflections. Such an expression for w of minimal length is called *reduced.* We write $\ell(w)$ for the length of a reduced expression for w. Typically an element w will have many reduced expressions.

For $w \in W$ and $s \in S$, if w has a reduced expression beginning with s, then it is easy to see that $\ell(sw) = \ell(w) - 1$. It turns out that if w does not have a reduced expression beginning with s, then $\ell(sw) = \ell(w) + 1$. (When I write "beginning with," I mean with respect to the usual left-to-right order in which English is read.)

Define $\Phi_{re} = \{we_i \mid w \in W, 1 \le i \le n\}$, and $\Phi^+_{re} = \Phi_{re} \cap \mathbb{Z}^n_{\ge 0}$. The elements of Φ^+_{re} are called the real roots for Q.

EXAMPLE 1.2. For type A_n, we find that Φ^+_{re} consists of the sums $e_i + e_{i+1} + \cdots + e_j$ for $1 \le i \le j \le n$.

Let $s_{i_1} \dots s_{i_r}$ be a reduced expression for $w \in W$. Define the inversion set corresponding to the reduced expression to be the following set of real roots: $\{e_{i_1}, s_{i_1}e_{i_2}, s_{i_1}s_{i_2}e_{i_3}, \dots, s_{i_1}s_{i_2}\dots s_{i_{r-1}}e_{i_r}\}$. This can also be viewed as the collection of positive roots which are sent to negative roots by w^{-1}. Note that this latter description does not depend on the choice of reduced expression for w; indeed, we have the following proposition:

PROPOSITION 1.1 ([**BB**, Corollary 1.4.4, Corollary 1.4.5, and Corollary 3.1.4]). *For $w \in W$, all the reduced words for w yield the same inversion set. Different elements of W have different inversion sets. The size of the inversion set of w is $\ell(w)$.*

We write $\mathrm{inv}(w)$ for the inversion set of w.

EXAMPLE 1.3. In type A_n, for $j \ge i$, we have $e_i + e_{i+1} + \cdots + e_j \in \mathrm{inv}(w)$ if and only if $j+1$ precedes i in the one-line notation for w.

In type A_2, the six elements of W, their expressions in one-line notation, and their corresponding inversion sets, are as follows:

e	123	$\emptyset$
s_1	213	$\{e_1\}$
s_2	132	$\{e_2\}$
s_1s_2	231	$\{e_1, e_1+e_2\}$
s_2s_1	312	$\{e_2, e_1+e_2\}$
$s_1s_2s_1$	321	$\{e_1, e_1+e_2, e_2\}$

We have listed the inversion sets in the order corresponding to the reduced expressions given above. If we take a different reduced word for the same element, considering $s_2s_1s_2$ instead of $s_1s_2s_1$, we get the roots e_2, e_1+e_2, e_1, in that order. Consistent with Proposition 1.1, the set of roots is the same as appears in the table for $s_1s_2s_1$, though the order in which they appear is different.

1.3. Finite versus infinite. There is a basic dichotomy in root system combinatorics. In our setting, it manifests in the following theorem:

THEOREM 1.3. *The following are equivalent:*

- *W is finite,*
- rep Q *has only finitely many indecomposables up to isomorphism (in which case we say Q is of* finite representation type*),*
- *Q consists of a finite number of Dynkin quivers, i.e., quivers of type A_n ($n \ge 1$), D_n ($n \ge 4$), E_6, E_7, or E_8.*

We have already met quivers of type A_n. A quiver of type D_n is a quiver with n vertices, obtained by orienting the following graph:

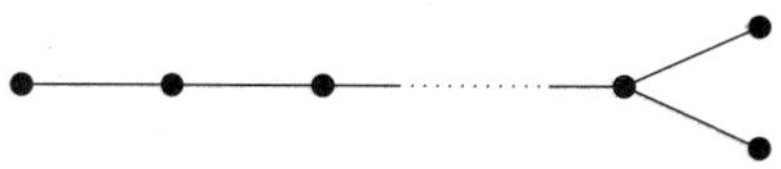

The quivers of type E_6, E_7, and E_8 are obtained by orienting the graphs shown below:

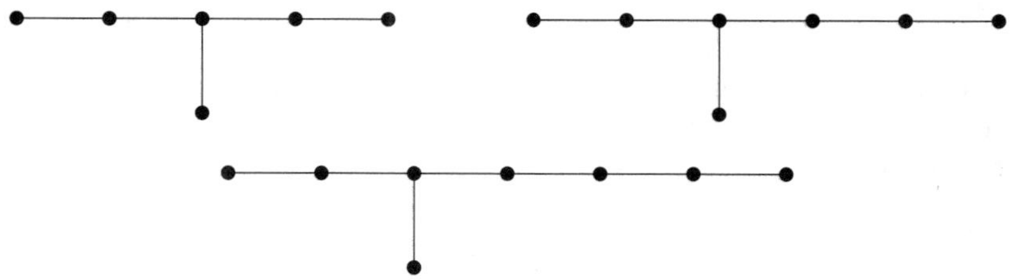

The equivalence of (ii) and (iii) is the celebrated theorem of Gabriel [**DDPW**, Theorem 1.23], [**Kir**, Theorem 3.3]. The equivalence of (i) and (iii) follow from the classification of finite Coxeter groups, see for example [**Kir**, Theorem A.11].

2. Real and imaginary roots

It turns out that Φ^+_{re}, as defined above, is intimately connected to the representation theory of Q.

THEOREM 2.1 ([**Kir**, Theorem 7.49], [**DDPW**, Remark 1.26]). *There is a unique indecomposable representation of Q whose dimension vector is α for each $\alpha \in \Phi^+_{re}$.*

This theorem was established under the assumption that k is algebraically closed by Kac [**Kac**]. As pointed out by Schofield [**Sch**], Kac's proof also applies to any field of characteristic p. The characteristic zero case is proved in [**Sch**]. Note that exceptionally [**DDPW**] only states the result for algebraically closed ground fields.

If Q is of finite representation type, then the representations of Theorem 2.1 exhaust the indecomposables. However, if Q is not of finite type, there is another class of roots, the *imaginary roots.*

Let M be the cone in $\mathbb{Z}^n_{\geq 0}$ consisting of vectors with connected support and such that for $\alpha \in M$, we have $(\alpha, e_i) \leq 0$. The imaginary roots are all elements of $\mathbb{Z}^n$ of the form $w\alpha$ for $w \in W$ and $\alpha \in M$. The following result is due to Kac [**Kac**]:

THEOREM 2.2 ([**Kir**, Theorem 7.49 and Theorem A.20]). *Let k be algebraically closed. The dimension vectors of indecomposable representations of Q are the positive real roots together with the imaginary roots. The positive real roots are those dimension vectors for which there exists a unique indecomposable representation, while the imaginary roots are those dimension vectors for which there are infinitely many non-isomorphic indecomposable representations.*

This theorem obviously requires some modification to accommodate finite fields; in fact, a full analogue of this theorem for general fields is not known.

3. Reflection functors

In this section we define the reflections functors, which play a decisive role in the representation theory of quivers of finite representation type, but which, as we will see, can also be important for general quiver representations. The recognition of their importance goes back to the classic paper of Bernstein-Gelfand-Ponomarev [**BGP**].

We say that a vertex i is a *sink* of Q if all the arrows incident to i point towards i. Dually, we say i is a *source* if all arrows incident to i point away from it.

If i is a sink or a source, we define $\mu_i(Q)$ to be the quiver obtained by changing the direction of all the arrows incident to i.

Suppose that i is a sink of Q. Let V be a representation of Q. We want to define a representation $R_i^+(V)$ of $Q' = \mu_i(Q)$.

First, we define the vector spaces:

$$R_i^+(V)_\ell = \begin{cases} V_\ell & \text{for } \ell \neq i \\ \ker\left(\displaystyle\bigoplus_{(a:j\to i)\in Q} V_j \right) \to V_i & \text{for } \ell = i \end{cases}$$

Note that the direct sum runs over all arrows pointing to i in Q. In the case that there is more than one arrow from j to i, the summand V_j will be repeated. The map whose kernel we take is the direct sum of the maps associated to the arrows indexing the sum.

We also have to define the linear maps associated to the arrows of $\mu_i(Q)$. For the arrows which are also in Q, we notice that neither the source nor the target vector spaces of the arrow have changed, so we can and do simply reuse the map from V. It remains to define the maps corresponding to the arrows in $\mu_i(Q)$ which point away from i. We use the maps coming from the natural inclusion of $R_i^+(V)_i$ into the sum over all the arrows from j to i in Q of V_j.

EXAMPLE 3.1. Let Q be the quiver of type A_3 shown below. Then $Q' = \mu_2(Q)$ is also as shown.

Below, in the lefthand column, we list the six indecomposable representations of Q, and in the righthand column, the representation of Q' to which it is sent by R_2^+ (which is indecomposable or, in the first case, the zero representation). We only give the vector spaces at each vertex; the maps between consecutive copies of k are non-zero.

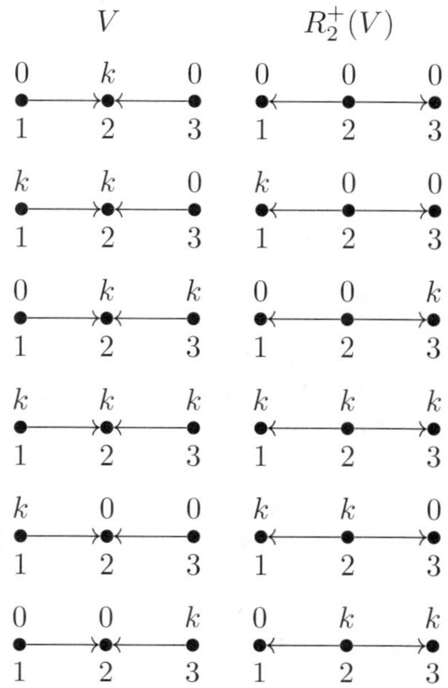

In order for R_i^+ to be a functor, for $f \in \operatorname{Hom}(V, W)$, we must define $R_i^+(f) \in \operatorname{Hom}(R_i^+(V), R_i^+(W))$. For $\ell \neq i$, we define $R_i^+(f)_\ell = f_\ell$. To define $R_i^+(f)_i$, we consider the following commutative diagram:

$$\begin{array}{ccccccc} 0 \longrightarrow & R_i^+(V)_i & \rightarrow & \bigoplus_{(a:j\to i)\in Q} V_j & \longrightarrow & V_i \\ & \vdots & & \downarrow & & \downarrow \\ 0 \longrightarrow & R_i^+(W)_i & \rightarrow & \bigoplus_{(a:j\to i)\in Q} W_j & \longrightarrow & W_i \end{array}$$

The rows of the diagram are the exact sequences which define $R_i^+(V)_i$ and $R_i^+(W)_i$. The vertical undotted arrows are maps defined by f; the dotted arrow is the arrow we want to define. We can see in the diagram a map from $R_i^+(V)_i$ to $\bigoplus_{(a:j\to i)} W_j$. We observe that if we compose it with the map from $\bigoplus_{(a:j\to i)} W_j$ to W_i, the result will be zero, which means that in fact the image of $R_i^+(V)_i$ falls in $R_i^+(W)_i$. We take this map from $R_i^+(V)_i$ to $R_i^+(W)_i$ to be the definition of $R_i^+(f)_i$.

We have gone into considerable detail describing R_i^+; the definition of R_i^- which is defined when i is a source is dual in the natural way. Let me mention only the key point that $R_i^-(V)_i$ is defined to be the cokernel of the natural map from V_i to $\bigoplus_{(a:i\to j)} V_j$. We then have the following theorem:

THEOREM 3.1 ([**DDPW**, Theorem 1.18], [**Kir**, Theorem 3.10]). *Let i be a sink in Q, and let $Q' = \mu_i(Q)$. Then R_i^+ is a functor from representations of Q to representations of Q', and R_i^- is a functor from representations of Q' to representations of Q.*

Let us note some important properties of R_i^+ and R_i^-.

PROPOSITION 3.1. *Let i be a sink in Q, and let $Q' = \mu_i(Q)$. We write S_i for the simple representation at i of Q, and S'_i for the simple representation at i of Q'.*

(1) *$R_i^+(S_i) = 0$ and $R_i^-(S'_i) = 0$.*
(2) *If $V \in \operatorname{rep} Q$ has no direct summand isomorphic to S_i, then*

$$\underline{\dim}\, R_i^+(V) = s_i(\underline{\dim}(V)).$$

Similarly if $V' \in \operatorname{rep} Q'$ has no direct summand isomorphic to S'_i then

$$\underline{\dim}\, R_i^-(V') = s_i(\underline{\dim}(V')).$$

(3) *R_i^+ is left exact; R_i^- is right exact.*
(4) *If $V \in \operatorname{rep} Q$, then $R_i^-(R_i^+(V))$ is isomorphic to the direct sum of the indecomposable summands of V which are not isomorphic to S_i (and a similar dual statement).*
(5) *The image of $R_i^+(\operatorname{rep} Q)$ consists of all representations of Q' not having S'_i as an indecomposable summand (and a similar dual statement).*

PROOF. (1) is immediate from the definition.

For (2), we prove on the first statement (about R_i^+). Suppose that V has no direct summands isomorphic to S_i. In this case, the map $\bigoplus_{(a:j\to i)} V_j \to V_i$ is surjective. This means that

$$\dim R_i^+(V)_i = \left(\sum_{(a:j\to i)\in Q} \dim V_j \right) - \dim V_i,$$

while $\dim R_i^+(V)_j = \dim V_j$ for $j \neq i$. This claim follows. Property (2) explains both the name ("reflection functor") and one of the reasons that this topic is included in this note.

For (3), the statement is clear except over vertex i. There, it follows from an application of the Snake Lemma. For another proof, see [**Kir**, Theorem 3.10] or [**DDPW**, Theorem 1.18].

(4) follows from applying (2) to the decomposition of V into indecomposables.

For (5), we check directly that for $M \in \operatorname{rep} Q$, we have that $R_i^+(M)$ has no S'_i direct summand. The fact that all representations without such a summand are in the image follows from (4). □

At the beginning of this section, we defined $\mu_i(Q)$ to be the result of reversing the arrows incident to i, provided that i was either a sink or a source. Clearly, this definition could still be applied if i is neither a sink nor a source, but it turns out that, from the point of view of representation theory, it is not the right thing to do: the right thing is to apply (Fomin-Zelevinsky) mutation, but the whole story becomes more complicated. See [**DWZ**] for more on this.

4. Torsion free classes and c-sortable elements

We are now ready to introduce c-sortable elements of W and finite torsion free classes in $\operatorname{rep} Q$. We will see that they have very similar inductive structures, which we will exploit to construct a bijection between them.

4.1. The correspondence between Coxeter elements and orientations of a quiver. Let $|Q|$ be the unoriented graph underlying the quiver Q. As was already mentioned, the Weyl group associated to Q only depends on $|Q|$. A Coxeter element for W is, by definition, a product of each of the Coxeter generators of W in some order. The following fact is easy but important.

LEMMA 4.1. *There is a bijection between the acyclic orientations of $|Q|$ and the Coxeter elements of W.*

PROOF. Associate to Q the product of the Coxeter generators in which s_i precedes s_j if there are one or more arrows from j to i in the quiver. This does not prescribe the order of all pairs of Coxeter generators, but any pair whose order is not prescribed is a pair whose corresponding vertices are not connected by an arrow, and which therefore commute. The fact that it is possible to find a linear order on the generators satisfying this prescription follows from the fact that the orientation of Q has no cycles.

Conversely, any Coxeter element determines an acyclic orientation of $|Q|$ by orienting any edges between i and j towards i if s_i precedes s_j in the Coxeter element. □

We write c_Q for the Coxeter element associated to Q.

EXAMPLE 4.1. Let Q_1 and Q_2 be as shown below:

In the Coxeter element associated to Q_1, we have s_3 before s_2 and s_2 before s_1, so $c_{Q_1} = s_3s_2s_1$. In the Coxeter element associated to Q_2, we have s_2 before s_1 and s_2 before s_3. We therefore could write $c_{Q_2} = s_2s_1s_3$ or $c_{Q_2} = s_2s_3s_1$; these are equal because $s_1s_3 = s_3s_1$.

4.2. Definition of c-sortable elements. Let $c = s_{i_1} \dots s_{i_n}$ be a Coxeter element for W. For $J \subseteq \{1, \dots, n\}$, we write c_J for the subword of c whose reflections are those s_j with $j \in J$.

An element $w \in W$ is called c-sortable if it has a reduced expression $w = c_{J_1}c_{J_2}\dots c_{J_k}$ with $J_1 \supseteq J_2 \supseteq \dots \supseteq J_k$.

EXAMPLE 4.2. In the Weyl group of type A_2, isomorphic to S_3, setting $c = s_1s_2$, five of the six group elements are c-sortable. The one which is not is s_2s_1. In the Weyl group of type A_3, for each choice of c there are 14 c-sortable elements. In general, the c-sortable elements in type A_n are counted by the n-th Catalan number, $\frac{1}{n+2}\binom{2n+2}{n+1}$. See [**Re**, Theorem 9.1], which gives complete information about enumeration in all finite types.

The c-sortable elements of a Coxeter group W were defined by Reading [**Re**] as a bridge between the combinatorics of finite-type cluster algebras and the noncrossing partitions of W. They play a central role in the Coxeter-theoretic approach to the combinatorics of cluster algebras, as developed by Reading and Speyer in a sequence of articles [**RS1, RS2, RS3**]; they have also turned out to have interesting applications in the representation theory of preprojective algebras, see [**AIRT**].

There is an extremely useful inductive characterization of c-sortable elements. In the theorem below, we use the notation $W_{\langle s_j \rangle}$ for the Coxeter group generated by the reflections s_i for $i \neq j$. If W is the Coxeter group corresponding to the quiver Q, then $W_{\langle s_j \rangle}$ is the Coxeter group for the quiver with vertex j *removed.* (While initially counter-intuitive, this notation is standard and very useful.)

THEOREM 4.1 ([**Re**, Lemma 2.4, Lemma 2.5, and Remark 2.7]). *Let $w \in W$, and let $c = s_{i_1} \dots s_{i_n}$.*

(1) *If $\ell(s_{i_1}w) > \ell(w)$, then w is c-sortable iff w is a $s_{i_2} \dots s_{i_n}$-sortable element in $W_{\langle s_{i_1} \rangle}$.*
(2) *If $\ell(s_{i_1}w) < \ell(w)$, then w is c-sortable iff $s_{i_1}w$ is $s_{i_2}s_{i_3} \dots s_{i_n}s_{i_1}$-sortable.*

PROOF. Suppose first that $\ell(s_{i_1}w) > \ell(w)$. This means that no reduced expression for w can begin with s_{i_1}. If w is c-sortable, then, since it does not use the s_{i_1} in the first copy of c, it can never use any s_{i_1}. Therefore it is in $W_{\langle s_{i_1} \rangle}$; the $s_{i_2} \dots s_{i_n}$-sortability there is immediate; the converse is also clear.

Now suppose that $\ell(s_{i_1}w) < \ell(w)$. This means that there is a reduced expression for w beginning with s_{i_1}. Suppose that w is c-sortable. Since $w \notin W_{\langle s_{i_1} \rangle}$, any expression for w must use s_{i_1} at least once, so the expression which manifests that w is c-sortable must begin with s_{i_1}. Therefore, it expresses w as s_{i_1} times a reduced word for $s_{i_1}w$, which, when we unwind the definition, must be $s_{i_2} \dots s_{i_n}s_{i_1}$-sortable. This is exactly what we wanted. The converse argument works the same way. □

EXAMPLE 4.3. In type A_2, with $c = s_1s_2$, the first case of Theorem 4.1 applies to the c-sortable elements e and s_2, while the second case applies to s_1, s_1s_2, and $s_1s_2s_1$, The non-c-sortable element s_2s_1 falls into the first case of the theorem, which says that it is not c-sortable because it is not an s_2-sortable element of $W_{\langle s_1 \rangle}$ (obviously true, since it is not an element of $W_{\langle s_1 \rangle}$ at all).

4.3. The correspondence between c-sortable elements and finite torsion free classes. A torsion free class in $\operatorname{rep} Q$ is a full additive subcategory closed under extensions and subrepresentations. That is to say, if $\mathcal{F}$ is a torsion free class and Y is in $\mathcal{F}$, so are all the subrepresentations of Y, and if $X, Z \in \mathcal{F}$, such that there is an extension

$$0 \to X \to Y \to Z \to 0,$$

then $Y \in \mathcal{F}$.

We say that a torsion free class is finite if it contains finitely many indecomposable representations up to isomorphism. If Q is a Dynkin quiver, then all torsion free classes will be finite. Conversely, if all torsion free classes in $\operatorname{rep} Q$ are finite, then Q is a Dynkin quiver, since the category of all representations is itself a torsion free class.

For w a c_Q-sortable element of W, define $\mathcal{F}(w)$ to be the full subcategory of $\operatorname{rep} Q$ consisting of direct sums of copies of the indecomposable representations whose dimension vectors are roots in $\operatorname{inv}(w)$.

We can now state the theorem whose proof occupies the remainder of the paper.

THEOREM 4.2. *The map $\mathcal{F}$ is a bijection from c_Q-sortable elements of W to finite torsion free classes in* $\operatorname{rep} Q$.

EXAMPLE 4.4. Let Q be the quiver $1 \leftarrow 2$. Then the correspondence between c_Q-sortable elements of W and torsion free classes is as follows:

w	$\mathcal{F}(w)$
e	0
s_1	$\operatorname{add} S_1$
s_2	$\operatorname{add} S_2$
s_1s_2	$\operatorname{add} S_1, P_2$
$s_1s_2s_1$	$\operatorname{add} S_1, P_2, S_2$

Here, we have written P_2 for the non-simple indecomposable representation, and we have written add for the additive category generated by the subsequent list of indecomposable representations. As is easly verified, the righthand column consists of the torsion free classes of $\operatorname{rep} Q$.

4.4. The inductive structure of torsion free classes. In order to prove Theorem 4.2, we will exhibit an inductive structure to torsion free classes which parallels the inductive structure of c-sortable elements. Suppose $\mathcal{F}$ is a torsion free class for $\operatorname{rep} Q$. Let i be a sink of Q, so S_i is a simple projective. We want to prove a structure theorem about $\mathcal{F}$, parallel to Theorem 4.1. We split into two cases, based on whether $S_i \in \mathcal{F}$.

PROPOSITION 4.1. *If $S_i \notin \mathcal{F}$, then $\mathcal{F}$ is supported on $Q \setminus \{i\}$, and thus defines a torsion free class in the representations of $Q \setminus \{i\}$.*

PROOF. Recall that for $M \in \operatorname{rep} Q$, we have that $\dim \operatorname{Hom}(S_i, M) = \dim M_i$, since S_i is projective. Thus, if $M_i \neq 0$, then $\operatorname{Hom}(S_i, M) \neq 0$. And note that in this case, the nonzero maps from S_i to M would have to be injections, because S_i has no non-zero subrepresentations. That means that if $M_i \neq 0$, we must have that $M \notin \mathcal{F}$, since otherwise the inclusion of S_i into M would contradict the fact that $S_i \notin \mathcal{F}$ but $M \in \mathcal{F}$. $\square$

PROPOSITION 4.2. *If $S_i \in \mathcal{F}$, then $R_i^+\mathcal{F}$ is a torsion free class for $\mu_i(Q)$ which does not contain S_i. If $\mathcal{F}$ contains finitely many indecomposable representations up to isomorphism, then $\mathcal{F}'$ contains one fewer.*

PROOF. Let us write Q' for $\mu_i(Q)$. We will write X', Y', etc. for representations of Q' and X, Y, etc. for representations of Q. Let $\mathcal{F}' = R_i^+\mathcal{F}$. We begin by proving that $\mathcal{F}'$ is closed under subrepresentations. Let $X' \in \mathcal{F}'$. Since $X' \in R_i^+(\operatorname{rep} Q)$, we know that X' has no direct summands isomorphic to S_i'. Let Y' be a subrepresentation of X'.

Let $Y = R_i^-(Y')$ and $X = R_i^-(X')$. Since $X' \in \mathcal{F}'$, we know that $X \in \mathcal{F}$. Since R_i^- is only right exact, it is not necessarily the case that Y is a subrepresentation of X. However, R_i^- applied to the inclusion from Y' to X' gives a map from Y to X. The image of this map is a subrepresentation of X, so lies in $\mathcal{F}$, and the kernel of this map is a sum of copies of S_i, which also lies in $\mathcal{F}$. Thus, since $\mathcal{F}$ is extension closed, it also contains Y, and thus $Y' \in \mathcal{F}'$, showing that $\mathcal{F}'$ is closed under subrepresentations.

We next check that $\mathcal{F}'$ is closed under extensions. Suppose that we have an extension of Q' representations:

$$0 \to Y' \to X' \to Z' \to 0$$

with Y' and Z' in $\mathcal{F}'$.

Any indecomposable representation of Q' that admits a nonzero morphism from S'_i is isomorphic to S'_i. By assumption, neither Y' nor Z' has an S'_i direct summand, so $\operatorname{Hom}(S'_i, Y') = 0 = \operatorname{Hom}(S'_i, Z')$. By the Hom long exact sequence, it follows that $\operatorname{Hom}(S'_i, X') = 0$, so X' does not have any direct summands isomorphic to S'_i either. Applying R_i^-, we get

$$R_i^-(Y') \to R_i^-(X') \to R_i^-(Z') \to 0, \tag{1}$$

since R_i^- is right exact. Since none of X', Y', Z' have any S'_i summands, the dimension vector of $R_i^-(X')$ is $s_i(\underline{\dim}(X'))$, and similarly for Y' and Z'. Since s_i is a linear map, $s_i\,\underline{\dim}(X') = s_i\,\underline{\dim}(Y') + s_i\,\underline{\dim}(Z')$, and it follows that (1) is exact on the left as well. Now $R_i^-(Y)$ and $R_i^-(Z)$ are in $\mathcal{F}$, so, since (1) is exact on the left as well, we can conclude that $R_i^-(X')$ is also in $\mathcal{F}$, and $X' \in \mathcal{F}'$, as desired.

The final statement follows from the fact that R_i^+ sends non-isomorphic indecompolsables other than S_i to non-isomorphic indecomposables, while killing S_i. □

The previous two propositions tell us that, given i a sink of Q and $\mathcal{F}$ a torsion free class in $\operatorname{rep} Q$, there is a way to associate to it a torsion free class that is "smaller" in a suitable sense: either smaller because it is defined on a smaller quiver (the case of Proposition 4.1) or smaller because it contains fewer indecomposables (the case of Proposition 4.2).

We now want to reverse the process and show that, given a candidate "smaller" torsion free class, it is possible to recover a torsion free class in $\operatorname{rep} Q$.

The first case, reversing Proposition 4.1, is obvious, but we state it to be able to refer to it later.

PROPOSITION 4.3. *If $\mathcal{F}$ is a torsion free class in the category of representations of $Q \setminus \{i\}$, then it is also a torsion free class in the category of representations of Q, via the inclusion of* $\operatorname{rep} Q \setminus \{i\}$ *into* $\operatorname{rep} Q$.

The second case, reversing Proposition 4.2, requires a proof, but it is similar to that of Proposition 4.2.

PROPOSITION 4.4. *Let $\mathcal{F}'$ be a torsion free class in* $\operatorname{rep} Q'$. *Let $\mathcal{F}$ be the additive hull of $R_i^- \mathcal{F}'$ and S_i. Then $\mathcal{F}$ is a torsion free class in* $\operatorname{rep} Q$.

PROOF. Let X be in $\mathcal{F}$, and let Y be a subrepresentation of X. We want to show that $Y \in \mathcal{F}$. Let $X' = R_i^+(X)$, and let $Y' = R_i^+(Y)$. Since R_i^+ kills S_i, $R_i^+\mathcal{F} = R_i^+(R_i^-\mathcal{F}') \subseteq \mathcal{F}'$. Since R_i^+ is left exact, Y' is a subrepresentation of X', and thus is in $\mathcal{F}'$. Now Y is isomorphic to the direct sum of $R_i^+(Y')$ with some copies of S_i, and both of these are in $\mathcal{F}$, so Y is in $\mathcal{F}$.

Now suppose that we have a short exact sequence

$$0 \to Y \to X \to Z \to 0$$

with Y and Z in $\mathcal{F}$. We want to show that $X \in \mathcal{F}$.

Applying R_i^+, we get an exact sequence:

$$0 \to R_i^+(Y) \to R_i^+(X) \to R_i^+(Z)$$

By our assumptions, $R_i^+(Y)$ and $R_i^+(Z)$ are both in $\mathcal{F}'$. In order to get exactness on the right, we can replace $R_i^+(Z)$ by the cokernel of the map from $R_i^+(Y)$ to $R_i^+(X)$. Note that it is isomorphic to a subrepresentation of $R_i^+(Z)$, and therefore it is also in $\mathcal{F}'$. It follows that $R_i^+(X) \in \mathcal{F}'$, and thus that X is in $\mathcal{F}$. □

Note that we actually only use the above proposition in the case that $\mathcal{F}'$ does not contain S_i', but we give the more general statement, since the proof is no different.

4.5. Proof of Theorem 4.2. We will prove Theorem 4.2 by establishing the next two propositions.

PROPOSITION 4.5. *Let $\mathcal{F}$ be a finite torsion free class in* $\operatorname{rep} Q$. *Then there is some c-sortable element w such that $\mathcal{F} = \mathcal{F}(w)$.*

PROOF. The proof is by induction on the number of vertices of Q and the number of indecomposable representations of $\mathcal{F}$.

Let the first reflection in c_Q be s_i. By the definition of c_Q, vertex i is a sink of Q. We split into two cases depending on whether or not $S_i \in \mathcal{F}$.

If $S_i \notin \mathcal{F}$, then by Proposition 4.1, $\mathcal{F}$ is a torsion free class in the representations of $Q \setminus \{i\}$. By induction, there is a $c_{Q\setminus\{i\}}$-sortable element w of $W_{\langle i \rangle}$ such that $\mathcal{F} = \mathcal{F}(w)$. This w is also c_Q-sortable in W by Theorem 4.1, and we are done.

If $S_i \in \mathcal{F}$, then by Proposition 4.2, $\mathcal{F}' = R_i^+\mathcal{F}$ is a torsion free class for $Q' = \mu_i(Q)$. By induction, $\mathcal{F}' = \mathcal{F}(w')$ for w' a $c_{Q'}$-sortable element. Since $\mathcal{F}'$ does not contain S_i', we know $\operatorname{inv}(w')$ does not contain e_i. Therefore, no reduced expression for w' begins with s_i, and $\ell(s_i w') = \ell(w') + 1$. Let $w = s_i w'$. Since $c_{Q'} = s_i c_Q s_i$, Theorem 4.1 tells us that w is c_Q-sortable. Now:

$$\begin{aligned} \operatorname{inv}(w) &= \{e_i\} \cup \{s_i\alpha \mid \alpha \in \operatorname{inv}(w')\} \\ &= \{\underline{\dim}\, S_i\} \cup \{s_i\, \underline{\dim}\, M' \mid M' \text{ indecomposable in } \mathcal{F}'\} \\ &= \{\underline{\dim}\, M \mid M \text{ indecomposable in } \mathcal{F}\}, \end{aligned}$$

and this shows that w is the desired c-sortable element of W. □

PROPOSITION 4.6. *Let w be a c_Q-sortable element of W. Then $\mathcal{F}(w)$ is a finite torsion free class.*

PROOF. The proof is by induction on the number of vertices of Q and the number of inversions of w. Let the first reflection of c_Q be s_i. Then vertex i is a sink of Q. We split into two cases depending on whether $\ell(s_i w) > \ell(w)$ or $\ell(s_i w) < \ell(w)$.

Suppose first that $\ell(s_i w) > \ell(w)$. Since w is c_Q-sortable, Theorem 4.1 tells us that w is $c_{Q\setminus i}$-sortable in $W_{\langle i \rangle}$. By induction, $\mathcal{F}(w)$ is a torsion free class in $\operatorname{rep} Q \setminus \{i\}$. Now Proposition 4.3 tells us that $F(w)$ is also a torsion free class in $\operatorname{rep} Q$.

Suppose next that $\ell(s_i w) < \ell(w)$. Theorem 4.1 tells us that $s_i w$ is $s_i c_Q s_i$-sortable. We note that this means that $s_i w$ is $c_{\mu_i(Q)}$-sortable, so, by induction, $\mathcal{F}(s_i w)$ is a torsion free class for $\mu_i(Q)$. We know that

$$\begin{aligned} \operatorname{inv}(w) &= \{e_i\} \cup \{s_i\alpha \mid \alpha \in \operatorname{inv}(s_i w)\} \\ &= \{\underline{\dim}\, S_i\} \cup \{\underline{\dim}\, R_i^- M \mid M \text{ indecomposable in } \mathcal{F}(s_i w)\} \end{aligned}$$

Proposition 4.4 now tells us that $\mathcal{F}(w)$ is a torsion free class, as desired. □

This completes the proof of Theorem 4.2.

References

[AIRT] Claire Amiot, Osamu Iyama, Idun Reiten, and Gordana Todorov, *Preprojective algebras and c-sortable words*, Proc. Lond. Math. Soc. (3) **104** (2012), no. 3, 513–539, DOI 10.1112/plms/pdr020. MR2900235

[BGP] I. N. Bernšteĭn, I. M. Gel′fand, and V. A. Ponomarev, *Coxeter functors, and Gabriel's theorem* (Russian), Uspehi Mat. Nauk **28** (1973), no. 2(170), 19–33. MR0393065

[BB] Anders Björner and Francesco Brenti, *Combinatorics of Coxeter groups*, Graduate Texts in Mathematics, vol. 231, Springer, New York, 2005. MR2133266

[DDPW] Bangming Deng, Jie Du, Brian Parshall, and Jianpan Wang, *Finite dimensional algebras and quantum groups*, Mathematical Surveys and Monographs, vol. 150, American Mathematical Society, Providence, RI, 2008. MR2457938

[DWZ] Harm Derksen, Jerzy Weyman, and Andrei Zelevinsky, *Quivers with potentials and their representations. I. Mutations*, Selecta Math. (N.S.) **14** (2008), no. 1, 59–119, DOI 10.1007/s00029-008-0057-9. MR2480710

[Hu] James E. Humphreys, *Reflection groups and Coxeter groups*, Cambridge Studies in Advanced Mathematics, vol. 29, Cambridge University Press, Cambridge, 1990. MR1066460

[IT] Colin Ingalls and Hugh Thomas, *Noncrossing partitions and representations of quivers*, Compos. Math. **145** (2009), no. 6, 1533–1562, DOI 10.1112/S0010437X09004023. MR2575093

[Kac] V. G. Kac, *Infinite root systems, representations of graphs and invariant theory*, Invent. Math. **56** (1980), no. 1, 57–92, DOI 10.1007/BF01403155. MR557581

[Kir] Alexander Kirillov Jr., *Quiver representations and quiver varieties*, Graduate Studies in Mathematics, vol. 174, American Mathematical Society, Providence, RI, 2016. MR3526103

[Sch] Aidan Schofield, *The field of definition of a real representation of a quiver Q*, Proc. Amer. Math. Soc. **116** (1992), no. 2, 293–295, DOI 10.2307/2159732. MR1072091

[Re] Nathan Reading, *Clusters, Coxeter-sortable elements and noncrossing partitions*, Trans. Amer. Math. Soc. **359** (2007), no. 12, 5931–5958, DOI 10.1090/S0002-9947-07-04319-X. MR2336311

[RS1] Nathan Reading and David E. Speyer, *Cambrian fans*, J. Eur. Math. Soc. (JEMS) **11** (2009), no. 2, 407–447, DOI 10.4171/JEMS/155. MR2486939

[RS2] Nathan Reading and David E. Speyer, *Sortable elements in infinite Coxeter groups*, Trans. Amer. Math. Soc. **363** (2011), no. 2, 699–761, DOI 10.1090/S0002-9947-2010-05050-0. MR2728584

[RS3] Nathan Reading and David E. Speyer, *Cambrian frameworks for cluster algebras of affine type*, Trans. Amer. Math. Soc. **370** (2018), no. 2, 1429–1468, DOI 10.1090/tran/7193. MR3729507

Département de mathématiques, Université du Québec à Montréal, Montréal, Québec, Canada

Email address: hugh.ross.thomas@gmail.com

Contemporary Mathematics
Volume **716**, 2018
https://doi.org/10.1090/conm/716/14430

Tree modules and limits of the approximation theory

Jan Trlifaj

ABSTRACT. In this expository paper, we present a construction of tree modules and combine it with (infinite dimensional) tilting theory and relative Mittag-Leffler conditions in order to explore limits of the approximation theory of modules. We also present a recent generalization of this construction due to Šaroch which applies to factorization properties of maps, and yields a solution of an old problem by Auslander concerning existence of almost split sequences.

Introduction

Relative homological algebra has gained momentum over the past two decades by the discovery of a number of new classes suitable for approximations (precovers and preenvelopes) by Enochs et al. [**16**], [**20**], and by the discovery of direct connections between complete cotorsion pairs and model category structures on complexes of modules by Hovey [**17**], [**23**]. The rich supply of approximations makes it possible to solve various particular problems of module theory by employing fitting approximations, while the model category structures enable us to compute morphisms between objects of the unbounded derived category of modules as morphisms between their cofibrant and fibrant replacements modulo chain homotopy.

Classes providing for approximations appeared ubiquitous in the early 2000s. A first warning sign came in the form of an independence result by Eklof and Shelah [**14**]: they proved that whether the class of all Whitehead groups is precovering or not depends on the extension of ZFC that we work in. Later on, the class of all $\aleph_1$-free groups was shown not to be precovering in ZFC (cf. [**18**, §5]). Since $\aleph_1$-free groups coincide with the flat Mittag-Leffler ones, the focus moved to Mitag-Leffler conditions and their relative versions defining various classes of locally free modules. Relative Mittag-Leffler conditions were known to be connected to (infinite dimensional) tilting theory by the work of Angeleri and Herbera [**1**]. However, even the basic case of (absolute) Mittag-Leffler modules resisted, with gradual improvements obtained in [**10**] and [**33**].

2010 *Mathematics Subject Classification.* Primary 16G70, 18G25; Secondary 03E75, 05C05, 16D70, 16D90, 16E30.

Key words and phrases. Almost split sequences, approximations of modules, infinite dimensional tilting theory, set-theoretic homological algebra, tree modules.

Research supported by GAČR 17-23112S.

The recent complete solution for the general case involving relative Mittag-Leffler conditions in [**4**] and [**32**] has revealed a surprising fact: despite the complexity of the set-theoretic homological algebra proofs, the modules witnessing non-existence of approximations can always be taken small, meaning countably generated. Their construction in the absolute case goes back to a classic work of Bass [**9**], where countably generated flat non-projective modules were constructed over any non-right perfect ring. The witnessing countably generated modules in the general setting have therefore been called *Bass modules*.

Tilting theory turned out to be helpful in passing from the absolute to the relative cases in [**4**] and [**32**]. If T is an (infinitely generated) tilting module, then the corresponding relative Mittag-Leffler modules are called *locally T-free*. Their class turns out to be precovering, iff T is Σ-pure split, that is, iff each pure submodule of a module in Add (T) splits. For example, each Σ-pure-injective tilting module is Σ-pure split, and hence so are all finitely generated tilting modules over any artin algebra. However, already for hereditary artin algebras of infinite representation type, there do exist infinitely generated tilting modules T that are not Σ-pure split [**2**]. So the non-precovering phenomenon for locally T-free modules does occur even in the artin algebra setting.

The key technical tool in proving non-existence of approximations (or factorization properties) of modules are the *tree modules*. Their construction, and its generalizations and applications, are the main topic of this expository paper. In Section 1, we define Bass modules and the tree modules induced by them first in the basic (absolute) case, and then in the general setting. We present their principal properties, and the limits that they impose on the approximation theory of modules. This part is based on [**4**], [**32**], and [**34**].

In Section 2, we present a recent generalization of the tree module construction due to Šaroch [**31**], used to answer a 40 years old question by Auslander on existence of almost split sequences from [**5**]. We will discuss his proof of the fact that for every associative unital ring R, a module M is the right term in an almost split sequence in Mod–R, only if M is finitely presented and has local endomorphism ring.

In what follows, R will denote an associative (but not necessarily commutative) ring with unit, and Mod–R the category of all (unitary right R-) modules. Moreover, given an infinite cardinal κ and a module M, we call M *κ-presented* (*$< \kappa$-presented*) provided that M is the direct limit of a direct system $\mathcal{D}$ of cardinality $\leq \kappa$ ($< \kappa$) such that $\mathcal{D}$ consists of finitely presented modules. For a class of modules $\mathcal{C}$, we will use the notation $\mathcal{C}^{<\kappa}$ and $\mathcal{C}^{\leq\kappa}$ to denote the subclass of $\mathcal{C}$ consisting of all $< \kappa$-presented modules, and κ-presented modules, respectively.

The notation mod–R stands for the class of all *strongly finitely presented* modules, i.e, the modules possessing a projective resolution consisting of finitely generated projective modules. For example, if R is a right coherent ring, then mod–$R =$ (Mod–$R)^{<\omega}$ is the class of all finitely presented modules.

1. Tree modules and their applications

1.1. Tree modules. The tree modules that we consider here were originally developed, in various mutations, for the setting of abelian groups by Eklof, Mekler and Shelah in order to study almost free groups with various additional properties (e.g., the $\aleph_1$-separable and the Whitehead ones, cf. [**12**, XVII.2] and [**13**]).

Our version here is based on the tree T_κ of all finite sequences of ordinals less than a given infinite cardinal κ. This version was employed e.g. in [**19**]. Our tree modules arise by a uniform decoration of the trees T_κ with Bass modules. In order to explain the basic idea in more detail, we need to introduce further notation.

Let κ be an infinite cardinal. The nodes of the *tree* T_κ are the maps (or sequences) of the form $\tau : n \to \kappa$ where $n < \omega$. Here, $n = \ell(\tau)$ is the *length* of τ. The partial order $\leq$ on T_κ is defined by $\tau' \leq \tau$, if $\ell(\tau') \leq \ell(\tau)$ and $\tau \restriction \ell(\tau') = \tau'$. In other words, $\tau' \leq \tau$, if the sequence τ' forms an initial segment of the sequence τ.

For example, $\tau' = (0,3,1) \leq \tau = (0,3,1,0)$, but $\tau' = (1) \nleq \tau = (3,3)$. Notice that the tree T_κ has cardinality κ, and each node of T_κ has κ immediate successors.

However, the branches of T_κ are short: they correspond 1-1 to countable sequences of ordinals $< \kappa$, that is, to the maps $\nu : \omega \to \kappa$. In particular, $\mathrm{Br}(T_\kappa)$, the set of all branches of T_κ, has cardinality κ^ω.

REMARK 1.1. Of course, $\kappa \leq \kappa^\omega$. We will especially be interested in the cardinals κ such that $\kappa < \kappa^\omega = 2^\kappa$. It is easy to see that there is a very good supply – a proper class – of such cardinals: for each cardinal μ, there is $\kappa \geq \mu$ such that $\kappa^\omega = 2^\kappa$ (see e.g. [**20**, 8.26]).

NOTATION 1.2. Let $\mathcal{F}$ be a set of countably presented modules. A *Bass module* for $\mathcal{F}$ is a module B which is a direct limit of a countably infinite direct system of modules from $\mathcal{F}$. Possibly taking a cofinal subset, we can w.l.o.g. express B as a direct limit $B = \varinjlim_{i<\omega} F_i$ where $\mathfrak{D} = (F_i, f_{ji} \mid i \leq j < \omega)$ is a direct system of modules from $\mathcal{F}$ indexed by ω. This is the direct system that will be used for decoration of branches of the tree T_κ.

The following particular example of a Bass module is a model one. Later on, we will see that the decoration of T_κ by this module always yields a flat Mittag-Leffler module:

EXAMPLE 1.3. Let $\mathcal{F} = \{R\}$, $a_i \in R$, and $f_i : R \to R$ be the left multiplication by a_i for each $i < \omega$. Then the chain $R \overset{f_0}{\to} R \overset{f_1}{\to} \dots R \overset{f_i}{\to} R \overset{f_{i+1}}{\to} \dots$ yields a direct system $\mathcal{D}$ (with $f_{ji} = f_{j,j-1} ... f_{i+1,i}$ for all $i < j < \omega$) whose direct limit is the countably presented module B with the presentation $0 \to R^{(\omega)} \overset{f}{\to} R^{(\omega)} \to B \to 0$ where f maps the ith term 1_i of the canonical basis of $R^{(\omega)}$ to $1_i - 1_{i+1}.a_i$. This is the *classic Bass module.*

By [**9**], B is a flat module of projective dimension at most 1, and if B is projective, then the chain of principal left ideals

$$Ra_0 \supseteq Ra_1a_0 \supseteq \dots Ra_n...a_0 \supseteq Ra_{n+1}a_n...a_0 \supseteq ...$$

stabilizes. In particular, if R is not right perfect, then there exists a classic Bass module which is not projective.

REMARK 1.4. If $\mathcal{F}$ is any class of countably generated projective modules, then the Bass modules over $\mathcal{F}$ are flat and countably presented, and hence of projective dimension at most 1. Conversely, each countably presented flat module is a Bass module for some class of finitely generated free modules, see e.g. [**20**, 2.23].

NOTATION 1.5. Now we turn to the construction of the tree modules. The idea is to combine the facts that all branches of the tree T_κ have length ω and the

Bass module B is a direct limit of the direct system $\mathfrak{D} = (F_i, f_{ji} \mid i \leq j < \omega)$ of modules indexed in ω, and uniformly decorate the branches of T_κ with the members of $\mathcal{D}$. The resulting tree module L is defined as a submodule of the product $P = \prod_{\tau \in T_\kappa} F_{\ell(\tau)}$ as follows:

For each $\nu \in \mathrm{Br}(T_\kappa)$, $i < \omega$, and $x \in F_i$, we define $x_{\nu i} \in P$ by

$$\pi_{\nu \restriction i}(x_{\nu i}) = x,$$

$$\pi_{\nu \restriction j}(x_{\nu i}) = f_{ji}(x) \text{ for each } i < j < \omega,$$

$$\pi_\tau(x_{\nu i}) = 0 \text{ otherwise},$$

where $\pi_\tau \in \mathrm{Hom}_R(P, F_{\ell(\tau)})$ denotes the τth projection for each $\tau \in T_\kappa$.

Let $X_{\nu i} = \{x_{\nu i} \mid x \in F_i\}$. Then $X_{\nu i}$ is a submodule of P isomorphic to F_i. Further, let $X_\nu = \sum_{i<\omega} X_{\nu i}$, and $L = \sum_{\nu \in \mathrm{Br}(T_\kappa)} X_\nu$. The module L is the *tree module* corresponding to κ and to the presentation of B as the direct limit of the direct system $\mathcal{D}$.

Here is our first observation concerning the module L:

LEMMA 1.6. (1) *$X_\nu \cong \bigoplus_{i<\omega} F_i$ for each $\nu \in Br(T_\kappa)$.*

(2) *Let $D = \bigoplus_{\tau \in T_\kappa} F_{\ell(\tau)}$. Then $D \subseteq L \subseteq P$, and we have the tree module exact sequence*

$$0 \to D \overset{\subseteq}{\to} L \to B^{(\mathrm{Br}(T_\kappa))} \to 0.$$

PROOF. 1. Let $Y_{\nu i} = \sum_{j \leq i} X_{\nu i}$. Then $Y_{\nu i} = \bigoplus_{j<i} F_j \oplus X_{\nu i}$, and the inclusion $Y_{\nu i} \subseteq Y_{\nu, i+1}$ splits, because there is a split exact sequence

$$0 \to X_{\nu i} \overset{p}{\hookrightarrow} F_i \oplus X_{\nu, i+1} \overset{q}{\twoheadrightarrow} F_{i+1} \to 0$$

where $p(x_{\nu i}) = x + (f_i(x))_{\nu, i+1}$, and $q(z + x_{\nu, i+1}) = x - f_i(z)$.

2. Let $\nu \in \mathrm{Br}(T_\kappa)$. Then $B \cong (X_\nu + D)/D$. Indeed, for each $i < \omega$, we can define $g_i : F_i \to (X_\nu + D)/D$ by $g_i(x) = x_{\nu i} + D$. Then $((X_\nu + D)/D, g_i \mid i \in I)$ is the direct limit of the direct system $\mathfrak{D}$.

Since each element of X_ν is a sequence in P whose τth component is zero for all $\tau \notin \{\nu \restriction i \mid i < \omega\}$, the modules $((X_\nu + D)/D \mid \nu \in \mathrm{Br}(T_\kappa))$ are independent. Thus $L/D = \bigoplus_{\nu \in \mathrm{Br}(T_\kappa)} (X_\nu + D)/D \cong B^{(\mathrm{Br}(T_\kappa))}$. $\square$

1.2. Locally $\mathcal{F}$-free modules. Tree modules can be used to construct locally free modules. Before introducing them, we recall several useful definitions.

DEFINITION 1.7. Let $\mathcal{C}$ be a class of modules. A module M is *$\mathcal{C}$-filtered* (or a *transfinite extension* of the modules in $\mathcal{C}$), provided that there exists an increasing chain $\mathcal{M} = (M_\alpha \mid \alpha \leq \sigma)$ of submodules of M with the following properties:

(1) $M_0 = 0$,
(2) $M_\alpha = \bigcup_{\beta<\alpha} M_\beta$ for each limit ordinal $\alpha \leq \sigma$,
(3) $M_{\alpha+1}/M_\alpha \cong C_\alpha$ for some $C_\alpha \in \mathcal{C}$, and
(4) $M_\sigma = M$.

The chain $\mathcal{M}$ is a *$\mathcal{C}$-filtration* (of length σ) of the module M. The class of all $\mathcal{C}$-filtered modules will be denoted by $\mathrm{Filt}(\mathcal{C})$.

A class $\mathcal{C} \subseteq \text{Mod–}R$ is *deconstructible* provided there exists a subset $\mathcal{C}' \subseteq \mathcal{C}$ such that $\mathcal{C} = \mathrm{Filt}(\mathcal{C}')$. If moreover $\mathcal{C}'$ can be chosen so that $\mathcal{C}' \subseteq \mathcal{C}^{<\kappa}$ for an infinite cardinal κ, then $\mathcal{C}$ is called *κ-deconstructible*.

Transfinite extensions include direct sums and extensions:

EXAMPLE 1.8. 1. Let $M = \bigoplus_{\alpha<\sigma} C_\alpha$ be a direct sum of copies of modules from $\mathcal{C}$. Then M is $\mathcal{C}$-filtered (just take $M_\alpha = \bigoplus_{\beta<\alpha} C_\beta$ for each $\alpha \leq \sigma$).

2. Let $0 \to C_1 \to M \to C_2 \to 0$ be an extension of the modules $C_1, C_2 \in \mathcal{C}$. Then M is $\mathcal{C}$-filtered (take $M_0 = 0$, $M_1 = C_1$, and $M_2 = M$).

The class $\operatorname{Filt}(\mathcal{C})$ is obviously closed under transfinite extensions (i.e., $\operatorname{Filt}(\operatorname{Filt}(\mathcal{C})) = \operatorname{Filt}(\mathcal{C})$), and hence under extensions and (arbitrary) direct sums.

However, $\operatorname{Filt}(\mathcal{C})$ is larger than the closure of $\mathcal{C}$ under extensions and direct sums in general. For example, if p is a prime integer and $\mathcal{C} = \{\mathbb{Z}_p\} \subseteq \text{Mod–}\mathbb{Z}$, then the latter closure does not contain any non-zero p-divisible abelian group, while $\operatorname{Filt}(\mathcal{C})$ is the class of all abelian p-groups.

Deconstructible classes provide for approximations, and make it possible to do relative homological algebra. We postpone discussing them in more detail after introducing basics of the approximation theory later in this Section.

We will now turn to locally free modules. We recall the notation of [**22**] and [**34**]:

DEFINITION 1.9. Let R be a ring, M a module, and κ an infinite regular cardinal.

A system $\mathcal{S}$ consisting of $< \kappa$-presented submodules of M satisfying the conditions

(1) $\mathcal{S}$ is closed under unions of well-ordered ascending chains of length $< \kappa$, and

(2) each subset $X \subseteq M$ such that $\operatorname{card} X < \kappa$ is contained in some $N \in \mathcal{S}$,

is called a *κ-dense system* of submodules of M.

Notice that in the setting of Definition 1.9, M is the directed union of the modules from $\mathcal{S}$.

In order to connect this notion with the setting of 1.2, we consider κ-dense systems consisting of countably $\mathcal{F}$-filtered modules:

DEFINITION 1.10. Let $\mathcal{F}$ be a set of countably presented modules. Denote by $\mathcal{C}$ the class of all modules possessing a countable $\mathcal{F}$-filtration. Let M be a module.

Then M is *locally $\mathcal{F}$-free* provided that M contains an $\aleph_1$-dense system $\mathcal{S}$ consisting of submodules from $\mathcal{C}$. The system $\mathcal{S}$ is said to *witness* the locally $\mathcal{F}$-freeness of M.

Notice that if M is countably generated, then M is locally $\mathcal{F}$-free, iff $M \in \mathcal{C}$. We also note the following result from [**34**]:

LEMMA 1.11. *Let $\mathcal{F}$ be a set of countably presented modules. Then the class of all locally $\mathcal{F}$-free modules is closed under transfinite extensions. In particular, it contains* $\operatorname{Filt}(\mathcal{F})$.

Let us pause to have another look at our basic setting:

EXAMPLE 1.12. Let $\mathcal{F}$ be the class of all countably presented projective modules. Then $\mathcal{C} = \mathcal{F}$, and $\mathcal{F}$-filtered modules coincide with the projective modules (by a classic result of Kaplansky).

The locally $\mathcal{F}$-free modules are called *$\aleph_1$-projective*, [**12**]. They are characterized by the existence of an $\aleph_1$-dense system consisting of countably generated projective modules.

$\aleph_1$-projective modules are nothing else than the well-known *flat Mittag-Leffler modules*, i.e., the flat modules M such that for each family $(N_i \mid i \in I)$ of left R-modules, the canonical map $M \otimes \prod_{i\in I} N_i \to \prod_{i\in I}(M \otimes_R N_i)$ is monic. In fact, we have (see [**22**] and [**28**]):

THEOREM 1.13. *The following are equivalent for a module M:*

(1) *M is $\aleph_1$-projective;*
(2) *M is flat Mittag-Leffler;*
(3) *M is flat, and for some (or any) presentation $M = \varinjlim_{i\in I} F_i$ of M as a direct limit of a direct system of finitely generated free modules $\mathcal{M} = (M_i \mid i \in I)$, and for each module N, the induced inverse system $\mathcal{H} = (\mathrm{Hom}_R(M_i, N) \mid i \in I)$ has the Mittag-Leffler property.*

Recall that an inverse system of modules $(H_i, h_{ij} \mid i \leq j \in I)$ has the *Mittag-Leffler property*, if for each $k \in I$ there exists $k \leq j \in I$, such that $\mathrm{Im}(h_{kj}) = \mathrm{Im}(h_{ki})$ for each $j \leq i \in I$, that is, the terms of the decreasing chain $(\mathrm{Im}(h_{ki}) \mid k \leq i \in I)$ of submodules of H_k stabilize.

We will denote by $\mathcal{FM}$ the class of all flat Mittag-Leffler modules. If R is a right perfect ring, then $\mathcal{P}_0 = \mathcal{FM} = \mathcal{F}_0$ [**9**], where $\mathcal{P}_0$ and $\mathcal{F}_0$ denotes the class of all projective and flat modules, respectively. However, if R is not right perfect, then $\mathcal{P}_0 \subsetneq \mathcal{FM} \subsetneq \mathcal{F}_0$, and we will see later that these three classes have rather different structural properties.

Now, we can prove that our general construction of the tree module L always yields a locally $\mathcal{F}$-free module. Of course, if B is $\mathcal{F}$-filtered, then so is L. But if B is not $\mathcal{F}$-filtered, the result is quite surprising: L has only a 'small' $\mathcal{F}$-filtered submodule D, while the 'big' quotient L/D is a direct sum of copies of B (cf. Lemma 1.6).

LEMMA 1.14. *The tree module L from Lemma* 1.6 *is locally $\mathcal{F}$-free.*

PROOF. For each countable subset $C = \{\nu_i \mid i < \omega\}$ of $\mathrm{Br}(T_\kappa)$, the module $X_C = \sum_{\nu\in C} X_\nu$ is isomorphic to a countable direct sum of the F_is. In fact, $X_C = \bigcup_{i<\omega} X_{C_i}$ where $X_{C_i} = \sum_{j\leq i} X_{\nu_j}$ is a direct summand in $X_{C_{i+1}}$ with a complement isomorphic to a countable direct sum of the F_is. So the local $\mathcal{F}$-freeness of L is witnessed by the set $\mathcal{S}$ of all X_C, where C runs over all countable subsets of $\mathrm{Br}(T_\kappa)$. □

1.3. Module approximations. In this section, we recall the relevant basic notions and results from the approximation theory of modules. For more details we refer to [**20**, Part II].

Approximations of modules were introduced by Auslander, Reiten and Smalø in the setting of finitely generated modules over artin algebras while Enochs and Xu studied them in the general setting of Mod–R, albeit using the different terminology of precovers and preenvelopes. We will primarily be interested in the general setting, so our terminology follows [**16**]:

DEFINITION 1.15. (i) A class of modules $\mathcal{A}$ is *precovering* if for each module M there is $f \in \mathrm{Hom}_R(A, M)$ with $A \in \mathcal{A}$ such that each

$f' \in \mathrm{Hom}_R(A', M)$ with $A' \in \mathcal{A}$ has a factorization through f:

$$\begin{array}{ccc} A & \xrightarrow{f} & M \\ {\scriptstyle g}\uparrow & \nearrow_{f'} & \\ A' & & \end{array}$$

The map f is called an $\mathcal{A}$-*precover* of M.

(ii) An $\mathcal{A}$-precover is *special* in case it is surjective, and its kernel K satisfies $\mathrm{Ext}^1_R(A, K) = 0$ for each $A \in \mathcal{A}$.

(iii) Let $\mathcal{A}$ be precovering. Assume that in the setting of (i), if $f' = f$ then each factorization g is an automorphism. Then f is called an $\mathcal{A}$-*cover* of M. The class $\mathcal{A}$ is *covering* in case each module has an $\mathcal{A}$-cover.

Dually, we define *(special) preenveloping* and *enveloping* classes of modules.

Precovering classes are ubiquitous because of the following basic facts due to Enochs and Šťovíček:

THEOREM 1.16. *Let $\mathcal{C}$ be a class of modules. Then the following implications hold:*

(1) *If $\mathcal{C}$ is deconstructible, then $\mathcal{C}$ is precovering.*
(2) *If $\mathcal{C}$ is precovering and closed under direct limits, then $\mathcal{C}$ is covering.*

We note that though verified in many particular instances, the validity of the converse implications in both statements of Theorem 1.16 remains an open problem in general.

One can obtain special precovering and special preenveloping classes by employing the notion of a complete cotorsion pair due to Salce [**30**]:

DEFINITION 1.17. A pair of classes of modules $\mathfrak{C} = (\mathcal{A}, \mathcal{B})$ is a *cotorsion pair* provided that

(1) $\mathcal{A} = {}^\perp\mathcal{B} = \{A \in \text{Mod-}R \mid \mathrm{Ext}^1_R(A, B) = 0 \text{ for all } B \in \mathcal{B}\}$, and
(2) $\mathcal{B} = \mathcal{A}^\perp = \{B \in \text{Mod-}R \mid \mathrm{Ext}^1_R(A, B) = 0 \text{ for all } A \in \mathcal{A}\}$.

In this case $\mathcal{A}$ is closed under transfinite extensions. If moreover

(3) For each module M, there exists an exact sequences $0 \to M \to B \to A \to 0$ with $A \in \mathcal{A}$ and $B \in \mathcal{B}$,

then $\mathfrak{C}$ is called *complete*. It that case, for each module M there also exists an exact sequences $0 \to B' \to A' \to M \to 0$ with $A' \in \mathcal{A}$ and $B' \in \mathcal{B}$, whence $\mathcal{A}$ is a special precovering class and $\mathcal{B}$ a special preenveloping class.

The ubiquity of special approximations comes from the following (cf. [**15**])

THEOREM 1.18. *Let $\mathcal{S}$ be any set of modules. Then the cotorsion pair $({}^\perp(\mathcal{S}^\perp), \mathcal{S}^\perp)$ is complete. In fact, for each module M there exists an exact sequence $0 \to M \to B \to A \to 0$ where $B \in \mathcal{S}^\perp$ and A is $\mathcal{S}$-filtered.*

Moreover, if $R \in \mathcal{S}$, then $\mathcal{A} = {}^\perp(\mathcal{S}^\perp)$ coincides with the class of all direct summands of $\mathcal{S}$-filtered modules, and $\mathcal{A}$ is deconstructible.

A further tool for constructing precovering classes comes from [**34**]:

THEOREM 1.19. *Assume that κ is an infinite cardinal such that each right ideal of R is $\leq \kappa$-generated. Let $0 \leq n < \omega$ and $\mathcal{C}$ be any κ^+-deconstructible class of* *modules. Then the class of all modules possessing a $\mathcal{C}$-resolution of length $\leq n$ is also κ^+-deconstructible.*

EXAMPLE 1.20. Since each projective module is a direct sum of countably generated modules, if $0 \leq n < \omega$ and κ is an infinite cardinal such that each right ideal of R is $\leq \kappa$-generated, then the class $\mathcal{P}_n$ of all modules of projective dimension at most n is κ^+-deconstructible.

Similarly, the class $\mathcal{F}_0$ of all flat modules over any ring is κ^+-deconstructible where κ is an infinite cardinal $\geq \operatorname{card} R$, cf. [**11**]. Hence so is the class $\mathcal{F}_n$ of all modules of projective dimension at most n, for each $n < \omega$.

1.4. Limits of the approximation theory. Theorems 1.16, 1.18 and 1.19 yield numerous approximation classes of modules. However, not all classes of modules closed under transfinite extensions are precovering. This surprising fact can be proved using locally free (tree) modules.

We will now present a full proof for the absolute case of $\aleph_1$-projective (= flat Mittag-Leffler) modules over any non-right perfect ring. The result was first proved in a different way in [**18**] for the particular case of $\aleph_1$-projective abelian groups. For countable non-right perfect rings, a proof was given in [**10**] (cf. also [**33**] and [**34**]). The following proof for general non-right perfect rings using tree modules is due to Šaroch:

THEOREM 1.21. *Assume that R is a non-right perfect ring and let $\mathcal{F} = \{R\}$. Let B be a Bass module for $\mathcal{F}$ which is not projective (see Example 1.3). Then B has no $\aleph_1$-projective precover.*

In particular, the class $\mathcal{FM}$ is not precovering, and hence not deconstructible.

PROOF. Assume there exists an $\mathcal{FM}$-precover $f : F \to B$ of B. Consider the short exact sequence $0 \to K \hookrightarrow F \xrightarrow{f} B \to 0$ where $K = \operatorname{Ker}(f)$. Let κ be an infinite cardinal such that $\operatorname{card} R \leq \kappa$ and $\operatorname{card} K \leq 2^\kappa = \kappa^\omega$ (see Remark 1.1).

Consider the tree module corresponding to T_κ, $\mathcal{F}$ and B. By Lemma 1.6, we have the tree module short exact sequence

$$0 \to D \hookrightarrow L \to B^{(2^\kappa)} \to 0,$$

where $L \in \mathcal{FM}$ by Lemma 1.14, and D is a free module of rank κ. Clearly, $L \in \mathcal{P}_1$.

Let $\eta : K \hookrightarrow E$ be a $\{L\}^\perp$-preenvelope of K with an $\{L\}$-filtered cokernel C (cf. Theorem 1.18). Consider the pushout

$$\begin{array}{ccccccccc}
 & & 0 & & 0 & & & & \\
 & & \downarrow & & \downarrow & & & & \\
0 & \longrightarrow & K & \xrightarrow{\subseteq} & F & \xrightarrow{f} & B & \longrightarrow & 0 \\
 & & {\scriptstyle\eta}\downarrow & & {\scriptstyle\varepsilon}\downarrow & & \| & & \\
0 & \longrightarrow & E & \xrightarrow{\subseteq} & P & \xrightarrow{g} & B & \longrightarrow & 0 \\
 & & \downarrow & & \downarrow & & & & \\
 & & C & \xrightarrow{\cong} & C & & & & \\
 & & \downarrow & & \downarrow & & & & \\
 & & 0 & & 0 & & & &
\end{array}$$

Then $P \in \mathcal{FM}$. Since f is an $\mathcal{FM}$-precover, there exists $h : P \to F$ such that $fh = g$. Then $f = g\varepsilon = fh\varepsilon$, whence $K + \mathrm{Im}(h) = F$. Let $h' = h \restriction E$. Then $h' : E \to K$ and $\mathrm{Im}(h') = K \cap \mathrm{Im}(h)$.

Consider the restricted short exact sequence

$$0 \longrightarrow \mathrm{Im}(h') \xrightarrow{\subseteq} \mathrm{Im}(h) \xrightarrow{f\restriction \mathrm{Im}(h)} B \longrightarrow 0.$$

As $E \in L^{\perp}$ and $L \in \mathcal{P}_1$, also $\mathrm{Im}(h') \in L^{\perp}$.

Applying $\mathrm{Hom}_R(-, \mathrm{Im}(h'))$ to the tree-module short exact sequence above, we obtain the exact sequence

$$\mathrm{Hom}_R(D, \mathrm{Im}(h')) \to \mathrm{Ext}^1_R(B, \mathrm{Im}(h'))^{2^{\kappa}} \to 0$$

where the first term has cardinality at most $(\mathrm{card}\, K)^{\kappa} \leq 2^{\kappa}$, so $\mathrm{Im}(h') \in B^{\perp}$ (otherwise, the second term would have cardinality at least $2^{2^{\kappa}}$).

Then $f \restriction \mathrm{Im}(h)$ splits, and so does the $\mathcal{FM}$-precover f. It follows that $B \in \mathcal{FM}$, whence B is projective, a contradiction. □

Theorem 1.21 is a special instance (for $\mathcal{F} = \{R\}$) of the following more general result proved in [**32**]. We will present its applications in the following section.

THEOREM 1.22. *Let $\mathcal{F}$ be a class of countably presented modules, and $\mathcal{L}$ the class of all locally $\mathcal{F}$-free modules. Let B be a Bass module for $\mathcal{F}$ such that B is not a direct summand in a module from $\mathcal{L}$.*

Then B has no $\mathcal{L}$-precover. In particular, the class $\mathcal{L}$ is not precovering, and hence not deconstructible.

1.5. Tilting approximations and locally T-free modules. The model case of flat Mittag-Leffler modules discussed in the previous section is actually a 0-dimensional instance of a more general phenomenon related to tilting. This relation was first noticed in [**34**], and fully developed and generalized in [**4**].

In order to formulate the key results, we recall the notion of an (infinitely generated) tilting module. For more details, we refer e.g. to [**20**, Part III].

For a module T, denote by $\mathrm{Add}(T)$ (resp. $\mathrm{add}(T)$) the class of all direct summands of arbitrary (resp. finite) direct sums of copies of T.

DEFINITION 1.23. A module T is *tilting* provided that

(T1) T has finite projective dimension,
(T2) $\mathrm{Ext}^i_R(T, T^{(\kappa)}) = 0$ for all $1 \leq i < \omega$ and all cardinals κ, and
(T3) there exist $r < \omega$ and an exact sequence $0 \to R \to T_0 \to \cdots \to T_r \to 0$ where $T_i \in \mathrm{Add}\, T$ for each $i \leq r$.

The class

$$\mathcal{T}_T = T^{\perp_\infty} = \{B \in \text{Mod-}R \mid \mathrm{Ext}^i_R(T, B) = 0 \text{ for all } 0 < i < \omega\}$$

is called the *(right) tilting class*, $\mathcal{A}_T = {}^{\perp}\mathcal{T}_T$ the *left tilting class*, and the (complete) cotorsion pair $\mathfrak{C}_T = (\mathcal{A}_T, \mathcal{T}_T)$ the *tilting cotorsion pair*, induced by T.

If T has projective dimension $\leq n$, then the tilting module T is called *n-tilting*, and similarly for $\mathcal{T}_T$, $\mathcal{A}_T$, and $\mathfrak{C}_T$.

It is easy to see that 0-tilting modules T coincide with (not necessarily finitely generated) projective generators, whence $\mathcal{A}_T = \mathcal{P}_0$ and $\mathcal{T}_T = \text{Mod–}R$.

Also, for each tilting module T, we have $\operatorname{Add} T = \mathcal{A}_T \cap \mathcal{B}_T$ (this is the *kernel* of the tilting cotorsion pair $\mathfrak{C}_T$), and the right tilting class $\mathcal{B}_T$ is *definable*, that is, $\mathcal{B}_T$ is closed under direct limits, products and pure submodules.

Tilting theory originated in the setting of finitely generated modules over finite dimensional algebras, but many of its aspects extend to the general setting of modules over arbitrary rings. Such extension is especially interesting for commutative rings, because all finitely generated tilting modules over a commutative ring are projective. For a recent classification of tilting classes over commutative rings, we refer to [**3**] and [**24**].

Though the left tilting class $\mathcal{A}_T$ is always special precovering and the right tilting class $\mathcal{T}_T$ special preenveloping, one can employ tilting modules to produce non-precovering classes of modules, namely the classes of locally T-free modules. The construction generalizes the base case of $T = R$, where locally T-free modules coincide with the flat Mittag-Leffler ones.

Before explaining the construction in more detail, we recall basic facts of infinite dimensional tilting theory over arbitrary rings:

THEOREM 1.24. (1) *Let $\mathfrak{C} = (\mathcal{A}, \mathcal{B})$ be a cotorsion pair. Then $\mathfrak{C}$ is tilting, iff $\mathcal{A} \subseteq \mathcal{P}_n$ for some $n < \omega$, and $\mathcal{B}$ is closed under arbitrary direct sums.*

(2) *Right tilting classes $\mathcal{T}$ in* Mod–R *coincide with the classes of finite type, i.e., the classes of the form $\mathcal{S}^{\perp_\infty}$ where $\mathcal{S}$ consists of strongly finitely presented modules of bounded projective dimension. Such class $\mathcal{T}$ is n-tilting, iff $\mathcal{S} \subseteq \mathcal{P}_n$.*

In particular, each left tilting class is $\aleph_1$-deconstructible.

The largest possible choice for the class $\mathcal{S}$ in Theorem 1.24.2 is $\mathcal{S} = \mathcal{A} \cap \text{mod–}R$ where $\mathcal{A} = {}^{\perp}\mathcal{T}$. Then $\mathcal{S} \subseteq \mathcal{A} \subseteq \varinjlim \mathcal{S}$ (where $\varinjlim \mathcal{S}$ denotes the class of all direct limits of modules from $\mathcal{S}$).

In the 0-tilting case, the largest choice is $\mathcal{S} = \mathcal{P}_0^{<\omega}$ (the class of all finitely generated projective modules), whence $\varinjlim \mathcal{S} = \mathcal{F}_0$ (the class of all flat modules). Of course, we also have $\mathcal{P}_0 \subseteq \mathcal{FM} \subseteq \mathcal{F}_0$, and the question is how to generalize the notion of a flat Mittag-Leffler module to the n-tilting setting for $n > 0$. The answer is given by the following definition:

DEFINITION 1.25. Consider the particular setting of Notation 1.5 when T is a tilting module and $\mathcal{F} = \mathcal{A}_T^{\leq\omega}$ (so $\mathcal{C} = \mathcal{F}$).

A module is *locally T-free* provided that M is locally $\mathcal{F}$-free, i.e., M admits a dense system of countably presented submodules from $\mathcal{A}$. We will denote by $\mathcal{L}_T$ the class of all locally T-free modules.

So if $T = R$, then the locally T-free modules coincide with the $\aleph_1$-projective (= flat Mittag-Leffler) modules. By Theorem 1.21, these modules form a (pre-) covering class, iff R is a right perfect ring, iff $\mathcal{P}_0 = \mathcal{F}_0$. This was substantially generalized in [**4**] as follows:

THEOREM 1.26. *Let T be a tilting module and $\mathcal{A}_T$ be the corresponding left tilting class (so $\mathcal{A}_T = \operatorname{Filt}((\mathcal{C}_T))$ where $\mathcal{C}_T = \mathcal{A}_T^{\leq\omega}$ by Theorem 1.24.2). Let $\mathcal{L}_T$ denote the class of all locally T-free modules. Then the following are equivalent:*

(1) *$\mathcal{L}_T$ is a (pre-) covering class;*
(2) *Each Bass module for $\mathcal{C}_T$ is contained in $\mathcal{C}_T$;*

(3) *The class $\mathcal{A}_T$ is closed under direct limits;*
(4) *T is $\sum$-pure split.*

Here, a module T is $\sum$*-pure split* provided that each pure embedding $N \hookrightarrow M$ with $M \in \operatorname{Add}(T)$, splits. For example, any $\sum$-pure injective module is $\sum$-pure split. Also, a ring R is right perfect, iff the regular module is $\sum$-pure split.

Since $\mathcal{S}_T = \mathcal{A}_T^{<\omega} \subseteq \mathcal{A}_T \subseteq \mathcal{L}_T \subseteq \varinjlim \mathcal{S}_T$, condition (3) above is further equivalent to $\mathcal{L}_T$ being closed under direct limits.

Theorem 1.26 indicates that the phenomenon of existence of non-precovering classes closed under transfinite extensions is much more widespread than originally expected. For example, though each finitely generated module over an artin algebra is $\sum$-pure injective (because it is endofinite), there do exist countably generated non-$\sum$-pure split tilting modules over each hereditary artin algebra of infinite representation type:

EXAMPLE 1.27. Let R be an indecomposable hereditary finite dimensional algebra of infinite representation type. Recall that there is a partition of the represetative set of all indecomposable finitely generated modules, ind-R, into three sets: q - the indecomposable preinjective modules, p - the indecomposable preprojective modules, and t the indecomposable regular modules.

By Theorem 1.24.2, $p^\perp$ is a right tilting class. The tilting module T inducing $p^\perp$ is called the *Lukas tilting module.* The left tilting class of T is the class $\mathcal{R}$ of all *Baer modules.* By [**2**], $\mathcal{R} = \operatorname{Filt}(p)$. The locally T-free modules are called *locally Baer modules.*

By [**2**] the Lukas tilting module is countably generated, but it has no finitely generated direct summands, and it is not Σ-pure split. Therefore, by Theorem 1.26, the class of all locally Baer modules is not precovering (and hence not deconstructible).

Of course, this means that there exist Bass modules for $\mathcal{R}^{<\omega}$ that are not Baer. Since $\mathcal{R}^{<\omega} = \operatorname{add}(p)$ is the class of all finitely generated preprojective modules, these Bass modules can be obtained as unions of the chains

$$P_0 \overset{f_0}{\hookrightarrow} P_1 \overset{f_1}{\hookrightarrow} \dots \overset{f_{i-1}}{\hookrightarrow} P_i \overset{f_i}{\hookrightarrow} P_{i+1} \overset{f_{i+1}}{\hookrightarrow} \dots$$

such that all the P_i are finitely generated preprojective, but the cokernels of all the f_i are regular (i.e., in $\operatorname{add}(t)$). Such chains exist for any hereditary finite dimensional algebra of infinite representation type, see [**4**].

Theorem 1.26 is proved in [**4**] as a corollary of a still more general result concerning cotorsion pairs:

THEOREM 1.28. *Let $\mathfrak{C} = (\mathcal{A}, \mathcal{B})$ be a cotorsion pair such that the class $\mathcal{B}$ is closed under direct limits. Then $\mathcal{A}$ is $\aleph_1$-deconstructible (whence $\mathfrak{C}$ is complete), and there is a module T such that $\mathcal{A} \cap \mathcal{B} = \operatorname{Add} T$.*

Let $\mathcal{F} = \mathcal{A}^{\leq\omega}$ and $\mathcal{L}$ be the class of all locally $\mathcal{F}$-free modules. Then the following are equivalent:

(1) *$\mathcal{L}$ is a (pre-) covering class;*
(2) *Each Bass module for $\mathcal{F}$ is contained in $\mathcal{F}$;*
(3) *The class $\mathcal{A}$ is closed under direct limits;*
(4) *T is $\sum$-pure split.*

The proof of Theorem 1.28 employs relative Mittag-Leffler conditions studied in [**1**], [**21**] and [**29**]. One of its interesting by-products is an alternative description of local $\mathcal{F}$-freeness using these conditions in the general setting of Theorem 1.28. First we recall the relevant definitions:

DEFINITION 1.29. Let $\mathcal{B}$ be a class of modules and $\mathcal{B}^* = \{\mathrm{Hom}_{\mathbb{Z}}(B, \mathbb{Q}/\mathbb{Z}) \mid B \in \mathcal{B}\}$ be the class of character left R-modules of the modules in $\mathcal{B}$.

A module M is *$\mathcal{B}$-stationary* provided that M can be expressed as the direct limit of a direct system $\mathcal{M}$ of finitely presented modules such that for each $B \in \mathcal{B}$, the induced inverse system $\mathcal{H}$ is Mittag-Leffler (see Theorem 1.13).

$\mathcal{B}$-stationarity is a relative Mittag-Leffler condition: by [**21**, 2.6(3)], M is $\mathcal{B}$-stationary, iff M is $\mathcal{B}^*$-Mittag-Leffler, where the latter means that the canonical map $M \otimes \prod_{i \in I} N_i \to \prod_{i \in I}(M \otimes_R N_i)$ is monic for each family $(N_i \mid i \in I)$ of left R-modules from $\mathcal{B}^*$.

THEOREM 1.30. *Let $M \in$ Mod–R and $(\mathcal{A}, \mathcal{B})$ be a cotorsion pair in* Mod–R *such that $\mathcal{B}$ is closed under direct limits. The the following are equivalent:*

(1) *M is locally $\mathcal{A}^{\leq \omega}$-free;*
(2) *M is $\mathcal{B}$-stationary, and M is a pure-epimorphic image of a module from $\mathcal{A}$.*

In particular, if $M \in \varinjlim \mathcal{A}$, then M is locally $\mathcal{A}^{\leq \omega}$-free, iff M is $\mathcal{B}$-stationary, iff M is $\mathcal{B}^*$-Mittag-Leffler.

REMARK 1.31. Lemma 1.22 is one of the key tools for proving Theorem 1.28. However, it can be used to construct non-precovering classes even outside the setting of 1.28: in [**35**], inspired by the ideas of Positselski from algebraic geometry [**26**], the authors studied very flat and locally very flat modules over commutative rings R. In analogy with our basic case of projective and flat Mittag-Leffler modules, they proved that for each noetherian domain, the class of all very flat modules is covering, iff the class of all locally very flat modules is precovering, iff the Zariski spectrum Spec(R) is finite.

2. Tree modules and almost split sequences

Almost split sequences play a central role in the representation theory of artin algebras. Since the founding work of Auslander and Reiten [**7**], they provide a crucial tool for understanding the category of finitely generated modules, notably for studying extensions of indecomposable finitely generated modules.

We will only briefly touch upon this vast topic. As in Section 1, we will stick to the general setting of (possibly infinitely generated) modules over arbitrary rings. After introducing the necessary basic definitions in this setting, we will directly concentrate on a major problem formulated by Auslander in 1975 in [**5**]: Auslander asked whether the notion of a right almost split map in Mod–R ending in a module M, available for any ring R and any indecomposable finitely presented module M with local endomorphism ring, does occur also for some infinitely generated modules. A negative solution has recently been obtained by Šaroch in [**31**] and it uses (generalized) tree modules. Before explaining the solution in more detail (following [**31**]), we recall the relevant definitions:

DEFINITION 2.1. Let R be a ring. A morphism $f \in \mathrm{Hom}_R(B, C)$ is said to be *right almost split* (in Mod–R) provided that the following two conditions are equivalent for each $B' \in \text{Mod–}R$ and $h \in \text{Hom}_R(B', C)$:

(1) h factorizes through f;
(2) h is not a split epimorphism.

Left almost split morphisms are defined dually. A short exact sequence $0 \to A \overset{g}{\to} B \overset{f}{\to} C \to 0$ in Mod–R is an *almost split sequence* provided that f is right almost split and g is left almost split.

If f is right almost split, then it is easy to see that f is not a split epimorphism, and the endomorphism ring of C is local (whence C is indecomposable). Auslander [**6**] proved that the converse holds true when C is finitely presented:

THEOREM 2.2. *Let C be a finitely presented module. Then there exists a right almost split morphism $f \in \text{Hom}_R(B, C)$ in* Mod–R, *iff the endomorphism ring of C is local. If this is the case and C is not projective, then there exists an almost split sequence $0 \to A \to B \to C \to 0$ in* Mod–R.

REMARK 2.3. In the particular case when R is an artin algebra, much more can be proved: the almost split sequences are unique and form the (simple) socles of the extension modules. More precisely, if $C \in \text{mod–}R$ is indecomposable and non-projective, then there exists a unique (up to equivalence of short exact sequences) almost split sequence $0 \to A \to B \to C \to 0$, and $A \in \text{mod–}R$ is indecomposable non-injective. The equivalence class of this almost split sequence forms the socle of the left S-module $\text{Ext}^1_R(C, A)$ (where S is the endomorphism ring of C), and this socle is a simple left S-module. Moreover, also the dual statement holds true — see e.g. [**8**, V.2] for more details.

In [**5**], Auslander asked whether given a ring R and a module C, there exists a right (left) almost split morphism in Mod–R ending (beginning) in C. A complete answer for right almost split morphisms has recently been given in [**31**]:

THEOREM 2.4. [**31**, Theorem 4.4] *Let R be a ring and $C \in$* Mod–R. *Then there exists a right almost split map $f \in \text{Hom}_R(B, C)$ in* Mod–R, *iff C is finitely presented and the endomorphism ring of C is local.*

The if-part of Theorem 2.4 follows from Auslander's Theorem 2.2. We will now proceed to indicate the modifications of the construction of tree modules T_κ from Section 1 that are needed for the proof of the only-if part.

First, instead of the Bass modules B, which are direct limits of the direct systems $\mathcal{D} = (F_i, f_{ji} \mid i \leq j < \omega)$ of small (= countably presented) modules F_i indexed in ω, we will need to consider direct limits of general well-ordered continuous direct systems $\mathcal{E} = (F_\alpha, f_{\beta\alpha} \mid \alpha \leq \beta < \text{cf}(\theta))$ of small (= $<\theta$-presented) modules F_α indexed by the regular infinite cardinal $\text{cf}(\theta)$. Here, $\text{cf}(\theta)$ denotes the cofinality of θ, and the term *continuous* means that for each limit ordinal $\alpha < \text{cf}(\theta)$, we have $F_\alpha = \varinjlim_{\beta<\alpha} F_\beta$.

These direct systems are important because of the following classic result by Iwamura [**25**] (see also [**31**, Lemma 2.3]):

LEMMA 2.5. *Let R be a ring, θ be an infinite cardinal, and C a θ-presented module. Then C is the direct limit of a well-ordered continuous direct system $\mathcal{E} = (F_\alpha, f_{\beta\alpha} \mid \alpha \leq \beta < \text{cf}(\theta))$ such that F_α is a $<\theta$-presented module for each $\alpha < \text{cf}(\theta)$.*

The second non-trivial modification concerns the trees T_κ. The modified trees will again be uniformly decorated (by the well-ordered continuous direct system $\mathcal{E}$).

But we now have two initial parameters: μ and κ, where $\mu = \mathrm{cf}(\theta)$ is a regular infinite cardinal, and κ is an infinite cardinal such that $\kappa^{<\mu} = \kappa$.

The latter equality is used in [**31**] to define the modified tree T so that the set of all its branches $\mathrm{Br}(T)$ is a certain subset of κ^μ of cardinality κ^μ such that any two distinct elements of $\mathrm{Br}(T)$ coincide on an initial segment of μ. The partial order on $T \subseteq \mu \times \kappa$ is defined by $(\alpha, \beta) \leq (\gamma, \delta)$, if $\alpha < \gamma$ and there exists $\eta \in \mathrm{Br}(T)$ such that $\eta(\alpha) = \beta$ and $\eta(\gamma) = \delta$.

Such construction of the tree T has the advantage that it still enables a presentation of the resulting tree module as a submodule of a product of copies of the modules F_α ($\alpha < \mu$). In order to explain the latter fact, we recall the following lemma relating well-ordered direct limits to (reduced) products (see e.g. [**27**, 3.3.2]):

LEMMA 2.6. *Let μ be a regular infinite cardinal and $\mathcal{E} = (F_\alpha, f_{\beta\alpha} \mid \alpha \leq \beta < \mu)$ be a well-ordered continuous direct system of modules. Then there is the following commutative diagram*

$$\begin{array}{ccccccccc}
0 & \longrightarrow & K & \overset{\subseteq}{\longrightarrow} & D & \overset{\pi}{\longrightarrow} & E & \longrightarrow & 0 \\
 & & \downarrow & & \downarrow{\scriptstyle \tau_0} & & \downarrow{\scriptstyle \sigma_0} & & \\
0 & \longrightarrow & Q & \overset{\subseteq}{\longrightarrow} & P & \longrightarrow & P/Q & \longrightarrow & 0
\end{array}$$

where $D = \bigoplus_{\alpha<\mu} F_\alpha$, $E = \varinjlim_{\alpha<\mu} F_\alpha$, $\pi : D \to E$ is the canonical pure epimorphism, $K = \mathrm{Ker}(\pi)$, $P = \prod_{\alpha<\mu} F_\alpha$, Q is the submodule of P consisting of the sequences with support of cardinality $< \mu$, and τ_0 is defined as follows: for each $\alpha < \mu$ and $x \in F_\alpha \subseteq D$, the βth component of the sequence $\tau_0(x)$ equals $f_{\beta\alpha}(x)$ in case $\alpha \leq \beta < \mu$, and it is 0 otherwise. Moreover, the induced map σ_0 is a pure monomorphism.

REMARK 2.7. While the first row of the diagram in Lemma 2.6 is the usual pure-exact sequence presenting a direct limit as the quotient of a direct sum, the monomorphism $\tau_0 : D \to P$ is not the usual pure embedding of the direct sum into the direct product. In fact, in the particular case of $\mu = \omega$, $\tau_0 \restriction F_i$ maps F_i into P the same way as the assignment $x \to x_{\nu i}$ in Notation 1.5.

Now, we will uniformly decorate the tree T by the direct system $\mathcal{E}$. First we use the tree T to have a multidimensional version of the commutative diagram from Lemma 2.6 as follows:

$$\begin{array}{ccccccccc}
0 & \longrightarrow & K^{(\mathrm{Br}(T))} & \overset{\subseteq}{\longrightarrow} & D^{(\mathrm{Br}(T))} & \longrightarrow & E^{(\mathrm{Br}(T))} & \longrightarrow & 0 \\
 & & \downarrow & & \downarrow{\scriptstyle \tau} & & \downarrow{\scriptstyle \sigma} & & \\
0 & \longrightarrow & Q' & \longrightarrow & P' & \overset{\rho}{\longrightarrow} & P'/Q' & \longrightarrow & 0
\end{array}$$

where $P' = \prod_{(\alpha,\beta)\in T} F_{\alpha\beta}$, Q' is the submodule of P' consisting of the sequences with support of cardinality $< \mu$, and $F_{\alpha\beta} = F_\alpha$ for all $(\alpha, \beta) \in T$.

Here, the upper exact sequence is just the direct sum of κ^μ copies of the canonical presentation of E as a pure-epimorphic image of D (i.e., of the first row of the diagram in Lemma 2.6), and τ restricted to the ηth component (for $\eta \in \mathrm{Br}(T)$) acts as τ_0 from Lemma 2.6 (again, compare with Notation 1.5).

Finally, we can define the generalized tree module:

LEMMA 2.8. *In the setting above, let* $L = \mathrm{Im}(\tau)$ *and* $H = \mathrm{Ker}(\rho \restriction L)$. *Then we have the tree module exact sequence*

$$0 \to H \to L \xrightarrow{g} C^{(\mathrm{Br}(T))} \to 0$$

where $C = \varinjlim F_\alpha$ *is the direct limit of the direct system* $\mathcal{E}$.

Following the pattern of Section 1, we should now proceed to showing that, in some sense, L is an almost free module:

LEMMA 2.9. *The module* L *from Lemma* 2.8 *is finitely* θ*-separable, that is,* M *is the directed union of a direct system consisting of* $< \theta$*-presented direct summands of* M.

Moreover, if e *is an idempotent endomorphism of* L *such that* $e \restriction H = 0$ *and the endomorphism ring of* $\mathrm{Im}\,(e)$ *is local, then* $\mathrm{Im}\,(e)$ *is* $< \theta$*-presented.*

The tree module from Lemma 1.6 was used in the proof of Theorem 1.21 to show that the Bass module B has no $\aleph_1$-projective precover. Our goal now is to use the tree module from Lemma 2.8 to show that almost split morphism cannot terminate in an infinitely generated module. The main tool is the following lemma from [**31**]:

LEMMA 2.10. *Let* θ *be an infinite cardinal,* C *be an* θ*-presented module and* $f : B \to C$ *be a non-split epimorphism. Let* κ *be an infinite cardinal such that* $\mathrm{card}\, R, \theta \leq \kappa$, $\kappa = \kappa^{<\mathrm{cf}(\theta)}$, *and* $\mathrm{card}\,\mathrm{Ker}(f) \leq \kappa^{\mathrm{cf}(\theta)} = 2^\kappa$.

Let L *be the tree module from Lemma* 2.8. *Then there exists* $h \in \mathrm{Hom}_R(C^{(\mathrm{Br}(T))}, C)$ *such that* hg *does not factorize through* f.

With all these tools at hand, we can prove the only-if part of Theorem 2.4:

PROOF. Assume there exists an infinite cardinal θ and a right almost split morphism $f : B \to C$ such that C is a θ-presented, but not $< \theta$-presented module. Since C is not finitely presented, each homomorphism from a finitely presented module to C factorizes through f, whence f is a (non-split) pure epimorphism. Moreover, the endomorphism ring of C is local.

The tree module L from Lemma 2.10 has the property that there exists $h \in \mathrm{Hom}_R(C^{(\mathrm{Br}(T))}, C)$ such that hg does not factorize through f. Then $hg : L \to C$ must be a split epimorphism, so there is $\iota : C \to L$ such that $hg\iota = 1_C$. Using Lemma 2.9 for the idempotent endomorphism $e = \iota hg$, we infer that $C \cong \mathrm{Im}(e)$ is $< \theta$-presented, a contradiction. □

REMARK 2.11. As noted in [**31**], Theorem 2.4 imposes strict limitations on the form of almost split sequences in Mod–R. Namely, it implies that if $0 \to A \xrightarrow{h} B \to C \to 0$ is an almost split sequence in Mod–R, then not only C is finitely presented with local endomorphism ring, but also A is pure-injective: otherwise, the non-split pure-embedding of A into its pure-injective envelope factors through h, whence h is a pure monomorphism and the almost split sequence splits, a contradiction.

Acknowledgment

The author thanks Jan Šaroch and the referee for valuable comments.

References

[1] Lidia Angeleri Hügel and Dolors Herbera, *Mittag-Leffler conditions on modules*, Indiana Univ. Math. J. **57** (2008), no. 5, 2459–2517, DOI 10.1512/iumj.2008.57.3325. MR2463975

[2] Lidia Angeleri Hügel, Otto Kerner, and Jan Trlifaj, *Large tilting modules and representation type*, Manuscripta Math. **132** (2010), no. 3-4, 483–499, DOI 10.1007/s00229-010-0356-2. MR2652443

[3] Lidia Angeleri Hügel, David Pospíšil, Jan Šťovíček, and Jan Trlifaj, *Tilting, cotilting, and spectra of commutative Noetherian rings*, Trans. Amer. Math. Soc. **366** (2014), no. 7, 3487–3517, DOI 10.1090/S0002-9947-2014-05904-7. MR3192604

[4] L. Angeleri Hügel, J. Šaroch, J. Trlifaj, *Approximations and Mittag-Leffler conditions – the applications*, to appear in Israel J. Math., arXiv:1612.01140v1.

[5] Maurice Auslander, *Existence theorems for almost split sequences*, Ring theory, II (Proc. Second Conf., Univ. Oklahoma, Norman, Okla., 1975), Dekker, New York, 1977, pp. 1–44. Lecture Notes in Pure and Appl. Math., Vol. 26. MR0439883

[6] Maurice Auslander, *A survey of existence theorems for almost split sequences*, Representations of algebras (Durham, 1985), London Math. Soc. Lecture Note Ser., vol. 116, Cambridge Univ. Press, Cambridge, 1986, pp. 81–89. MR897320

[7] Maurice Auslander and Idun Reiten, *Representation theory of Artin algebras. III. Almost split sequences*, Comm. Algebra **3** (1975), 239–294, DOI 10.1080/00927877508822046. MR0379599

[8] Maurice Auslander, Idun Reiten, and Sverre O. Smalø, *Representation theory of Artin algebras*, Cambridge Studies in Advanced Mathematics, vol. 36, Cambridge University Press, Cambridge, 1995. MR1314422

[9] Hyman Bass, *Finitistic dimension and a homological generalization of semi-primary rings*, Trans. Amer. Math. Soc. **95** (1960), 466–488, DOI 10.2307/1993568. MR0157984

[10] Silvana Bazzoni and Jan Šťovíček, *Flat Mittag-Leffler modules over countable rings*, Proc. Amer. Math. Soc. **140** (2012), no. 5, 1527–1533, DOI 10.1090/S0002-9939-2011-11070-0. MR2869137

[11] L. Bican, R. El Bashir, and E. Enochs, *All modules have flat covers*, Bull. London Math. Soc. **33** (2001), no. 4, 385–390, DOI 10.1017/S0024609301008104. MR1832549

[12] Paul C. Eklof and Alan H. Mekler, *Almost free modules*, Revised edition, North-Holland Mathematical Library, vol. 65, North-Holland Publishing Co., Amsterdam, 2002. Set-theoretic methods. MR1914985

[13] Paul C. Eklof and Saharon Shelah, *On Whitehead modules*, J. Algebra **142** (1991), no. 2, 492–510, DOI 10.1016/0021-8693(91)90321-X. MR1127077

[14] Paul C. Eklof and Saharon Shelah, *On the existence of precovers*, Illinois J. Math. **47** (2003), no. 1-2, 173–188. Special issue in honor of Reinhold Baer (1902–1979). MR2031314

[15] Paul C. Eklof and Jan Trlifaj, *How to make Ext vanish*, Bull. London Math. Soc. **33** (2001), no. 1, 41–51, DOI 10.1112/blms/33.1.41. MR1798574

[16] E. E. Enochs, O. M. G. Jenda, *Relative Homological Algebra 1*, 2nd rev. ext. ed., GEM 30, W. de Gruyter, Berlin 2011.

[17] E. E. Enochs, O. M. G. Jenda, *Relative Homological Algebra 2*, GEM 54, W. de Gruyter, Berlin 2011.

[18] Sergio Estrada, Pedro A. Guil Asensio, Mike Prest, and Jan Trlifaj, *Model category structures arising from Drinfeld vector bundles*, Adv. Math. **231** (2012), no. 3-4, 1417–1438, DOI 10.1016/j.aim.2012.06.011. MR2964610

[19] Rüdiger Göbel and Jan Trlifaj, *Cotilting and a hierarchy of almost cotorsion groups*, J. Algebra **224** (2000), no. 1, 110–122, DOI 10.1006/jabr.1999.8103. MR1736696

[20] R. Göbel, J. Trlifaj, *Approximations and Endomorphism Algebras of Modules*, 2nd rev. ext. ed., Vols. 1 and 2, GEM 41, W. de Gruyter, Berlin 2012.

[21] Dolors Herbera, *Definable classes and Mittag-Leffler conditions*, Ring theory and its applications, Contemp. Math., vol. 609, Amer. Math. Soc., Providence, RI, 2014, pp. 137–166, DOI 10.1090/conm/609/12123. MR3204357

[22] Dolors Herbera and Jan Trlifaj, *Almost free modules and Mittag-Leffler conditions*, Adv. Math. **229** (2012), no. 6, 3436–3467, DOI 10.1016/j.aim.2012.02.013. MR2900444

[23] Mark Hovey, *Cotorsion pairs, model category structures, and representation theory*, Math. Z. **241** (2002), no. 3, 553–592, DOI 10.1007/s00209-002-0431-9. MR1938704

[24] M. Hrbek, J. Šťovíček, *Tilting classes over commutative rings*, preprint, arXiv:1701.05534v1.

[25] T. Iwamura, *A lemma on directed sets*, Zenkoku Shijo Sugaku Danwakai 262(1944), 107-111, (in Japanese).

[26] L.Positselski, *Contraherent cosheaves*, preprint, arXiv:1209.2995v5.

[27] Mike Prest, *Purity, spectra and localisation*, Encyclopedia of Mathematics and its Applications, vol. 121, Cambridge University Press, Cambridge, 2009. MR2530988

[28] Michel Raynaud and Laurent Gruson, *Critères de platitude et de projectivité. Techniques de "platification" d'un module* (French), Invent. Math. **13** (1971), 1–89, DOI 10.1007/BF01390094. MR0308104

[29] P. Rothmaler, *Mittag-Leffler modules and positive atomicity*, Habilitationsschrift, Christian-Albrechts-Universität zu Kiel 1994.

[30] Luigi Salce, *Cotorsion theories for abelian groups*, Symposia Mathematica, Vol. XXIII (Conf. Abelian Groups and their Relationship to the Theory of Modules, INDAM, Rome, 1977), Academic Press, London-New York, 1979, pp. 11–32. MR565595

[31] Jan Šaroch, *On the non-existence of right almost split maps*, Invent. Math. **209** (2017), no. 2, 463–479, DOI 10.1007/s00222-016-0712-2. MR3674220

[32] J. Šaroch, *Approximations and Mittag-Leffler conditions – the tools*, to appear in Israel J. Math., arXiv:1612.01138v1.

[33] Jan Šaroch and Jan Trlifaj, *Kaplansky classes, finite character and $\aleph_1$-projectivity*, Forum Math. **24** (2012), no. 5, 1091–1109, DOI 10.1515/form.2011.101. MR2988573

[34] Alexander Slávik and Jan Trlifaj, *Approximations and locally free modules*, Bull. Lond. Math. Soc. **46** (2014), no. 1, 76–90, DOI 10.1112/blms/bdt069. MR3161764

[35] Alexander Slávik and Jan Trlifaj, *Very flat, locally very flat, and contraadjusted modules*, J. Pure Appl. Algebra **220** (2016), no. 12, 3910–3926, DOI 10.1016/j.jpaa.2016.05.020. MR3517562

Faculty of Mathematics and Physics, Department of Algebra, Charles University, Sokolovská 83, 186 75 Prague 8, Czech Republic

Email address: `trlifaj@karlin.mff.cuni.cz`

Selected Published Titles in This Series

716 **Alex Martsinkovsky, Kiyoshi Igusa, and Gordana Todorov, Editors,** Surveys in Representation Theory of Algebras, 2018

710 **Raphaël Danchin, Reinhard Farwig, Jiří Neustupa, and Patrick Penel, Editors,** Mathematical Analysis in Fluid Mechanics, 2018

709 **Fernando Galaz-García, Juan Carlos Pardo Millán, and Pedro Solórzano, Editors,** Contributions of Mexican Mathematicians Abroad in Pure and Applied Mathematics, 2018

708 **Christian Ausoni, Kathryn Hess, Brenda Johnson, Ieke Moerdijk, and Jérôme Scherer, Editors,** An Alpine Bouquet of Algebraic Topology, 2018

707 **Nitya Kitchloo, Mona Merling, Jack Morava, Emily Riehl, and W. Stephen Wilson, Editors,** New Directions in Homotopy Theory, 2018

706 **Yeonhyang Kim, Sivaram K. Narayan, Gabriel Picioroaga, and Eric S. Weber, Editors,** Frames and Harmonic Analysis, 2018

705 **Graham J. Leuschke, Frauke Bleher, Ralf Schiffler, and Dan Zacharia, Editors,** Representations of Algebras, 2018

704 **Alain Escassut, Cristina Perez-Garcia, and Khodr Shamseddine, Editors,** Advances in Ultrametric Analysis, 2018

703 **Andreas Malmendier and Tony Shaska, Editors,** Higher Genus Curves in Mathematical Physics and Arithmetic Geometry, 2018

702 **Mark Grant, Gregory Lupton, and Lucile Vandembroucq, Editors,** Topological Complexity and Related Topics, 2018

701 **Joan-Carles Lario and V. Kumar Murty, Editors,** Number Theory Related to Modular Curves, 2018

700 **Alexandre Girouard, Dmitry Jakobson, Michael Levitin, Nilima Nigam, Iosif Polterovich, and Frédéric Rochon, Editors,** Geometric and Computational Spectral Theory, 2017

699 **Mark L. Agranovsky, Matania Ben-Artzi, Catherine Bénéteau, Lavi Karp, Dmitry Khavinson, Simeon Reich, David Shoikhet, Gilbert Weinstein, and Lawrence Zalcman, Editors,** Complex Analysis and Dynamical Systems VII, 2017

698 **Alexander M. Blokh, Leonid A. Bunimovich, Paul H. Jung, Lex G. Oversteegen, and Yakov G. Sinai, Editors,** Dynamical Systems, Ergodic Theory, and Probability: in Memory of Kolya Chernov, 2017

697 **Fabrizio Broglia, Françoise Delon, Max Dickmann, Danielle Gondard-Cozette, and Victoria Ann Powers, Editors,** Ordered Algebraic Structures and Related Topics, 2017

696 **Ara S. Basmajian, Yair N. Minsky, and Alan W. Reid, Editors,** In the Tradition of Ahlfors–Bers, VII, 2017

695 **Katrina Barron, Elizabeth Jurisich, Antun Milas, and Kailash Misra, Editors,** Lie Algebras, Vertex Operator Algebras, and Related Topics, 2017

694 **Manjul Bhargava, Robert Guralnick, Gerhard Hiss, Klaus Lux, and Pham Huu Tiep, Editors,** Finite Simple Groups: Thirty Years of the Atlas and Beyond, 2017

693 **Michael Cwikel and Mario Milman, Editors,** Functional Analysis, Harmonic Analysis, and Image Processing, 2017

692 **Anatole Katok, Yakov Pesin, and Federico Rodriguez Hertz, Editors,** Modern Theory of Dynamical Systems, 2017

691 **Farrell Brumley, Maria Paula Gómez Aparicio, and Alberto Mínguez, Editors,** Around Langlands Correspondences, 2017

690 **Andrés Eduardo Caicedo, James Cummings, Peter Koellner, and Paul B. Larson, Editors,** Foundations of Mathematics, 2017

For a complete list of titles in this series, visit the
AMS Bookstore at **www.ams.org/bookstore/conmseries/**.